Evolutionary Biology

VOLUME 26

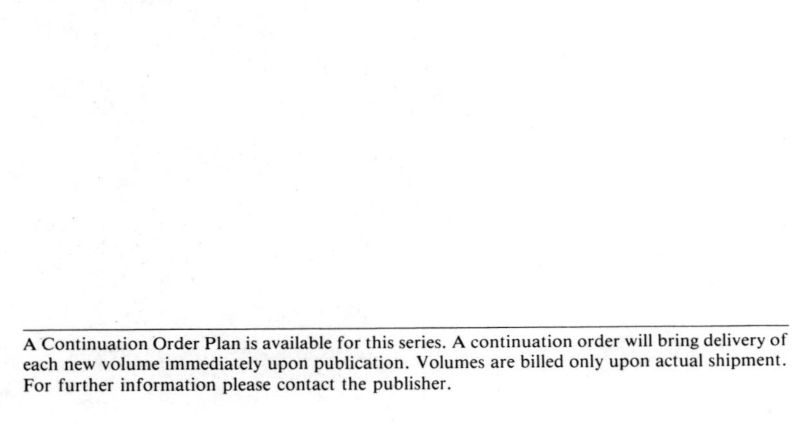

A Continuation Order Plan is available for this series. A continuation order will bring delivery of each new volume immediately upon publication. Volumes are billed only upon actual shipment. For further information please contact the publisher.

Evolutionary Biology

VOLUME 26

Edited by

MAX K. HECHT
Queens College of the City University of New York
Flushing, New York

BRUCE WALLACE
Virginia Polytechnic Institute and State University
Blacksburg, Virginia

and

ROSS J. MACINTYRE
Cornell University
Ithaca, New York

PLENUM PRESS • NEW YORK AND LONDON

The Library of Congress cataloged the first volume of this title as follows:

Evolutionary biology. v. 1- 1967-
New York, Appleton-Century-Crofts.
v. illus. 24 cm annual.
Editors: 1967- T. Dobzhansky and others.
1. Evolution—Period. 2. Biology—Period. I. Dobzhansky, Theodosius Grigorievich, 1900-
QH366.A1E9 575'.005 67-11961

ISBN 0-306-44154-3

A Division of Plenum Publishing Corporation
233 Spring Street, New York, N.Y. 10013

Printed in the United States of America

Contributors

John C. Avise • *Department of Genetics, University of Georgia, Athens, Georgia 30602*

Nicolas Borot • *CRPG/CNRS, UPR 8291, CHU Purpan, F-31300 Toulouse, Cedex, France*

Anne Cambon-Thomsen • *CRPG/CNRS, UPR 8291, and INSERM U100, CHU Purpan, F-31052 Toulouse, Cedex, France*

David K. Jacobs • *Department of Invertebrates, American Museum of Natural History, New York, New York 10024*

William Klitz • *Department of Integrative Biology, University of California, Berkeley, California 94720*

David McElroy • *Section of Biochemistry, Molecular and Cellular Biology, Cornell University, Ithaca, New York 14853*

Horst Nöthel • *Institut für Genetik, Freie Universität Berlin, D 1000 Berlin-33 (Dahlem), Germany*

P. A. Parsons • *Waite Institute, University of Adelaide, Glen Osmond, S.A. 5064, Australia*

Joseph M. Quattro • *Center for Theoretical and Applied Genetics, Rutgers University, New Brunswick, New Jersey 08903-0231; present address: Hopkins Marine Station, Stanford University, Pacific Grove, California 93950*

Kimberly S. Reece • *Department of Biology, College of William and Mary, Williamsburg, Virginia 23185*

Jan Salick • *Department of Botany, Ohio University, Athens, Ohio 45701*

David O. F. Skibinski • *School of Biological Sciences, University College of Swansea, Swansea, SA2 8PP, United Kingdom*

Glenys Thomson • *Department of Integrative Biology, University of California, Berkeley, California 94720*

Robert C. Vrijenhoek • *Center for Theoretical and Applied Genetics, Rutgers University, New Brunswick, New Jersey 08903-0231*

Robert D. Ward • *Department of Human Sciences, Loughborough University of Technology, Loughborough, Leicestershire, LE11 3TU, United Kingdom; present address: CSIRO Division of Fisheries, GPO Box 1538, Hobart, Tasmania 7001, Australia*

Florence Wasserman • *Department of Natural Sciences, York College of the City University of New York, Jamaica, New York 11451*

Marvin Wasserman • *Biology Department, Queens College of the City University of New York, Flushing, New York 11367*

Mathew Woodwark • *School of Biological Sciences, University College of Swansea, Swansea, SA2 8PP, United Kingdom*

Ray Wu • *Section of Biochemistry, Molecular and Cellular Biology, Cornell University, Ithaca, New York 14853*

Preface

The first volume of *Evolutionary Biology* was published twenty-five years ago. Since that time, twenty-five volumes and one supplement have appeared. As stated in earlier volumes, we are continuing to focus this series on critical reviews, commentaries, original papers, and controversies in evolutionary biology. The topics of the reviews range from anthropology to molecular evolution and from population biology to paleobiology.

Recent volumes have presented a broad spectrum of chapters on such subjects as molecular phylogenetics, evolutionary ecology, population genetics, paleobiology, comparative morphology, developmental biology, and the history and philosophy of evolutionary biology. With this volume, Bruce Wallace has decided to retire from the editorial direction of *Evolutionary Biology.* His co-editors and the publisher thank him for his dedicated service to the series. Michael T. Clegg will replace Wallace for forthcoming volumes.

The editors continue to solicit manuscripts in all areas of evolutionary biology. Manuscripts should be sent to any one of the following: Max K. Hecht, Department of Biology, Queens College of the City University of New York, Flushing, New York 11367; Ross J. MacIntyre, Department of Genetics and Development, Cornell University, Ithaca, New York 14853; or Michael T. Clegg, Department of Botany, University of California, Riverside, California 92521.

Contents

Evolutionary Biology

1

Function and Evolution of Actins

KIMBERLY S. REECE, DAVID McELROY, and RAY WU

INTRODUCTION

In this review, evolutionary relationships of actin gene and protein sequences, as well as intron positions within actin genes, are analyzed both within and between eukaryotic kingdoms.

With the exception of fungi (e.g., *Thermomyces lanuginosus, Saccharomyces cerevisiae,* and *Aspergillus nidulans*) and some protists (e.g., *Tetrahymena thermophila* and *Chlamydomonas reinhardtii*), actin is coded for by gene families in all eukaryotic organisms studied to date (Table I). Actin protein isoforms in vertebrates include skeletal muscle α actins, cardiac and smooth-muscle actins, as well as cytoplasmic β and γ actins. In organisms where actin is encoded by a gene family, the various actin isoforms appear to be involved in different cellular functions, such as maintenance of cell architecture or cellular motility. Studies in *Dictyostelium* (McKeown and Firtel, 1981), *Drosophila* (Fyrberg *et al.,* 1983), *Stronglocentrotus* (sea urchin) (Cox *et al.,* 1986; J. J. Lee *et al.,* 1986), *Xenopus* (Gurdon *et al.,* 1985; Mohun *et al.,* 1986; Sargent *et al.,* 1986; Wilson *et al.,* 1986), and *Mus* (mouse) (Shani, 1986) have demonstrated patterns of tissue-specific and/or developmental stage-specific expression of actin genes. Eventually we hope to understand the relationship of functionally distinct actin protein isoforms to the evolutionary development of actin gene structure.

KIMBERLY S. REECE • Department of Biology, College of William and Mary, Williamsburg, Virginia 23185. DAVID McELROY and RAY WU • Section of Biochemistry, Molecular and Cellular Biology, Cornell University, Ithaca, New York 14853.

Evolutionary Biology, Volume 26, edited by Max K. Hecht *et al.* Plenum Press, New York, 1992.

TABLE I. The Number of Actinlike Sequences Found in the Organisms of Different Eukaryotic Kingdoms and the Abbreviations Used for the Actin Genes

Organism	Kingdom	Number of actin genes	Gene	Abbreviation	Reference
Human (*Homo sapiens*)	Animalia	20–30	β Cytoplasmic	HuCy	Hamada *et al.* (1982), Ng *et al.* (1985)
			α Cardiac	HuCa	
Mouse (*Mus musca*)	Animalia	>20	—	—	Minty *et al.* (1983)
Chicken (*Gallus gallus*)	Animalia	8–10	β-Cytoplasmic type 5	ChCy	Fornwald *et al.* (1982), Bergsma *et al.* (1985)
		—	α Skeletal	ChM	
Frog (*Xenopus leavis*)	Animalia	—	α Cardiac	Xeno	Mohun *et al.* (1986)
Fruit fly (*Drosophila melanogaster*)	Animalia	6	79B	Dros	Fyrberg *et al.* (1981), Sanchez *et al.* (1983)
Sea urchin (*Stronglocentrotus purpuratus*)	Animalia	>10	—	SeUr	Scheller *et al.* (1981), Schuler *et al.* (1983)
Nematode (*Caenorhabditis elegans*)	Animalia	4	—	—	Files *et al.* (1983)
Rice (*Oryza sativa*)	Plantae	8–10	1	RAc1	Reece (1988), Reece *et al.* (1990), McElroy *et al.* (1990)
		—	2	RAc2	
		—	3	RAc3	
		—	7	RAc7	
Maize (*Zea mays*)	Plantae	6–8	1	MAc1	Shah *et al.* (1983)
Soybean (*Glycine max*)	Plantae	6–8	1	SAc1	Shah *et al.* (1982), Shah *et al.* (1983)
		—	3	SAc3	
Arabidopsis thaliana	Plantae	3	1	AAc1	Nairn *et al.* (1988)
Potato (*Solanum tuberasum*)	Plantae	5	—	—	Drouin and Dover (1987)
Tomato (*Lycopersicon esculentum*)	Plantae	>10	1	—	Bernatzky and Tanksley (1986)
Petunia hybrida	Plantae	>100	—	—	Baird and Meagher (1987)
Pine (*Pinus contorta*)	Plantae	>2	—	PAc1-A	Kenny *et al.* (1988)
Chlamydomonas reinhardii	Plantae	1	—	—	K. Kindle (personal communication)
Dictyostelium discoideum	Protista	17	A8	Dict	McKeown *et al.* (1978), Romans and Firtel (1985)
Physarum polycephalum	Protista	5	C	Phys	Schedl and Dove (1982), Gonzalez-y-Mechand and Cox (1988)
Tetrahymena pyriformis	Protista	1	Actin	Tetr	Cupples and Pearlman (1986), Hirono *et al.* (1987)
Trypanosoma brucei	Protista	2–4	Actin	Tryp	Ben Amar *et al.* (1988)
Thermomyces lanuginosus	Fungi	1	Actin	Ther	Wildeman (1988)
Saccharomyces cerevisiae	Fungi	1	Actin	Yeas	Ng and Abelson (1980), Gallwitz and Sures (1980)
Aspergillus nidulans	Fungi	1	Actin	Aspe	Fidel *et al.* (1988)

The first of the three main sections of this chapter is a review of actin protein functions in different eukaryotic groups, with special attention to plants and algae, whose actin gene evolution is of particular interest to us. The next section is an analysis of representative actin sequences from the four eukaryotic kingdoms in an attempt to better understand evolutionary relationships among the various actin genes. The last section is an analysis of intron number and location in actin genes (and four other eukaryotic genes) with a view to shedding some light on the significance of introns in the structural evolution of eukaryotic genes.

FUNCTIONS OF ACTIN

Actin in Animals

Actin in Vertebrates

Actin was first identified for its role in muscle contraction (Straub, 1941), but has since been implicated in many other cellular functions, including cell motility (Wehland *et al.,* 1977), cytoskeletal formation, and intracellular transport (Stossel, 1984). Actin has been found associated with contractile rings in dividing cells (Lazarides and Weber, 1974) and may play a role in cytokinesis (Pollard, 1986), chromosomal condensation, and mitosis (Maupin and Pollard, 1986; Crowley and Brasch, 1987).

In the animal kingdom actin is highly conserved both among species and between various actin isoforms. Classically, vertebrate actins have been divided into three categories (1) nonmuscle or cytoplasmic (β and γ) types, (2) smooth-muscle actins, and (3) skeletal and cardiac (α) muscle forms. Although muscle actins from several different species, including human, rat, and chicken, show little variation in their amino acid sequences, these isoforms can be distinguished by amino acid sequence differences, especially in the amino-terminal region (Vandekerckhove and Weber, 1984). Cytoplasmic actins are also highly conserved among species, and in fact show greater similarity across species than to muscle actins within a species. Even so, muscle and cytoplasmic actins normally show at least 90% amino acid sequence similarity. Two new actin isoforms recently isolated from vertebrate intestinal brush border cells (Sawtell *et al.,* 1988) have an amino-terminal region similar to that of cytoplasmic γ and β actins, yet they contain an epitope in common with muscle actins. These isoforms may represent a functional and evolutionary intermediate of the actin protein.

Actin in Invertebrates

Essentially, actin performs the same function in invertebrates as in vertebrates. Interestingly, however, all actin genes in invertebrates, regardless of their function, appear to be more closely related in amino acid sequence to vertebrate cytoplasmic actins. Even actins found in the muscle tissue of *Caenorhabditis elegans* (Files *et al.*, 1983), *Drosophila* (Fyrberg *et al.*, 1981), and sea urchin (Cooper and Crain, 1982) display greater similarity to vertebrate nonmuscle actin isoforms than they do to vertebrate muscle actins. Although invertebrate muscle actins are all "cytoplasmic" in their primary sequence, they perform many of the same functions as vertebrate muscle actins, and tissue-dependent expression of actin genes, differentiating between muscle and nonmuscle cells, has been observed in *Drosophila* (Fyrberg *et al.*, 1983). These findings suggest that vertebrate muscle actins did not develop until after the divergence of vertebrates and invertebrates (Vandekerckhove and Weber, 1984).

Actin in Fungi and Protists

Actin proteins and genes isolated from fungi and protists are also more similar in sequence to vertebrate cytoplasmic than to vertebrate muscle actins. The functions of actin in these organisms are less well studied than in animals. However, in several genera of fungi [*Saccharomyces* (Adams and Pringle, 1984), *Candida* (Anderson and Soll, 1986), and *Uromyces* (Hoch and Staples, 1983)] actin has been found associated with hyphal apical growth zones or buds. In addition, some fungal systems appear to require actin for organelle movements (McKerracher and Heath, 1987), although there is evidence against such a function in the fungus *Saprolegnia* (Heath, 1988).

In the protist *Physarum,* the force-generating system responsible for organelle and nutrient movement through the cytoplasm, as well as locomotion and chemotaxis during starvation, requires actinomyosin (Wohlfarth-Bottermann, 1986). Actin filaments similar to those in animal cells have been detected in contractile rings of *Tetrahymena* cells (R. D. Allen, 1978). Although all of the actins of *Dictyostelium* are similar to vertebrate cytoplasmic actins, the synthesis of actin isotypes is differentially regulated both spatially and temporally during development (MacLeod *et al.*, 1980; Romans *et al.*, 1985), suggesting that actin may play a variety of roles in this organism.

Extensive studies have been conducted on actin function in some green algae [see Staiger and Schliwa (1987) for review]. Actin has been found in the cytoplasm (La Claire, 1984) and associated with the nucleus (Tischendorf *et al.*, 1987). Many of the algal studies have focused on Characean algae, most notably *Nitella* and *Chara*, which are thought to be ancestral to higher plants (Graham, 1982). Examination of *Chara* using the light microscope has shown actin filaments (Williamson and Toh, 1979). Actin isolated from *Nitella* has solubility properties, ultrastructure, and electrophoretic mobility similar to those of animal actins (N. S. Allen and Allen, 1978). *Mougeotia*, an organism in the same class, Charophyceae, as *Nitella* and *Chara*, has actinlike filaments which can be decorated with the myosin derivative, heavy meromyosin (HMM) (Wagner and Klein, 1981).

Actin appears to be involved in cytoplasmic streaming and organelle movement, including chloroplast aggregation and rotation in some green algae. When actin filaments from *Nitella* are decorated with HMM they are found to have polarity and are oriented in the direction opposite that of cytoplasmic streaming (N. S. Allen, 1980). In these cells the cytoplasm can be seen to move in a helical path around a large internal vacuole, suggesting that the movement of myosin along polar actin bundles could produce the observed directional streaming. *Mougeotia* contains one large chloroplast which rotates in response to variations in light intensity. This movement is inhibited by cytochalasin B (which breaks down actin filaments), but is not affected by colchicine, which inhibits microtubule formation (Klein *et al.*, 1980). Mechanically dissected *Chara* cells contain actin cables which can support the movement of organelles (Williamson and Toh, 1979) and myosin-coated beads (Scheetz and Spudich, 1983). Recent studies in another class of green algae, Chlorophyceae, have demonstrated that the clustering of chloroplasts at wounds following treatment by UV light is inhibited by cytochalasin B, implicating a role for actin filaments in this movement (La Claire, 1989; Menzel and Elsner-Menzel, 1989). Although much work has been done on actin in protists, in particular algae, most has concentrated on only a few genera from a few phyla within this very large heterogeneous assemblage of organisms.

Actin in Higher Plants

Analysis of actin protein function in higher plants is still rudimentary and is less well characterized than in algal systems. Unlike Characean algae, where *in vitro* analysis of actin function can be done relatively easily, satisfactory *in vitro* systems have not yet been obtained for higher plant cells. Preparation of plant cells for actin localization studies has been difficult. Many of the typical preparatory agents for electron microscopic visualization, such as osmium tetroxide, will, under some conditions, destroy or cause rearrangements of actin filaments (Maupin-Szamier and Pollard, 1978). Antibodies tagged with fluorescent molecules have often been used to visualize actin filaments using fluorescence microscopy (Pesacreta *et al.,* 1982; Clayton and Lloyd, 1985; Marc and Gunning, 1986; Schmit and Lambert, 1987; Seagull *et al.,* 1987), but the resulting staining patterns are usually diffuse and difficult to interpret (Staiger and Schliwa, 1987).

Under the transmission electron microscope, 5- to 7-nm filaments seen in *Vallisneria* cells which morphologically resembled F-actin were found to bind reversibly with HMM or myosin-S1 fragments (Seagull and Heath, 1979). Similar filaments isolated from *Amaryllis* also specifically bound to HMM with a 35-nm periodicity (Condeelis, 1974). Biochemical techniques have been used to isolate actin from stringbean (*Phaseolus*) root tips (Jackson and Doyle, 1977), wheat (*Triticum*) germ (Ilker *et al.,* 1979), soybean (*Glycine*) seedlings (Metcalf *et al.,* 1980, 1984), and tomato (*Lycopersion*) (Vahey and Scordilis, 1980). Recently an actinlike protein was found associated with the outer envelope of pea (*Pisum*) chloroplasts (McCurdy and Williamson, 1987). As in all other organisms studied, these actins from higher plants have solubility, ultrastructure, and electrophoretic mobility properties similar to those of animal actins.

Actin has been found associated with the endoplasmic reticulum of plant cells (Quader *et al.,* 1987) and, as in animal cells, there is some evidence that actin may play a role in mitosis and cell division (Clayton and Lloyd, 1985; Schmit and Lambert, 1987; Seagull *et al.,* 1987; Traas *et al.,* 1987) and with the growth of pollen tubes (Heslop-Harrison *et al.,* 1986; Pierson *et al.,* 1986) and root hairs (Herth *et al.,* 1972). In addition, actin may be involved in organelle movement (Klein *et al.,* 1980; McCurdy and Williamson, 1987) and cytoplasmic streaming (Pesacreta *et al.,* 1982; Lloyd, 1983). In actively streaming cells of plant vascular tissue, cellular filaments are oriented parallel to the direction of streaming (Parthasarathy *et al.,* 1985). This streaming can be disrupted by cytochalasins or phalloidins (Wessels *et al.,* 1971; Bradley, 1973; Palevitz, 1980; Seagull and Heath, 1979; Parthasarathy *et al.,* 1985), as can chloroplast migration in several plant species (Staiger and Schliwa, 1987), implicating a role for actin filaments in these processes. In pollen tubes of tobacco and lily, microtubules and actin were found to colocalize (Pierson *et al.,* 1989), suggesting that, as in other cells, actin functions in cytoplasmic and cell surface movements in plants.

As in animal cells, the role of actin filaments in karyokinesis and cell division in plants is somewhat controversial. However, nuclear migration after mitosis appears to require intact actin filaments, since this movement is inhibited by treatments with either phalloidins or cytochalasins in *Allium* guard cells (Palevitz, 1980). Also, a recent study in endosperm cells showed that during cytokinesis actin colocalized with microtubules and appeared to contribute to cell plate formation (Schmit and Lambert, 1988).

ANALYSIS AND EVOLUTIONARY RELATIONSHIPS OF ACTIN GENE AND PROTEIN SEQUENCES

Sequence Analysis of Actin Genes

Sequence analysis of 24 actin genes from all four eukaryotic kingdoms, protists, fungi, plants, and animals (Whittaker, 1969), was done to determine the evolutionary relationships between these genes. Since, in most organisms, actin is encoded by a multigene family resulting in proteins with different functions, it is important to understand the function of each actin. It would be more meaningful in an evolutionary analysis of actins if the actin genes coding for proteins that perform the same function in different organisms could be compared. At this time there is not sufficient understanding of the differential expression of actin genes in many organisms or enough sequence data available for a comprehensive analysis. Therefore, what we present are comparisons between mostly paralogous, not orthologous genes.

Deduced amino acid sequences were compared using the Beckman Microgenie™ program (Queen and Korn, 1984), which produces an optimal alignment of two sequences to determine matches/length. This analysis counts similar amino acids as a partial match but does not correct for multiple hits. Pairwise comparisons of coding-region nucleotide sequences was done to determine the percentage of replacement substitution using the program of Li *et al.* (1985). This analysis takes into account nonrandom patterns of nucleotide substitution and considers all possible paths between two codons.

Results of the pairwise sequence analyses are presented in Table II. In general, similar relationships among the actin sequences were observed for the comparisons made at either the amino acid level, for percentage of similarity, or at the nucleotide level, to determine percentage replacement substitution. The range in percentage replacement substitution observed between the genes from different groups of organisms is displayed graphically in Fig. 1. Results of the pairwise comparisons were used to construct gene phenograms with the SAHN clustering program in the NTSYS-PC package (Rohlf, 1989). The phenogram generated based upon the percent similarity is shown in Fig. 2a. One of the four alternative trees constructed based upon the percentage replacement substitution is shown in Fig. 2b. The goodness of fit for the trees was determined by first calculating a cophenetic value matrix using the COPH program in the NTSYS-PC package. This value was compared to the original tree matrix generated by the SAHN program. Each of the cluster analyses generated represents a good fit based on this comparison.

Actin genes within the animal kingdom, as a group, display the highest degree of intrakingdom sequence conservation (0.7–6.4% replacement sub-

TABLE II. Coding-Region Amino Acid Sequence Comparisons (Percent Similarity)[a] and Percentage Replacement Substitution[b] in Pairwise Comparisons between Actin Genes[c]

	HuCy	HuCa	ChCy	ChM	Xeno	Dros	SeUr	RAc1	RAc2	RAc3	RAc7	MAc1	SAc1	SAc3	PAc1	AAc1	Dict	Phys	Tetr	Tryp	Ther	Yeas	Aspe
HuCy	—	93.9	98.9	92.8	93.9	95.2	94.7	88.0	84.4	84.9	83.6	85.1	81.2	85.1	82.6	88.8	94.7	94.1	76.8	71.8	88.2	88.5	88.5
HuCa	4.6	—	93.9	98.4	99.5	92.8	91.8	86.7	83.6	84.4	82.0	83.0	79.6	84.4	82.6	88.0	91.5	90.4	74.5	71.6	84.8	86.7	85.9
ChCy	1.4	5.3	—	92.3	93.9	95.5	93.9	88.0	84.7	85.1	83.6	85.1	81.2	85.1	82.0	88.8	94.1	93.6	76.3	71.8	88.3	88.5	88.5
ChM	5.3	1.0	5.7	—	97.9	91.8	90.7	85.6	82.3	83.2	81.5	82.2	78.5	83.0	81.4	87.0	90.2	89.4	74.5	71.1	84.6	86.1	85.6
Xeno	5.1	0.7	5.5	1.8	—	92.8	91.8	86.7	83.6	84.4	82.0	83.0	79.6	84.4	82.6	88.0	91.5	90.4	74.5	71.6	84.8	86.7	85.9
Dros	4.4	5.7	4.8	6.4	6.4	—	93.1	86.4	83.1	83.3	82.5	83.5	79.8	83.5	79.5	87.5	92.3	91.7	74.2	70.7	86.4	87.2	86.7
SeUr	4.4	5.6	4.6	6.1	5.5	5.8	—	86.7	83.1	82.5	82.5	83.5	79.6	84.0	80.7	87.8	93.3	91.4	75.2	71.2	86.4	87.4	85.9
RAc1	9.8	10.9	10.6	11.7	11.1	12.9	10.7	—	90.5	89.1	86.0	87.0	87.3	89.6	87.6	93.9	89.1	88.3	74.7	70.3	80.6	82.2	80.9
RAc2	13.7	14.3	14.2	15.1	14.7	16.0	14.8	5.7	—	87.4	84.8	86.8	83.7	87.9	88.3	90.2	85.5	85.0	73.1	68.2	78.1	79.1	78.6
RAc3	13.6	14.6	13.8	14.9	14.5	16.2	14.4	6.5	8.4	—	83.4	85.4	82.8	86.5	85.2	90.7	85.7	85.1	73.2	67.4	79.0	79.3	78.8
RAc7	17.8	18.9	18.5	19.1	19.2	19.8	18.7	11.4	12.1	13.3	—	87.0	80.7	83.9	84.5	87.0	84.4	84.4	71.1	66.1	77.7	79.3	78.5
MAc1	12.9	14.4	13.4	14.6	15.0	15.4	14.0	7.2	8.1	8.7	10.6	—	82.8	86.2	84.5	89.1	85.9	86.1	72.3	69.1	77.3	79.2	78.1
SAc1	12.5	14.5	13.3	15.2	14.9	15.5	14.0	7.3	10.1	10.8	16.3	10.3	—	85.1	85.1	88.1	82.0	81.4	69.5	65.6	75.3	76.4	75.6
SAc3	11.9	13.7	12.5	13.9	13.8	14.9	12.7	6.0	8.2	7.8	13.3	7.7	8.7	—	90.7	89.9	85.9	85.4	72.7	69.2	78.7	80.3	80.1
PAc1	15.4	16.8	15.9	16.8	16.8	19.1	16.9	6.7	7.8	9.6	20.4	9.0	8.9	5.9	—	90.7	83.9	84.5	76.4	65.4	77.6	77.5	78.3
AAc1	9.7	10.4	10.4	11.0	10.5	12.2	10.4	3.6	7.2	6.9	12.3	7.2	7.0	6.5	7.2	—	89.6	89.1	76.1	70.8	82.4	83.5	82.4
Dict	6.2	7.5	6.6	8.6	8.1	7.0	7.4	10.6	14.1	13.8	18.5	12.8	13.1	12.9	16.5	10.2	—	97.6	76.3	70.7	87.2	88.0	86.7
Phys	4.7	6.9	5.4	7.7	7.6	5.6	5.0	9.3	12.7	12.6	16.6	11.9	12.0	11.2	14.9	9.5	2.9	—	76.3	71.2	87.2	87.2	86.7
Tetr	18.5	20.4	18.4	20.7	20.2	20.0	18.8	21.1	23.3	21.5	26.6	23.2	23.9	23.0	22.2	20.2	18.5	20.0	—	70.5	75.2	75.2	73.9
Tryp	22.6	23.2	23.2	23.4	22.7	22.1	22.8	24.8	27.1	26.9	29.8	26.9	27.2	26.0	31.0	26.1	23.6	23.1	25.2	—	70.5	72.6	71.0
Ther	8.9	11.9	9.9	11.6	12.5	9.2	10.6	16.0	19.5	19.0	23.0	19.2	18.7	17.5	21.8	14.5	10.3	9.0	19.7	23.6	—	89.6	94.1
Yeas	9.4	11.4	9.8	11.8	11.5	10.1	10.9	16.2	18.1	18.8	22.2	17.3	18.0	17.8	20.7	15.6	9.5	10.1	19.1	25.0	8.1	—	91.7
Aspe	8.9	11.3	10.0	11.1	11.6	9.3	10.2	15.7	18.9	19.4	23.1	18.4	18.2	16.8	21.4	14.2	10.7	9.1	21.2	23.7	3.4	8.3	—

[a] Comparisons of the deduced protein sequences (upper portion of matrix) using the Beckman Microgenie™ program to determine percent similarity based upon Dayhoff's algorithm (Dayhoff and Eck, 1978) as modified by Queen and Korn (1984). In determining percent similarity, the following pairs of amino acids are considered to be the same: Asp versus Glu, Asn versus Gln, Lys versus Arg, Leu versus Ile, Leu versus Val, and Ser versus Thr.

[b] The percentage replacement substitution (bottom portion of matrix) as determined by the program of Li *et al.* (1985).

[c] The actin gene sequences are abbreviated as indicated in Table I.

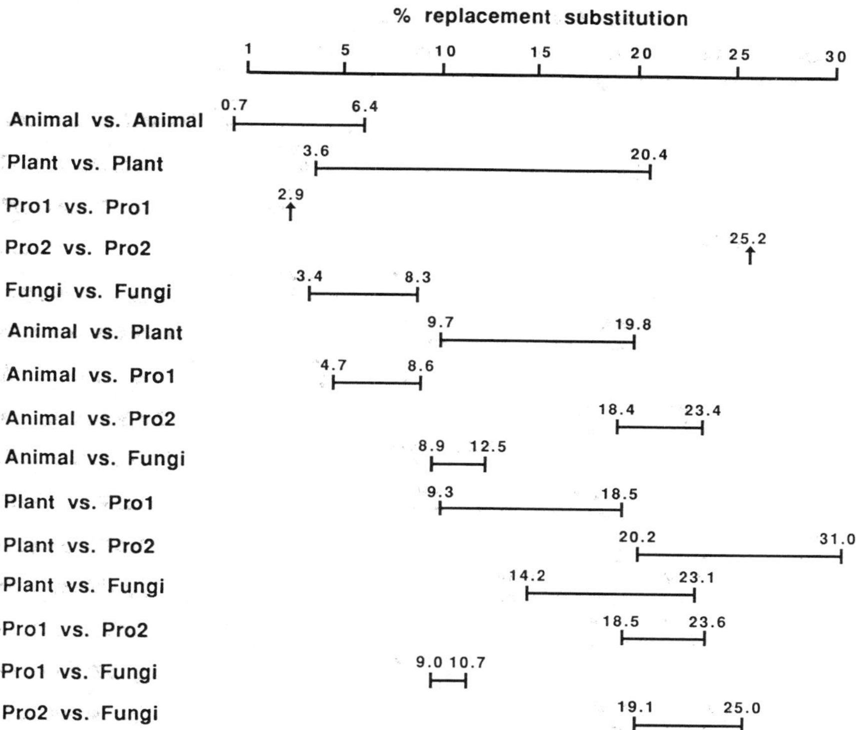

FIG. 1. Bar graph displaying the range of percentage replacement substitution between different groups of actin genes. Group 1 protists (Pro1) are represented by *Dictyostelium* and *Physarum*. Group 2 protists (Pro2) are represented by *Tetrahymena* and *Trypanosoma*. The fungi include *Thermomyces, Aspergillus,* and *Saccharomyces*.

stitution) (Fig. 1). Vertebrate sequences coding for muscle α-actin isoforms [represented by human cardiac (HuCa) and chicken muscle (ChM)] show 4.6–5.7% replacement substitution (Table II) when compared with vertebrate genes coding for cytoplasmic nonmuscle β-actin protein [represented by human cytoplasmic (HuCy) and chicken cytoplasmic (ChCy)]. The invertebrate actins, *Drosophila* (Dros) and sea urchin (SeUr), are somewhat more highly diverged from vertebrate α-actin genes (5.6–6.4% replacement substitution) than they are from vertebrate β-actin cytoplasmic gene sequences (4.4–4.8% replacement substitution). However, the largest replacement substitution amounts to only 6.4% (between Dros and ChM). As a rule, actin gene sequences of plants, protists, and fungi showed slightly more similarity

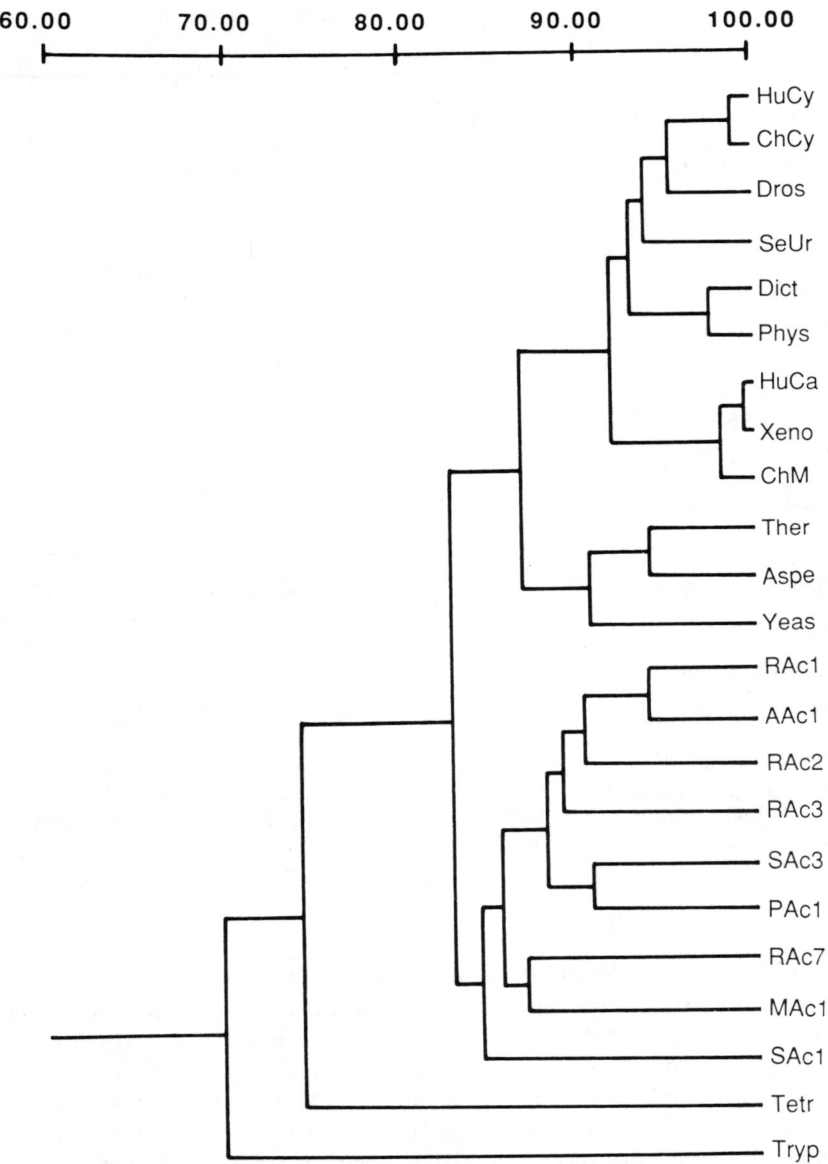

FIG. 2. Phenetic trees for actin genes. The SAHN (sequential, agglomerative, hierarchical, and nested) clustering program of the NTSYS (numerical taxonomy systematics) package was used to construct these unweighted pair-group, unrooted phenetic trees (Rohlf, 1989). (a) The scale on top correlates branch length with the percent similarity of the actin amino acid sequences.

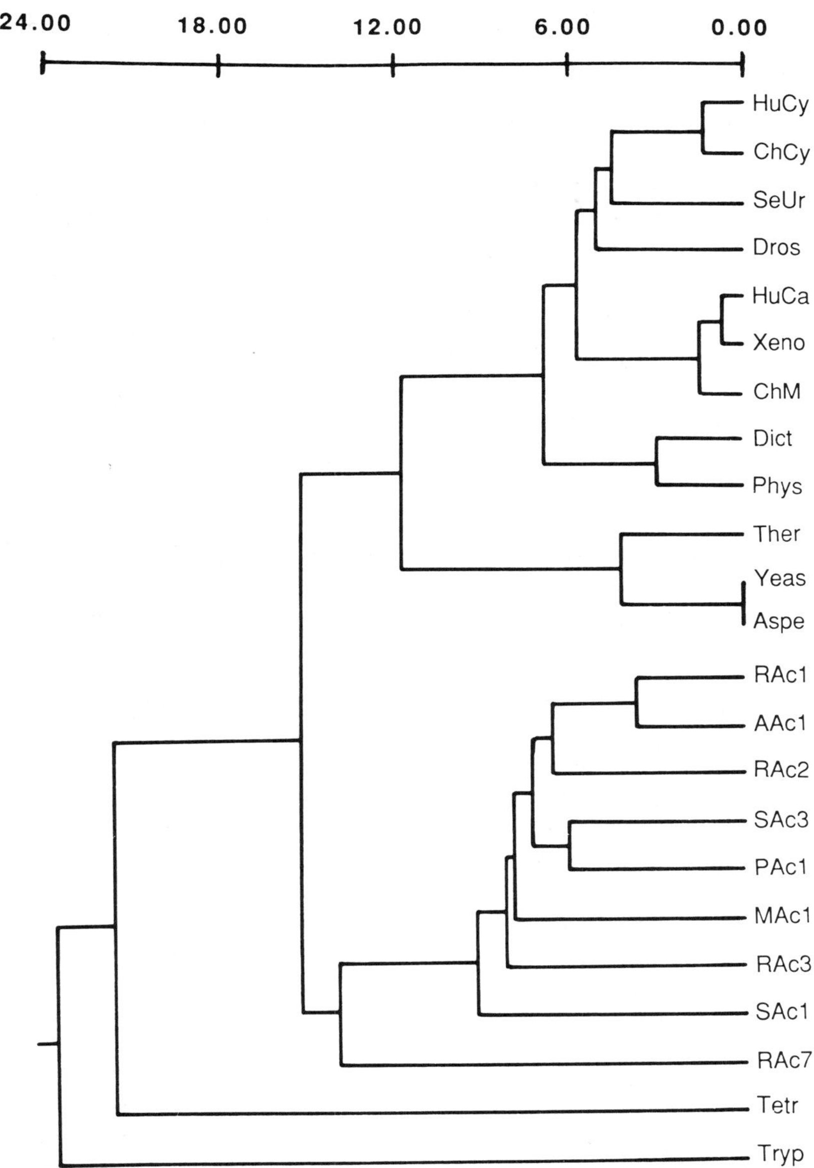

FIG. 2 (*Continued*). (b) The scale on top correlates branch length with percent replacement substitution in the nucleotide sequences of the coding region (data from Table II).

to those of vertebrate cytoplasmic actins (such as HuCy) than to muscle actins (such as HuCa).

As shown in Fig. 1, replacement substitution among plant actins generally ranged from 3.6 to 20.4%, indicating that, as a group, they are as diverged from each other as they are from animal actins (animal versus plant, 9.7–19.8%). Interestingly, the comparison between rice (a monocot) actin gene RAc1 and *Arabidopsis* (a dicot) actin gene AAc1 showed only 3.6% replacement substitution between these two sequences (Table II). This difference is less than that for any other plant versus plant comparison. In addition, these two plant actin sequences are more similar to each other than vertebrate nonmuscle and muscle actins are to each other. Both phenograms illustrate the similarity between RAc1 and AAc1, suggesting that these genes are coding for functionally homologous actins in these two species and may be orthologous. Also, the analyses suggest that soybean actin gene 3 (SAC3) and the pine actin gene (PAc1) are quite similar and might be orthologous (see Figs. 2a and b). The other plant actin genes are probably paralogous.

Actin gene sequences of the four protists showed great heterogeneity. Two of the protists (*Dictyostelium* and *Physarum*) showed a replacement substitution value of only 2.9%. These organisms are referred to as group 1 protists (Pro1) in Fig. 1. The other two protists (*Tetrahymena* and *Trypanosoma*) were highly diverged from all other organisms (18.4–31.0% replacement substitution) and from each other (25.2%). *Tetrahymena* and *Trypanosoma* are referred to as group 2 protists (Pro2) in Fig. 1 simply for the ease of data presentation, although this designation does not suggest a close relationship between the two organisms. All four of these organisms have been placed into the protist kingdom (Whittaker, 1969; Margulis and Schwartz, 1988), but these data lend strong support to the concept that the kingdom Protista is a taxonomic but not phylogenetic group. It is interesting to note that the extent of replacement substitution between animals and group 1 protists (4.7–8.6%) is as low as that observed within the fungi (3.4–8.3%), and much lower than that observed between animals and plants (9.7–19.8%). This result suggests that *Dictyostelium* and *Physarum* actin genes are much more closely related to those of animals than they are to either those of other protists or fungi.

Animal genes show a higher degree of similarity to group 1 protist (4.7–8.6%) and to fungal (8.9–12.5%) than to plant genes (9.7–19.8%). In addition, the fungal and group 1 protist genes are more similar to animal (4.7–18.5%) than they are to plant actin genes (9.3–23.1%). In general, actin genes of group 2 protists and plants show the greatest divergence. This suggest that the duplication and divergence of actin genes began earlier in plants than in animals and fungi. Hightower and Meagher (1986) determined that soybean actin genes are diverging at the same rate as other "nonplant" actin genes and determined that the time of radiance in plants was at least 330 million years ago. If the same rate of divergence is occurring in *Tetrahymena* and *Trypanosoma,* then we must assume that they diverged from each other and other organisms even earlier than the plant actin genes duplicated and diverged from each other. *Physarum* and *Dictyostelium* actin genes are closely related to each other and to animal actin genes, indicating a later time of divergence from animals and each other. Protists are largely defined by exclusion and placed into the kingdom Protista because they do not clearly belong in any of the other three eukaryotic kingdoms. This criterion for classification can lead to the unfortunate circumstance of grouping together organisms which are clearly less related to each other than they are to organisms in other kingdoms. The actin gene sequence data indicate that this may indeed be the case for *Dictyostelium* and *Physarum,* whose genes are more similar to those of animals. *Tetrahymena* and *Trypanosoma* cannot be easily associated with each other or with another kingdom based upon actin gene data. Further studies at the molecular level of these organisms presently classified into Protista may help elucidate the evolution of and relationships within this "catch-all" kingdom.

Conservation of Structural Domains within Actin Proteins

From studies of animal systems, it is known that actin has specific interaction sites within its sequence. Actin is known to have interaction sites for various actin-binding proteins, and for some nucleotides and cations as well as for the actin monomers themselves during actin polymerization. The binding proteins are known to regulate actin polymerization as well as actin interaction with other components of the cell cytoskeleton (Korn, 1982; Pollard and Cooper, 1986). It is thought that animal actin-binding proteins, which include severin, vinculin, gelsolin, fragmin, and villin (Fig. 3a), contain amino acid sequences homologous to actin itself and act to regulate actin structure and function by binding to sites of actin–actin interaction (Tellam *et al.,* 1989). For example, actin "capping" proteins replace actin monomers and prevent the addition of new monomers.

As mentioned previously, the precise functions of plant actin isoforms have yet to be determined. However, it is believed that these isoforms play different roles in the plant cell. Consequently plant actins should also contain a number of protein–protein interaction sites whose structural requirements would act to constrain sequence divergence of functionally equivalent plant actin isoforms. Plant actins do contain stretches of amino acids which are conserved both within plants and between plant and animal actin proteins (Fig. 3b) despite a high degree of intra- and interspecific divergence.

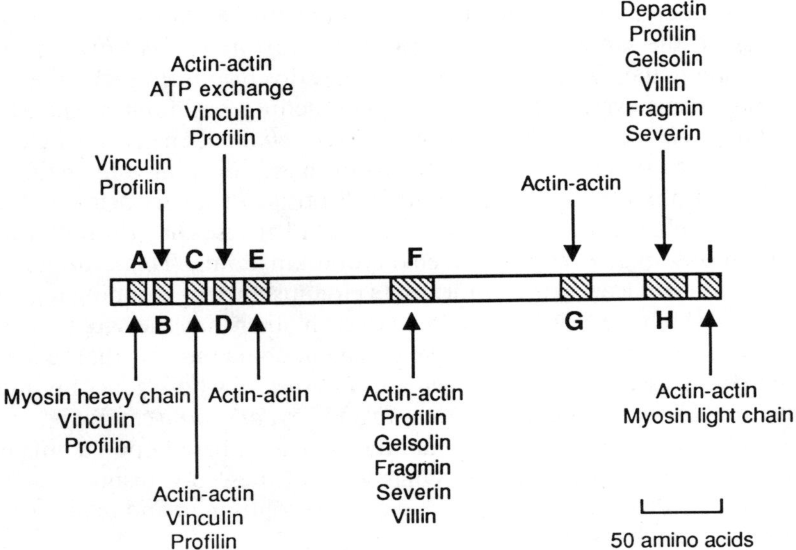

FIG. 3. Comparisons between plant and animal actin proteins. (a) Graphic representation indicating the position of conserved amino acid regions (striped boxes) within plant and animal actin proteins. The sites of actin–actin and actin–actin-binding-protein interactions inferred from studies of animal systems are indicated by vertical arrows. The equivalent regions in a and b are indicated by letters.

These conserved regions within plant actin proteins often coincide with sites which are believed to function in actin–actin and actin–actin-binding-protein interactions in animal systems (Fig. 3a). Therefore, it is likely that many of the actin interactions found in animal cells will also be found in plants, and that plants will have equivalents to the actin-binding proteins found in animals. Furthermore, these actin interaction sites should act to restrain the extent of sequence divergence observed within plant actin genes, leading to a saturation of nonsynonymous nucleotide substitution within such conserved regions.

Analysis of the Amino-Terminal Region of Actin Proteins

Vertebrate α actins are quite different from β and γ actins in the 17–18 amino acids of their respective amino-terminal regions. In fact, these amino-terminal sequences have been used to differentiate between different vertebrate actin isotypes. This region was also used to determine that both inver-

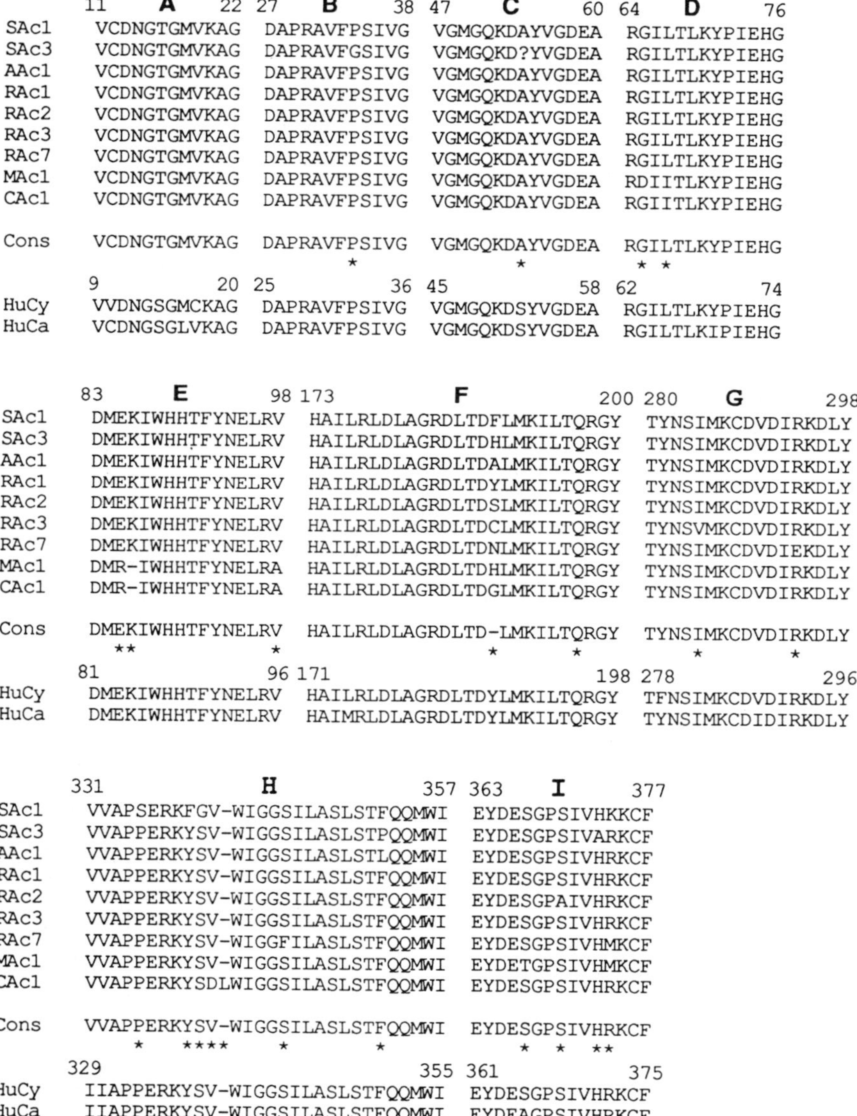

FIG. 3. (*Continued*). (b) Regions of amino acid sequence similarity between plant and animal actin proteins. The human β-cytoplasmic (HuCy) and α-cardiac (HuCa) actin sequences are shown as representative animal nonmuscle and muscle actin protein isoforms, respectively. Plant actin sequence abbreviations are as in Table I. A consensus sequence (Cons*) is indicated below each conserved plant actin region.

TABLE III. The Percentage Similarity[a] Observed in Pairwise Comparisons between the Amino-Terminal 18–20[b] Amino Acids in a Number of Actin Proteins[c]

	18 HuCy	20 HuCa	19 ChCy	20 ChM	20 Xeno	19 Dros	19 SeUr	20 RAc1	20 RAc2	20 RAc3	19 RAc7	19 MAc1	20 SAc1	20 SAc3	20 AAc1	19 Dict	19 Phys	19 Tetr	19 Tryp	18 Ther	18 Yeas	18 Aspe
HuCy	—	60.0	84.2	60.0	60.0	73.7	78.9	60.0	60.0	55.0	47.4	57.9	60.0	60.0	60.0	73.7	57.9	68.4	52.6	72.7	77.8	72.2
HuCa		—	65.0	90.0	100	60.0	70.0	60.0	60.0	60.0	45.0	55.0	60.0	60.0	60.0	60.0	40.0	45.0	60.0	55.0	60.0	55.0
ChCy			—	55.0	65.0	84.2	68.4	65.0	65.0	65.0	52.6	63.2	65.0	65.0	65.0	68.4	52.6	63.2	57.9	78.9	78.9	78.9
ChM				—	90.0	60.0	70.0	55.0	55.0	55.0	50.0	60.0	55.0	55.0	55.0	55.0	40.0	50.0	55.0	55.0	55.0	55.0
Xeno					—	60.0	70.0	60.0	60.0	60.0	45.0	55.0	60.0	60.0	60.0	60.0	40.0	45.0	60.0	55.0	60.0	55.0
Dros						—	68.4	55.0	55.0	55.0	42.1	52.6	55.0	55.0	55.0	68.4	52.6	63.2	57.9	73.7	73.7	73.7
SeUr							—	60.0	60.0	55.0	47.4	57.9	60.0	60.0	60.0	73.7	63.2	68.4	57.9	73.7	78.9	73.7
RAc1								—	100	90.0	80.0	90.0	100	95.0	95.0	65.0	45.0	45.0	55.0	50.0	55.0	50.0
RAc2									—	90.0	80.0	90.0	100	95.0	95.0	65.0	45.0	45.0	55.0	50.0	55.0	50.0
RAc3										—	70.0	80.0	90.0	85.0	95.0	65.0	45.0	45.0	55.0	50.0	55.0	50.0
RAc7											—	84.2	80.0	75.0	80.0	52.6	42.1	42.1	42.1	47.4	36.8	47.4
MAc1												—	90.0	85.0	90.0	57.9	52.6	57.9	57.9	47.4	47.4	47.4
SAc1													—	95.0	95.0	65.0	45.0	45.0	55.0	50.0	55.0	50.0
SAc3														—	90.0	60.0	40.0	45.0	55.0	50.0	55.0	50.0
AAc1															—	70.0	50.0	45.0	55.0	50.0	55.0	50.0
Dict																—	78.9	68.4	47.4	78.9	84.2	78.9
Phys																	—	68.4	52.6	68.4	63.2	68.4
Tetr																		—	57.9	63.2	73.7	63.2
Tryp																			—	52.6	57.9	52.6
Ther																				—	88.9	100
Yeas																					—	88.9

[a] Percentage similarity determined using Beckman's Microgenie™ program based upon Dayhoff's algorithm (Dayhoff and Eck, 1978).
[b] Alignment of the sequences accounted for gaps in some relative to other sequences.
[c] The actin gene sequences are abbreviated as indicated in Table I.

tebrate muscle and nonmuscle actins were more similar to vertebrate cytoplasmic than to vertebrate muscle actin proteins (Vandekerckhove and Weber, 1984). The amino-terminal 18–20 residues from all of the actin sequences were aligned and the percent similarity determined (Table III). In both the fungal and plant kingdoms actins show a relatively high degree of conservation in their respective amino-terminal regions. In general, the plant actins show 80.0–100% amino acid similarity to each other and the fungal sequences display 88.9–100% amino acid similarity in the amino-terminal 18–20 residues. These similarities are in contrast to the situation in animals, where the nonmuscle and muscle actin proteins show only 55.0–65.0% similarity in their respective amino-terminal regions, yet show a comparatively high degree of similarity when the entire sequences of their respective proteins are compared (90.7–98.9%).

Protist actins show relatively little conservation of the amino-terminal region when compared to each other or other organisms (Table III). Interestingly, the *Physarum* actin protein, which is quite similar to animal actins when their entire amino acid sequences are compared, shows little similarity (40.0–63.2%) to animal actins in its amino-terminal region. The conservation of the amino-terminal region within the plants and within the fungi may indicate that this region constitutes a domain of the protein which is important for actin function and/or regulation within each of these kingdoms.

Analysis of amino-terminal amino acids of animal actins has identified some amino acids which are characteristically located at specific positions in vertebrate muscle-actin proteins and other residues which are consistently found in vertebrate cytoplasmic (Vandekerckhove and Weber, 1978) actin sequences. By determining the amino acids located at these characteristic positions, invertebrate (sea urchin, scallop, and *Drosophila*) actin isoforms were found to resemble vertebrate cytoplasmic actin isoforms (Vandekerckhove and Weber, 1984). Vertebrate muscle actins generally begin with four acidic residues, while vertebrate cytoplasmic and invertebrate actin isoforms start with three acidic amino acids (data not shown). Alignment of a number of actin protein sequences (Table IV) shows that Ile and Ala are often found at positions 5 and 6, respectively, in vertebrate cytoplasmic and invertebrate proteins, while vertebrate muscle actins have Thr residues at these positions. Position 10 is occupied by a Cys residue and position 17 by a Val in muscle actin isoforms, while cytoplasmic actins have a Val (or Ile) at position 10 and a Cys at position 17. Fungal actins also resemble invertebrate and vertebrate cytoplasmic actins at these amino acid positions. Plant actins have amino acids at these positions which are characteristic of both the cytoplasmic and muscle isoforms. They have an Ile at position 5, as do the vertebrate cytoplasmic isoforms. Position 6 is usually occupied by a Gln in plant actins (an exception is in SAc1, where it is Glu), which is not seen in animal or fungal

TABLE IV. NH_2-Terminal Amino Acid Residues Found to Be Characteristic of Genes Encoding Actin Proteins of Particular Isotypes and/or Groups of Organisms

Organism and gene	Residue at given amino acid position number			
	5	6	10	17
Human cardiac α	Thr	Thr	Cys	Val
Human skeletal α	Thr	Thr	Cys	Val
Xenopus cardiac α	Thr	Thr	Cys	Val
Rabbit skeletal α	*Asp*	Thr	Cys	Val
Human cytoplasmic β	Ile	Ala	Val	Cys
Bovine cytoplasmic β	Ile	Ala	Val	Cys
Human cytoplasmic γ	Ile	Ala	Ile	Cys
Bovine cytoplasmic γ	Ile	Ala	Ile	Cys
Sea urchin	Val	Ala	Ile	Cys
Scallop	Val	Ala	Val	Cys
Drosophila 79B	*Glu*	Ala	Val	Cys
Rice 1	Ile	Gln	Cys	Val
Rice 2	Ile	Gln	Cys	Val
Rice 3	Ile	Gln	Cys	Val
Rice 7	Ile	Gln	Cys	Val
Maize 1	Ile	Gln	Cys	Val
Soybean 1	Ile	*Glu*	Cys	Val
Soybean 3	Ile	Gln	Cys	Val
Arabidopsis 1	Ile	Gln	Cys	Val
Dictyostelium	Val	Gln	Ile	Cys
Physarum	Val	Gln	Ile	Cys
Tetrahymena	*Ser*	*Pro*	Ile	Cys
Trypanosoma	*Gln*	Thr	Cys	Val
Aspergillus	Val	Ala	Ile	Cys
Saccharomyces	Val	Ala	Ile	Cys
Thermomyces	Val	Ala	Ile	Cys

proteins. As with vertebrate muscle actins, Cys and Val are found at positions 10 and 17 in the plant actins. Protist actins do not show any consistency of the amino acids occupying these positions. The hypothesis that plant, fungal, and protist actins are "cytoplasmic" in nature is further supported by the observation that (with the exception of the maize actin 1 gene) they contain a DESG motif (Fig. 3b) between residues 365 and 368 of their respective COOH-terminal sequences as do vertebrate nonmuscle actins, while a DEAG motif is associated with muscle actins. These findings call into question the practice of using specific amino-terminal residues in the characterization of actin isoforms.

CONSERVATION OF INTRON POSITIONS IN EUKARYOTIC GENES

Conservation of Intron Positions in Actin Genes

All of the eight plant actin genomic sequences published have three coding-region introns at precisely the same locations (Table V). This intron conservation is especially interesting, since these plants include two dicots (soybean and *Arabidopsis*) and two monocots (maize and rice) which are believed to have diverged approximately 140–200 million years ago (Baba *et al.*, 1981). Shah *et al.* (1983) have stated that within the animal kingdom the intron positions of actin genes vary, while in plants the positions are conserved. There are seven introns in the human aortic and chicken α-smooth genes, five in the mouse and *Xenopus laevis* α-actin genes, and only four in the human and *Xenopus borealis* β-actin genes. The positions of these introns, when present, are well conserved (Table V). More variations in intron number and location occurs among invertebrate actin genes. The small number of protists and fungal actin genes examined did not allow productive comment on the nature of actin gene introns in these two groups. However, it is important that three of the four protists examined had no introns, while two of the fungi, *Aspergillus* and *Thermomyces,* had identical intron numbers and locations in their respective actin genes, only two of which were in a similar position to those of other organisms.

As discussed earlier, plant actin genes show less conservation in nucleotide sequence than animal actin genes, suggesting that the ancestral plant actin gene duplicated and began diverging at an earlier time than the ancestral animal actin gene. Yet during the evolution of animal actin genes, which has presumably been for a shorter time than plant actin gene evolution, the number of introns and their positions have varied. This observation leads to the question of whether plant actin gene introns are highly conserved because the mechanism for intron insertion and/or deletion is no longer functional or less active in plants, or because there are functional constraints which prevent the removal of actin gene introns.

Conservation of Intron Positions in Nonactin Genes

To determine whether the observed conservation of coding-region intron positions was a feature common to plant genes in general or unique to plant actin genes alone, the structure of four other homologous nuclear genes (those for α tubulin, β tubulin, alcohol dehydrogenase, and globin and its

TABLE V. Intron Positions of Five Nuclear Genes[a]: Actin, α Tubulin, β Tubulin, Globin (leghemoglobin), and Alcohol Dehydrogenase

Organism	Gene	Number of introns	Intron positions	Reference
Intron position in actin genes				
Human	Aortic	7	41, 84, 121, 150, 204, 267, 327	Ueyama *et al.* (1984)
Chicken	α smooth	7	41, 84, 121, 150, 204, 267, 327	Carroll *et al.* (1986)
Mouse	α	5	41, 150, 204, 267, 327	Hu *et al.* (1986)
Chicken	α skeletal	5	41, 150, 204, 267, 327	Fornwald *et al.* (1982)
X. laevis	α	5	41, 150, 204, 267, 327	Mohun *et al.* (1986)
Human	β	4	41, 121, 267, 327	Ng *et al.* (1985)
X. borealis	β	4	41, 121, 267, 327	Cross *et al.* (1988)
S. purpuratus	J	4	41, 121, 204, 267	Cooper and Crain (1982)
S. purpuratus	15B	2	121, 204	Schuler *et al.* (1983)
S. franciscanus	15A	2	121, 204	Foran *et al.* (1985)
C. elegans	IV	2	19, 322	Files *et al.* (1983)
C. elegans	I	1	64	Files *et al.* (1983)
C. elegans	II	1	64	Files *et al.* (1983)
C. elegans	III	1	64	Files *et al.* (1983)
Drosophila	57A	1	13	Fyrberg *et al.* (1981)
Drosophila	79B	1	307	Sanchez *et al.* (1983)
Drosophila	5C	0		Sanchez *et al.* (1983)
Soybean	SAc3	3	19, 150, 355	Shah *et al.* (1982)
Soybean	SAc1	3	19, 150, 355	Shah *et al.* (1983)
Maize	1	3	19, 150, 355	Shah *et al.* (1983)
Arabidopsis		3	19, 150, 355	Nairn *et al.* (1988)
Rice	RAc1	3	19, 150, 355	Reece *et al.* (1990)

Rice	RAc2	3	19, 150, 355	Reece *et al.* (1990)
Rice	RAc3	3	19, 150, 355	Reece *et al.* (1990)
Rice	RAc7	3	19, 150, 355	Reece *et al.* (1990)
Physarum		4	26, 94 214, 322	Gonzalez-y-Merchand and Cox (1988)
Dictyostelium		0		Romans and Firtel (1985)
Trypanosoma		0		Ben Amar *et al.* (1988)
Tetrahymena		0		Hirono *et al.* (1987)
Aspergillus		5	3, 13, 32, 41, 299	Fidel *et al.* (1988)
Thermomyces		5	3, 13, 32, 41, 299	Wildeman (1988)
S. cerevisiae		1	4	Galliwitz and Sures (1980)
Intron positions in globin and leghemoglobin genes				
Human	α	2	30, 104	Orkin (1978)
Orangutan	α	2	30, 104	Marks *et al.* (1986)
Equine	α	2	30, 104	Clegg *et al.* (1984)
Goat	α	2	30, 104	Schon *et al.* (1982)
Mouse	α	2	30, 104	Nishioka and Leder (1979)
Chicken	α	2	30, 104	Dodgson and Engel (1983)
Duck	α	2	30, 104	Erbil and Niessing (1982)
Xenopus	α	2	30, 104	Patient *et al.* (1980)
Human	β	2	30, 104	Poncz *et al.* (1983)
Goat	β	2	30, 104	Shapiro *et al.* (1983)
Rabbit	β	2	30, 104	Hardison *et al.* (1979)
Mouse	β	2	30, 104	Nishioka and Leder (1979)
Chicken	β	2	30, 104	Dolan *et al.* (1983)
Xenopus	β	2	30, 104	Meyerof *et al.* (1984)

(continued)

TABLE V. Continued

Organism	Gene	Number of introns	Intron positions	Reference
Soybean	1	3	30, 68, 104	Wiborg *et al.* (1982)
Soybean	2	3	30, 68, 104	Wiborg *et al.* (1982)
Kidney bean		3	30, 68, 104	J. S. Lee and Verma (1984)
Parasponia (leghemoglobinlike sequence)		3	30, 68, 104	Landsmann *et al.* (1986)
Intron positions in α-tubulin genes				
Human	α	3	1, 76, 125	Hall and Cowan (1985)
Rat	α	3	1, 76, 125	Lemischka and Sharp (1982)
Chicken	cα3	3	1, 76, 125	Pratt and Cleveland (1988)
Chicken	cα4	4	1, 76, 125, 352	Pratt and Cleveland (1988)
Chicken	cα8	4	1, 76, 125, 352	Pratt and Cleveland (1988)
Chicken	cα5/6	3	1, 76, 125	Pratt and Cleveland (1988)
Drosophila	α1	1	1	Theurkauf *et al.* (1986)
Drosophila	α2	2	1, 176	Theurkauf *et al.* (1986)
Drosophila	α3	0		Theurkauf *et al.* (1986)
Drosophila	α4	1	1	Theurkauf *et al.* (1986)
Arabidopsis	α1	4	38, 109, 176, 346	Ludwig *et al.* (1988)
Arabidopsis	α3	4	38, 109, 176, 346	Ludwig *et al.* (1987)
Volvox	α1	3	15, 90, 211	Mages *et al.* (1988)
Chlamydomonas	α1	2	15, 90	Brunke *et al.* (1984)
Chlamydomonas	α2	2	15, 90	Brunke *et al.* (1984)

Physarum	α	7	4, 40, 90, 176, 256, 326, 412	Monteiro and Cox (1987)
S. lemnae	α	0		Helftenbein (1985)
Trypanosoma	α	0		Kimmel *et al.* (1985)
S. pombe	$\alpha 1$	1	19	Toda *et al.* (1984)
S. pombe	$\alpha 2$	0		Toda *et al.* (1984)
S. cerevisiae	$\alpha 1$	1	9	Schatz *et al.* (1986)
S. cerevisiae	$\alpha 3$	1	9	Schatz *et al.* (1986)
Intron positions in β-tubulin genes				
Human	β	3	19, 56, 93	Lewis *et al.* (1985)
Chicken	$\beta 2$	3	19, 56, 93	Sullivan *et al.* (1986*b*)
Chicken	$\beta 3$	3	19, 56, 93	Sullivan *et al.* (1986*a*)
Chicken	$\beta 5$	3	19, 56, 93	Sullivan *et al.* (1986*a*
Drosophila	$\beta 2$	3	19, 56, 131	Rudolph *et al.* (1987)
Drosophila	$\beta 3$	3	19, 56, 131	Rudolph *et al.* (1987)
Arabidopsis	$\beta 1$	2	131, 222	Oppenheimer *et al.* (1988)
Arabidopsis	$\beta 4$	2	131, 222	Marks *et al.* (1987)
Arabidopsis	$\beta 5$	2	131, 222	Silflow *et al.* (1987)
Maize	$\beta 10$	1	131	Silflow *et al.* (1987)
Chlamydomonas	$\beta 1$	3	8, 56, 131	Youngblom *et al.* (1984)
Chlamydomonas	$\beta 2$	3	8, 56, 131	Youngblom *et al.* (1984)
Tetrahymena	β	0		May *et al.* (1987)
Aspergillus	βA	8	4, 12, 19, 35, 56, 205, 317, 437	May *et al.* (1987)
Aspergillus	βC	5	4, 12, 19, 35, 317	May *et al.* (1987)

(continued)

TABLE V. Continued

Organism	Gene	Number of introns	Intron positions	Reference
Neurospora	β	6	4, 12, 19, 35, 56, 317	Orbach *et al.* (1986)
S. pombe	β	5	4, 19, 35, 56, 349	Toda *et al.* (1984)
S. cerevisiae	β	0		Neff *et al.* (1983)
Intron positions in alcohol dehydrogenase genes				
Human		8	5, 39, 86, 115, 188, 275, 321, 367	Duester *et al.* (1986)
Drosophila		2	32, 167	Benyajati *et al.* (1981)
Maize	1	9	11, 56, 71, 179, 206, 232, 252, 284, 338	Dennis *et al.* (1984)
Maize	2	9	Same as maize 1	Dennis *et al.* (1985)
Rice		9	Same as maize 1	Xie and Wu (1989)
Pea	1	9	Same as maize 1	Llewellyn *et al.* (1987)
Barley	2	8	11, 56, 71, 179, 206, 232, 252, 284	Trick *et al.* (1988)
Barley	3	8	Same as barley 1	Trick *et al.* (1988)
Arabidopsis		6	11, 56, 71, 232, 284, 338	Chang and Meyerowitz (1986)
Aspergillus	A	2	252, 367	Gwynne *et al.* (1987)
S. cerevisiae	2	0		Russell *et al.* (1983)
S. cerevisiae	1	0		Bennetzen and Hall (1982)

[a] Intron positions of the five nuclear genes were compared within and between kingdoms. The positions are given relative to the encoded amino acid position in the protein. In some cases alignment was necessary due to deletions or insertions and the position numbers are given relative to the homologous amino acid in the human sequence.

plant relative leghemoglobin) from several different organisms was examined (see Table V).

Animal globin and plant leghemoglobin genes show conservation of both intron number and position within each kingdom (Table V). Among animals, all 14 globin genes (α and β) have two introns within their coding regions; one intron is located at position 30, and the other at position 104. All four plant leghemoglobin genes examined have three introns which are located at identical positions, two of which are common to the animal intron positions.

Intron number but not location of α-tubulin genes varies within the animal kingdom (Table V). Intron number ranges from zero to four among the six animal genes examined, but these introns, when present, all occur at conserved positions. α-Tubulin gene data was only available for one plant species, *Arabidopsis.* Both *Arabidopsis* genes sequenced have four introns at the same positions. The number of α-tubulin introns and their locations vary among the three protists examined (*Physarum, Stylonychia lemnae,* and *Trypanosoma*), and among the four yeast α-tubulin genes as well as the two algae.

The number of introns and their locations within β-tubulin genes is relatively well conserved within the animal and plant kingdoms (Table V). The number of β-tubulin introns, however, is not well conserved within the protists or fungi, although some positions are conserved both within and between kingdoms. However, intron position is highly conserved among the six plant alcohol dehydrogenase genes examined, with the number of introns ranging from six in *Arabidopsis* to nine in maize, rice, and pea.

In summary, intron numbers and locations for five genes have been analyzed among organisms in the four eukaryotic kingdoms. In general, the intron number and location are highly conserved among plants, moderately well conserved among animals, and not well conserved among fungi and among protists. The most interesting observation here is that, although the nucleotide sequences of the plant actin genes are less conserved than those of the animal or fungal genes, intron number and location are more highly conserved. This suggests that intron insertion and/or deletion is less frequent in relation to the rate of sequence divergence for plants than for fungi or animals. The analysis, however, is based on a small number of genes with relatively few representatives of each kingdom. The small sample size warrants caution in making conclusions.

CONCLUSIONS

Actin function needs to be studied further, especially in plants, fungi, and protists, so that the effect of functional constraints on the evolution of the actin gene sequences can be better determined. Based upon the sequence analyses presented here, as expected, the actin sequences are highly conserved within the animal kingdom, while the plant actin genes are quite divergent. The ancestral plant actin gene probably duplicated and began diverging before the ancestral animal actin gene. Fungal actins show a higher degree of intrakingdom conservation than plant actins and are more closely related to animal than to plant actin genes. Actin genes from the protists *Physarum* and *Dictyostelium* are closely related to the animal actin genes, while *Trypanosoma* and *Tetrahymena* genes are very highly diverged from each other and other organisms. Ribosomal RNA genes have also been found to show a high degree of sequence diversity within the kingdom Protista (Sogin *et al.,* 1986; Cedergren *et al.,* 1988). With the continued collection of molecular data it is becoming increasingly evident that the protists are a somewhat haphazard collection of organisms which display a much higher level of genetic diversity than is observed within the other kingdoms. The Protista is likely to be continually reorganized as more molecular and morphological data become available.

We see a clear pattern of coding-sequence similarity of actin genes among different organisms in the four eukaryotic kingdoms as evidenced by the conservation of certain domains. This conservation is likely to be related to the functions of actin, many of which are conserved across all four eukaryotic kingdoms. In animals, however, the amino-terminal region of the protein is relatively highly diverged between the actin isoforms. This divergence may be related to functional differences between the muscle and cytoplasmic forms of the protein. The amino-terminal region of plant and fungal actins is well conserved, suggesting that this region may have an important and possibly unique function in these organisms.

Finally, this preliminary analysis of intron positions indicates that plant genes have highly conserved intron positions, relative to the degree of sequence conservation. Intron positions of animal genes are fairly well conserved, while the intron positions of fungal and protist genes vary. Determinations of evolutionary relationship based upon gene structures should be made cautiously, since the rate of insertion and deletion of introns may not be constant among organisms or even among different genes. Introns may be less frequently inserted into or deleted from plant genes than in those of the animal, fungal, or protist kingdoms.

Acknowledgments

We want to thank Elizabeth Kemmerer for running the NTSYS-PC computer analysis to generate the phenogram. We also thank Sharon T.

Broadwater for a careful reading of the manuscript and helpful suggestions. This work was supported by research grants GM29179 from the National Institutes of Health, U.S. Public Health Service, and RF84066, Allocation No. 3, from the Rockefeller Foundation.

REFERENCES

Adams, A. E. M., and Pringle, J. R., 1984, Relationship of actin and tubulin distribution to bud growth in wild-type and morphogenetic mutant *Saccharomyces cerevisiae*, *J. Cell Biol.* **98:**934–945.

Allen, N. S., 1980, Cytoplasmic streaming and transport in the Characean algae *Nitella*, *Can. J. Bot.* **58:**786–796.

Allen, N. S., and Allen, R. D., 1978, Cytoplasmic streaming in green plants, *Annu. Rev. Biophys. Bioeng.* **7:**497–526.

Allen, R. D., 1978, Membranes of ciliates: Biochemistry and fusion, in: *Cell Surface Reviews, Membrane Fusion* (G. Poste and G. L. Nicholson, eds.), Vol. 5, pp. 657–763, Elsevier/North-Holland, New York.

Anderson, J. M., and Soll, D. R., 1986, Differences in actin localization during bud and hypha formation in the yeast *Candida albicans*, *J. Gen. Microbiol.* **132:**2035–2047.

Baba, M. L., Darga, L. L., Goodman, M., and Czelusniak, J., 1981, Evolution of cytochrome c investigated by the maximum parsimony method, *J. Mol. Biol.* **17:**197–213.

Baird, W. V., and Meagher, R. B., 1987, A complex gene superfamily encodes actin in petunia, *EMBO J.* **6:**3223–3234.

Ben Amar, M. F., Pays, A., Tebabi, P., Dero, B., Seebeck, T., Steinert, M., and Pays, E., 1988, Structure and transcription of the actin gene of *Trypanosoma brucei*, *Mol. Cell. Biol.* **8:**2166–2176.

Bennetzen, J. L., and Hall, B. D., 1982, The primary structure of the *Saccharomyces cerevisiae* gene for alcohol dehydrogenase I, *J. Biol. Chem.* **257:**3018–3035.

Benyajati, C., Place, A. R., Powers, D. A., and Sofer, W., 1981, Alcohol dehydrogenase gene of *Drosophila melanogaster*. Relationship of intervening sequences to functional domains in the protein, *Proc. Natl. Acad. Sci. USA* **78:**2717–2721.

Bergsma, D. J., Chang, K. S., and Schwartz, R. J., 1985, Novel chicken actin gene: Third cytoplasmic isoform, *Mol. Cell. Biol.* **5:**1151–1162.

Bernatzky, R., and Tanksley, S. D., 1986, Genetics of actin-related sequences in tomato, *Theor. Appl. Genet.* **72:**314–321.

Bradley, M. O., 1973, Microfilaments and cytoplasmic streaming: Inhibition of streaming with cytochalasin, *J. Cell Sci.* **12:**327–343.

Brunke, K. J., Anthony, A. G., Sternberg, E. J., and Weeks, D. P., 1984, Repeated consensus sequence and pseudopromoters in the four coordinately regulated tubulin genes of *Chlamydomonas reinhardtii*, *Mol. Cell. Biol.* **4:**1115–1124.

Carroll, S. L., Bergsma, D. J., and Schwartz, R. J., 1986, Structure and complete nucleotide sequence of the chicken α-smooth muscle (aortic) actin gene, *J. Biol. Chem.* **261:**8965–8976.

Cedergren, R., Gray, M. W., Abel, Y., and Sankoff, D., 1988, The evolutionary relationships among known life forms, *J. Mol. Evol.* **28:**98–112.

Chang, C., and Meyerowitz, E. M., 1986, Molecular cloning and DNA sequence of the *Arabidopsis thaliana* alcohol dehydrogenase gene, *Proc. Natl. Acad. Sci. USA* **83:**1408–1412.

Clayton, L., and Lloyd, C. W., 1985, Actin organization during the cell cycle in meristematic plant cells, *Exp. Cell Res.* **156:**231–238.

Clegg, J. B., Goodbourn, S. E. Y., and Braend, M., 1984, Genetic organization of the polymorphic equine a globin locus and sequences of the BII a1 gene, *Nucleic Acids Res.* **12:**7847–7858.

Condeelis, J. S., 1974, The identification of F-actin in the pollen tube and protoplast of *Amaryllis belladonna, Exp. Cell Res.* **88:**435–439.

Cooper, A. D., and Crain, W. R., Jr., 1982, Complete nucleotide sequence of a sea urchin actin gene, *Nucleic Acids Res.* **10:**4081–4092.

Cox, K. H., Angerer, L. M., Lee, J. J., Davidson, E. H., and Angerer, R. C., 1986, Cell lineage-specific programs of expression of multiple actin genes during sea urchin embryogenesis, *J. Mol. Biol.* **188:**159–172.

Cross, G. S., Wilson, C., Erba, H. P., and Woodland, H. R., 1988, Cytoskeletal actin gene families of *Xenopus borealis* and *Xenopus laevis, J. Mol. Evol.* **27:**17–28.

Crowley, K. S., and Brasch, K., 1987, Does the interchromatin compartment contain actin?, *Cell Biol. Int. Rep.* **11:**537–546.

Cupples, C., and Pearlman, R., 1986, Isolation and characterization of the actin gene from *Tetrahymena thermophila, Proc. Natl. Acad. Sci. USA* **83:**5160–5164.

Dayhoff, M. O., Schwartz, R. M., and Orcutt, B. C., 1978, A model of evolutionary change in proteins, *Atlas of Protein Sequence and Structure,* Vol. 5, Suppl. 3, pp. 345–352, National Biomedical Research Foundation, Silver Springs, Maryland.

Dennis, E. S., Gerlach, W. L., Pryor, A. J., Bennetzen, J. L., Inglis, A., Llewellyn, D., Sachs, M. M., Ferl, R. J., and Peacock, W. J., 1984, Molecular analysis of the alcohol dehydrogenase (Adh1) gene of maize, *Nucleic Acids Res.* **12:**3983–4000.

Dennis, E. S., Sachs, M. M., Gerlach, W. L., Finnegan, E. J., and Peacock, W. J., 1985, Molecular analysis of the alcohol dehydrogenase 2 (Adh2) gene of maize, *Nucleic Acids Res.* **13:**727–743.

Dodgson, J. B., and Engels, J. D., 1983, The nucleotide sequence of the adult chicken α-globin genes, *J. Biol. Chem.* **258:**4623–4629.

Dolan, M., Dodgson, J. B., and Engel, J. D., 1983, Analysis of the adult chicken β-globin gene, *J. Biol. Chem.* **258:**3983–3990.

Drouin, G., and Dover, G. A., 1987, A plant processed pseudogene, *Nature* **328:**557–558.

Duester, G., Smith, M., Bilanchone, V., and Hatfield, G. W., 1986, Molecular analysis of the human class I alcohol dehydrogenase gene family and nucleotide sequence of the gene encoding the β subunit, *J. Biol. Chem.* **261:**2027–2033.

Erbil, C., and Niessing, J., 1982, The complete nucleotide sequence of the duck α^A-globin gene, *Gene* **20:**211–217.

Fidel, S., Doonan, J. H., and Morris, N. R., 1988, *Aspergillus nidulans* contains a single actin gene which has unique intron location and encodes a γ-actin, *Gene* **70:**283–293.

Files, J. G., Carr, S., and Hirsh, D., 1983, The actin gene family of *Caenorhabditis elegans, J. Mol. Biol.* **164:**355–375.

Foran, D. R., Johnson, P. J., and Moore, G. P., 1985, Evolution of two actin genes in the sea urchin *Strongylocentrotus franciscanus, J. Mol. Evol.* **22:**108–116.

Fornwald, J. A., Kuncio, G., Peng, I., and Ordahl, C. P., 1982, The complete nucleotide sequence of the chick α-actin gene and its evolutionary relationship to the actin gene family, *Nucleic Acids Res.* **10:**3861–3876.

Fyrberg, E. A., Bond, J. B., Hershey, N. D., Mixter, K. S., and Davidson, N., 1981, The actin genes of *Drosophila:* Protein coding regions are highly conserved but intron positions are not, *Cell* **24:**107–116.

Fyrberg, E. A., Mahaffey, J. W., Bond, B. J., and Davidson, N., 1983, Transcripts of the six *Drosophila* actin genes accumulate in a stage- and tissue-specific manner, *Cell* **33:**115–123.

Gallwitz, D., and Sures, I., 1980, Structure of a split yeast gene: Complete nucleotide sequence of the actin gene in *Saccharomyces cerevisiae, Proc. Natl. Acad. Sci. USA* **77**:2546–2550.

Gonzalez-y-Merchand, J. A., and Cox, R. A., 1988, Structure and expression of an actin gene of *Physarum polycephalum, J. Mol. Biol.* **202**:161–168.

Graham, L. E., 1982, *Coleochaete* and the origin of land plants, *Am. J. Bot.* **71**:603–608.

Gurdon, J. B., Fairman, S., Mohun, T. J., and Brennan, S., 1985, Activation of muscle-specific actin genes in *Xenopus* development by an induction between animal and vegetal cells of a blastula, *Cell* **41**:913–922.

Gwynne, D. I., Buxton, F. P., Sibley, S., Davies, R. W., Lockington, R. A., Scazzocchio, C., and Sealy-Lewis, H. M., 1987, Comparison of the *cis*-acting control regions of two coordinately controlled genes involved in ethanol utilization in *Aspergillus nidulans, Gene* **51**:205–216.

Hall, J. L., and Cowan, N. J., 1985, Structural features and restricted expression of a human α-tubulin gene, *Nucleic Acids Res.* **13**:207–223.

Hamada, H., Petrino, M. G., and Kakunaga, T., 1982, Molecular structure and evolutionary origin of human cardiac muscle actin gene, *Proc. Natl. Acad. Sci. USA* **79**:5901–5905.

Hardison, R. C., Butler, E. T., Lacy, E., Maniatis, T., Rosenthal, N., and Efstratiadis, A., 1979, The structure and transcription of four linked rabbit β-like globin genes, *Cell* **18**:1285–1297.

Heath, I. B., 1988, Evidence against a direct role for cortical actin arrays in saltatory organelle motility in hyphae of the fungus *Saprolegnia ferax, J. Cell Sci.* **91**:41–47.

Helftenbein, E., 1985, Nucleotide sequence of the macronuclear DNA molecule coding for α-tubulin from the ciliate *Stylonychia lemnae.* Special codon usage: TAA is not a translation termination codon, *Nucleic Acids Res.* **13**:415–433.

Herth, W., Franke, W. W., and Vanderwoude, W. J., 1972, Cytochalsin stops tip growth in plants, *Naturwissenschaften* **59**:38–39.

Heslop-Harrison, J., Heslop-Harrison, Y., Cesti, M., Tiezzi, A., and Ciampolini, F., 1986, Actin during pollen germination, *J. Cell Sci.* **86**:1–8.

Hightower, R. C., and Meagher, R. B., 1986, The molecular evolution of actin, *Genetics* **114**:315–332.

Hirono, M., Endoh, H., Okada, N., Numata, O., and Watanabe, Y., 1987, *Tetrahymena* actin gene and identification of its gene product, *J. Mol. Biol.* **194**:181–192.

Hoch, H. C., and Staples, R. C., 1983, Visualization of actin *in situ* by rhodamine-conjugated phalloin in the fungus *Uromyces phaseoli, Eur. J. Cell Biol.* **32**:52–58.

Hu, M. C.-T., Sharp, S. B., and Davidson, N., 1986, The complete sequence of the mouse skeletal α-actin gene reveals several conserved and inverted repeat sequences outside of the protein-coding region, *Mol. Cell. Biol.* **6**:15–25.

Ilker, R. A., Breidenbach, R. W., and Murphy, T. M., 1979, Partial purification of actin from wheat germ, *Phytochemistry* **18**:1781–1783.

Jackson, W. T., and Doyle, B. G., 1977, Characterization of actin from root tips of *Phaseolus vulgaris, J. Cell Biol.* **75**:268a.

Kenny, J. R., Dancik, B. P., and Florence, L. Z., 1988, Nucleotide sequence of the carboxy-terminal portion of a lodgepole pine actin gene, *Can. J. For. Res.* **18**:1595–1602.

Kimmel, B. E., Samson, S., Wu, J., Hirschberg, R., and Yarbrough, L. R., 1985, Tubulin genes of the African trypanosoma *Brucei rhodesiense:* Nucleotide sequence of a 3.7 kb fragment containing genes for alpha and beta tubulins, *Gene* **35**:237–248.

Klein, K., Wagner, G., and Blatt, M. R., 1980, Heavy-meromyosin decoration of microfilaments from *Mougeotia* protoplasts, *Planta* **150**:354–356.

Korn, E. D., 1982, Actin polymerization and its regulation by proteins from nonmuscle cells, *Physiol. Rev.* **62**:672–737.

La Claire, J. W., II, 1984, Actin is present in a green alga that lacks cytoplasmic streaming, *Protoplasma* **120:**242–244.

La Claire, J. W., II, 1989, Actin cytoskeleton in intact and wounded coenocytic green algae, *Planta* **177:**47–57.

Landsmann, J., Dennis, E. S., Higgins, T. J. V., Appleby, C. A., Kortt, A. A., and Peacock, W. J., 1986, Common evolutionary origin of legume and non-legume plant haemoglobins, *Nature* **324:**166–168.

Lazarides, E., and Weber, K., 1974, Actin antibody: The specific visualization of actin filaments in non-muscle cells, *Proc. Natl. Acad. Sci. USA* **71:**2268–2274.

Lee, J. J., Calzone, F. J., Britten, R. J., Angerer, R. C., and Davidson, E. H., 1986, Activation of sea urchin actin genes during embryogenesis, *J. Mol. Biol.* **188:**173–183.

Lee, J. S., and Verma, D. P. S., 1984, Structure and chromosomal arrangement of leghemoglobin genes in kidney bean suggest divergence in soybean leghemoglobin gene loci following tetraploidation, *EMBO J.* **3:**2745–2752.

Lemischka, I., and Sharp, P. A., 1982, The sequences of an expressed rat α-tubulin gene and a pseudogene with an inserted repetitive element, *Nature* **300:**330–335.

Lewis, S. A., Gilmartin, M. E., Hall, J. L., and Cowan, N. J., 1985, Three expressed sequences within the β-tubulin multigene family each define a distinct isotype, *J. Mol. Biol.* **182:**11–20.

Li, W.-H., Wu, C.-I., and Luo, C.-C., 1985, A new method for estimating synonymous and nonsynonymous rates of nucleotide substitution considering the relative likelihood of nucleotide and codon changes, *Mol. Biol. Evol.* **2:**150–174.

Llewellyn, D. J., Finnegan, E. J., Ellis, J. G., Dennis, E. S., and Peacock, W. J., 1987, Structure and expression of an alcohol dehydrogenase 1 gene from *Pisum sativum* (cv. "greenfeast"), *J. Mol. Biol.* **195:**115–123.

Lloyd, C. W., ed., 1983, *The Cytoskeleton in Plant Growth and Development,* Academic Press, London.

Ludwig, S. R., Oppenheimer, D. G., Silflow, C. D., and Snustad, D. P., 1987, Characterization of the α-tubulin gene family of *Arabidopsis thaliana, Proc. Natl. Acad. Sci. USA* **84:**5833–5837.

Ludwig, S. R., Oppenheimer, D. G., Silflow, C. D., and Snustad, D. P., 1988, The α-tubulin of *Arabidopsis thaliana:* Primary structure and preferential expression in flowers, *Plant Mol. Biol.* **10:**311–321.

MacLeod, C. L., Firtel, R. A., and Papkoff, J., 1980, Regulation of actin gene expression during spore germination in *Dictyostelium discoideum, Dev. Biol.* **76:**263–274.

Mages, W., Salbaum, J. M., Harper, J. F., and Schmitt, R., 1988, Organization and structure of *Volvox* α-tubulin genes, *Mol. Gen. Genet.* **213:**449–458.

Marc, J., and Gunning, B. E. S., 1986, Immunofluorescent localization of cytoskeletal tubulin and actin during spermatogenesis in *Pterisdium aquilinum* (L.) Kuhn, *Protoplasma* **134:**163–177.

Margulis, L., and Schwartz, K. V., 1988, *Five Kingdoms,* 2nd ed., W. H. Freeman, New York.

Marks, J., Shaw, J.-P., and Shen, C. K. J., 1986, The orangutan adult α-globin gene locus: Duplicated functional genes and a newly detected member of the primate α-globin gene family, *Proc. Natl. Acad. Sci. USA* **83:**1413–1417.

Marks, M. D., West, J., and Weeks, D. P., 1987, The relatively large beta-tubulin gene family of *Arabidopsis* contains a member with an unusual transcribed 5′ noncoding sequence, *Plant Mol. Biol.* **10:**91–104.

Maupin, P., and Pollard, T. D., 1986, Arrangement of actin filaments and myosin-like filaments in the contractile ring of actin-like filaments in the mitotic spindle of dividing HeLa cells, *J. Ultrastruct. Mol. Struct.* **94:**92–103.

Maupin-Szamier, P., and Pollard, T. D., 1978, Actin filament destruction by osmium tetroxide, *J. Cell Biol.* **77**:837–852.

May, G. S., Tsang, M. L.-S., Smith, H., Fidel, S., and Morris, N. R., 1987, *Aspergillus nidulans* β-tubulin genes are unusually divergent, *Gene* **55**:231–243.

McCurdy, D. W., and Williamson, R. E., 1987, An actin related protein inside pea chloroplasts, *J. Cell Sci.* **87**:449–456.

McElroy, D., Rothenberg, M., and Wu, R., 1990, Structural characterization of a rice actin gene, *Plant Mol. Biol.* **14**:163–172.

McKeown, M., and Firtel, R. A., 1981, Differential expression and 5′ end mapping of actin genes in *Dictyostelium, Cell* **24**:799–807.

McKeown, M., Taylor, W., Kindle, K., Firtel, R., Bender, W., and Davidson, N., 1978, Multiple heterogeneous actin genes in *Dictyostelium, Cell* **15**:789–800.

McKerracher, L. J., and Heath, I. B., 1987, Cytoplasmic migration and intracellular movements during tip growth in fungal hyphae, *Exp. Mycol.* **11**:79–100.

Menzel, D., and Elsner-Menzel, C., 1989, Actin-based chloroplast rearrangements in the cortex of the giant coenocytic green alga *Caulerpa, Protoplasma* **150**:1–8.

Metcalf, T. N., Szabo, L. J., Schubert, K. R., and Wang, J. L., 1980, Immunochemical identification of an actin-like protein from soybean seedlings, *Nature* **285**:171–172.

Metcalf, T. N., Szabo, L. J., Schubert, K. R., and Wang, J. L., 1984, Ultrastructure and immunocytochemical analyses of the distribution of microfilaments in seedlings and plants of *Glycine max, Protoplasma* **120**:91–99.

Meyerhof, W., Klinger-Mitropoulos, S., Stadler, J., Weber, R., and Knochel, W., 1984, The primary structure of the larval β-globin gene of *Xenopus laevis* and its flanking regions, *Nucleic Acids Res.* **12**:7705–7719.

Minty, A. J., Alonso, S., Guenet, J. L., and Buckingham, M., 1983, Number and organization of actin-related sequences in the mouse genome, *J. Mol. Biol.* **167**:77–101.

Mohun, T. J., Garrett, N., and Gurdon, J. B., 1986, Upstream sequences required for tissue-specific activation of the cardiac actin gene in *Xenopus laevis* embryos, *EMBO J.* **5**:3185–3193.

Monteiro, M. J., and Cox, R. A., 1987, Primary structure of an α-tubulin gene of *Physarum polycephalum, J. Mol. Biol.* **193**:427–438.

Nairn, C. J., Winesett, L., and Ferl, R. J., 1988, Nucleotide sequence of an actin gene from *Arabidopsis thaliana, Gene* **65**:247–257.

Neff, N. F., Thomas, J. H., Grisafi, P., and Botstein, P., 1983, Isolation of the β-tubulin gene from yeast and demonstration of its essential function *in vivo, Cell* **33**:211–219.

Ng, R., and Abelson, J., 1980, Isolation and sequence of the gene for actin in *Saccharomyces cerevisiae, Proc. Natl. Acad. Sci. USA* **77**:3912–3916.

Ng, S.-Y., Gunning, P., Eddy, R., Ponte, P., Leavitt, J., Shows, T., and Kedes, L., 1985, Evolution of the functional human β-actin gene and its multi-pseudogene family: Conservation of noncoding regions and chromosomal dispersion of pseudogenes, *Mol. Cell. Biol.* **5**:2720–2723.

Nishioka, Y., and Leder, P., 1979, The complete sequence of a chromosomal mouse α-globin gene reveals elements conserved throughout vertebrate evolution, *Cell* **18**:875–882.

Ohrbach, M. J., Porro, E. B., and Yanofsky, C., 1986, Cloning and characterization of the gene for β-tubulin from a benomyl resistant mutant of *Neurospora crassa* and its use as a dominant selectable marker, *Mol. Cell. Biol.* **6**:2452–2461.

Orkin, S. H., 1978, The duplicated human α-globin genes lie close together in cellular DNA, *Proc. Natl. Acad. Sci. USA* **75**:5950–5954.

Oppenheimer, D. G., Haas, N., Silflow, C. D., and Snustad, D. P., 1988, The β-tubulin gene family of *Arabidopsis thaliana:* Preferential accumulation of the $\beta1$ transcript in roots, *Gene* **63:**87–102.

Palevitz, B. A., 1980, Comparative effects of phalloidin and cytochalasin B on motility and morphogenesis in *Allium, Can. J. Bot.* **58:**773–785.

Parthasarathy, M. V., Perdue, T. D., Witzmun, A., and Alvernaz, J., 1985, Actin network as a normal component of the cytoskeleton in many vascular plant cells, *Am. J. Bot.* **72:**1318–1323.

Patient, R. K., Elkington, J. A., Kay, R. M., and Williams, J. G., 1980, Internal organization of the major adult α- and β-globin genes of *X. laevis, Cell* **21:**565–573.

Pesacreta, T. C., Carly, W. W., Webb, W. W., and Parthasarathy, M., 1982, F-actin in conifer roots, *Proc. Natl. Acad. Sci. USA* **79:**2898–2901.

Pierson, E. S., Kengen, H. M. P., and Derksen, J., 1989, Microtubules and actin filaments co-localize in pollen tubes of *Nicotiana tabacum* L. and *Lilium longiflorum* Thunb, *Protoplasma* **150:**75–77.

Pollard, T. D., 1986, Assembly and dynamics of the actin filament system in nonmuscle cells, *J. Cell. Biochem.* **31:**87–95.

Pollard, T. D., and Cooper, J. A., 1986, Actin and actin-binding proteins. A critical evaluation of mechanisms and functions, *Annu. Rev. Biochem.* **55:**987–1035.

Poncz, M., Schwartz, E., Ballantine, M., and Surrey, S., 1983, Nucleotide sequence analysis of the $\gamma\beta$-globin gene region in humans, *J. Biol. Chem.* **258:**11599–11609.

Pratt, L. F., and Cleveland, D. W., 1988, A survey of the α-tubulin gene family in chicken: Unexpected sequence heterogeneity in the polypeptides encoded by five expressed genes, *EMBO J.* **7:**931–940.

Quader, H., Hofmann, A., and Schnepf, E., 1987, Shape and movement of the endoplasmic reticulum in onion bulb epidermis cells: Possible involvement of actin, *Eur. J. Cell Biol.* **44:**17–26.

Queen, C., and Korn, L. J., 1984, A comprehensive analysis program for the IBM personal computer, *Nucleic Acids Res.* **12:**581–599.

Reece, K. S., 1988, The actin gene family of rice (Oryza sativa L), Ph.D. thesis, Cornell University, Ithaca, New York.

Reece, K. S., McElroy, D., and Wu, R., 1990, Genomic nucleotide sequence of four rice (*Oryza sativa*) actin genes, *Plant Mol. Biol.* **14:**621–624.

Rohlf, F. J., 1989, *NTSYS-pc: Numerical Taxonomy and Multivariate Analysis System,* Exeter, Setauket, New York.

Romans, P., and Firtel, R. A., 1985, Organization of the actin multigene family of *Dictyostelium discoideum* and analysis of variability in the protein coding regions, *J. Mol. Biol.* **186:**321–335.

Romans, P., Firtel, R. A., and Saxe, C. L., III, 1985, Gene-specific expression of the actin multigene family of *Dictyostelium discoideum, J. Mol. Biol.* **186:**337–355.

Rudolph, J. E., Kimble, M., Hoyle, H. D., Subler, M. A., and Raff, E. C., 1987, Three *Drosophila* beta-tubulin sequences: A developmentally regulated isoform ($\beta3$), the testis-specific isoform ($\beta2$), and an assembly-defective mutation of the testis-specific isoform ($\beta2t^8$) reveal both an ancient divergence in metazoan isotypes and structural constraints for beta-tubulin function, *Mol. Cell. Biol.* **7:**2231–2242.

Russell, D. W., Smith, M., Williamson, V. M., and Young, E. T., 1983, Nucleotide sequence of the yeast alcohol dehydrogenase II gene, *J. Biol. Chem.* **258:**2674–2682.

Sanchez, F., Tobin, S. L., Rdent, U., Zulauf, E., and McCarthy, B. J., 1983, Two *Drosophila* actin genes in detail, *J. Mol. Biol.* **163:**533–551.

Sargent, T. D., Jamrich, M., and Dawid, I. B., 1986, Cell interaction and the control of gene activity during early development of *Xenopus laevis, Dev. Biol.* **114:**238–246.

Sawtell, N. M., Hartman, A. L., and Lessard, J. L., 1988, Unique isoactins in the brush border of rat intestinal epithelial cells, *Cell Motil. Cytoskeleton* **11:**318–325.

Schatz, P. J., Pillus, L., Grisafi, P., Solomon, F., and Botstein, D., 1986, Two functional α-tubulin genes of the yeast *Saccharomyces cerevisiae* encode divergent proteins, *Mol. Cell. Biol.* **6**:3711–3721.

Schedl, T., and Dove, W. F., 1982, Mendelian analysis of the organization of actin sequences in *Physarum polycephalum, J. Mol. Biol.* **159**:41–57.

Scheetz, M. P., and Spudich, J. A., 1983, Movement of myosin-coated fluorescent beads on actin cables *in vitro, Nature* **305**:31–35.

Scheller, R. H., McAllister, L. B., Crain, W. R., Durica, D. S., Posakony, J. W., Thomas, T. L., Britten, R. J., and Davidson, E. H., 1981, Organization and expression of multiple actin genes in the sea urchin, *Mol. Cell. Biol.* **1**:609–628.

Schmit, A.-C., and Lambert, A.-M., 1987, Characterization and dynamics of cytoplasmic F-actin, *J. Cell Biol.* **105**:2157–2166.

Schmit, A.-C., and Lambert, A.-M., 1988, Plant actin filament and microtubule interactions during anaphase–telophase transition: Effects of antagonist drugs, *Biol. Cell* **64**:309–318.

Schon, E. A., Wernke, S. M., and Lingre, J. B., 1982, Gene conversion of two functional goat α-globin preserves only minimal flanking sequences, *J. Biol. Chem.* **257**:6825–6835.

Schuler, M. A., McOsker, P., and Keller, E. B., 1983, DNA sequence of two linked actin genes of sea urchin, *Mol. Cell. Biol.* **3**:448–456.

Seagull, R. W., and Heath, I. B., 1979, The effects of tannic acid on the *in vivo* preservation of microfilaments, *Eur. J. Cell. Biol.* **20**:184–188.

Seagull, R. W., Falconer, M. M., and Weedenburg, C. A., 1987, Microfilaments: Dynamic arrays in higher plant cells, *J. Cell Biol.* **104**:995–1000.

Shah, D. M., Hightower, R. C., and Meagher, R. B., 1982, Complete nucleotide sequence of a soybean actin gene, *Proc. Natl. Acad. Sci. USA* **79**:1022–1026.

Shah, D. M., Hightower, R. C., and Meagher, R. B., 1983, Genes encoding actin in higher plants: Intron positions are highly conserved but the coding sequences are not, *J. Mol. Appl. Genet.* **2**:111–126.

Shani, M., 1986, Tissue-specific and developmentally regulated expression of a chimeric actin–globin gene in transgenic mice, *Mol. Cell. Biol.* **6**:2624–2631.

Shapiro, S. G., Schon, E. A., Townes, T. M., and Lingrel, J. B., 1983, Sequence and linkage of the goat e^{I} and e^{II} β-globin genes, *J. Mol. Biol.* **169**:31–52.

Silflow, C. D., Oppenheimer, D. G., Kopczak, S. D., Ploense, S. E., Ludwig, N. H., and Snustad, D. P., 1987, Plant tubulin genes: Structure and differential expression during development, *Dev. Genet.* **8**:435–460.

Sogin, M. L., Elwood, H. J., and Gunderson, J. H., 1986, Evolutionary diversity of small-subunit rRNA genes, *Proc. Natl. Acad. Sci. USA* **83**:1383–1387.

Staiger, C. J., and Schliwa, M., 1987, Actin localization and function in higher plants, *Protoplasma* **141**:1–12.

Stossel, T. P., 1984, Contribution of actin to the structure of the cytoplasmic matrix, *J. Cell Biol.* **99**:15s–21s.

Straub, F. B., 1941, *Studies from the Institute of Medical Chemistry* (A. Szent-Gyorgy, ed.), Vol. 1A, University of Szeged, Szeged, Hungary.

Sullivan, K. F., Havecroft, J. C., Machlin, P. S., and Cleveland, D. W., 1986*a,* Sequence and expression of the chicken β5- and β4-tubulin genes define a pair of divergent β-tubulins with complementary patterns of expression, *Mol. Cell. Biol.* **6**:4409–4418.

Sullivan, K. F., Machlin, P. S., Ratrie, H., and Cleveland, D. W., 1986*b,* Sequence and expression of the chicken β3 tubulin gene, *J. Biol. Chem.* **261**:13317–13322.

Tellam, R. L., Morton, D. J., and Clarke, F. M., 1989, A common theme in the amino acid sequences of actin and many actin-binding proteins?, *TIBS* **14**:130–133.

Theurkauf, W. E., Baum, H., Bo, J., and Wensink, P. C., 1986, Tissue-specific and constitutive α-tubulin genes of *Drosophila melanogaster* code for structurally distinct proteins, *Proc. Natl. Acad. Sci. USA* **83**:8477–8481.

Tischendorf, G., Sawitzky, D., and Werz, G., 1987, Antibodies specific for vertebrate actin, myosin, actinin or vinculin recognize epitopes in the giant nucleus of the marine green alga, *Acetabularia, Cell Motil. Cytoskel.* **7**:78–86.

Toda, T., Adachi, Y., Yasushi, H., and Yanagida, M., 1984, Identification of the pleiotropic cell division cycle gene NDA2 as one of two different α-tubulin genes in *Schizosaccharomyces pombe, Cell* **37**:233–242.

Traas, J. A., Doonan, J. H., Rawlins, D. J., Shaw, P. J., Watts, J., and Lloyd, C. W., 1987, An actin network is present in the cytoplasm throughout the cell cycle of carrot cells and associated with the dividing nucleus, *J. Cell Biol.* **105**:387–395.

Trick, M., Dennis, E. S., Edwards, K. J. R., and Peacock, W. J., 1988, Molecular analysis of the alcohol dehydrogenase gene family of barley, *Plant Mol. Biol.* **11**:147–160.

Ueyama, H., Hamada, H., Battula, N., and Kakunaga, T., 1984, Structure of a human smooth muscle actin gene (aortic type) with a unique intron site, *Mol. Cell. Biol.* **4**:1073–1078.

Vahey, M., and Scordilis, S. P., 1980, Contractile proteins from tomato, *Can. J. Bot.* **58**:797–801.

Vandekerckhove, J., and Weber, K., 1978, Mammalian cytoplasmic actins are the products of at least two genes and differ in primary structure in at least 25 identified positions from skeletal muscle actins, *Proc. Natl. Acad. Sci. USA* **75**:1106–1110.

Vandekerckhove, J., and Weber, K., 1984, Chordate muscle actins differ distinctly from invertebrate muscle actins, *J. Mol. Biol.* **179**:391–413.

Wagner, G., and Klein, K., 1981, Mechanism of chloroplast movement in *Mougeotia, Protoplasma* **109**:169–185.

Wehland, J., Osborn, M., and Weber, K., 1977, Phalloidin-induced polymerization in the cytoplasm of cultured cells interfered with cell locomotion and growth, *Proc. Natl. Acad. Sci. USA* **74**:5613–5617.

Wessels, N. K., Spooner, B. S., Ash, J. F., Bradley, M. O., Luduena, M. A., Taylor, E. L., Wrenn, J. T., and Yamada, K. M., 1971, Microfilaments in cellular and developmental processes, *Science* **171**:135–143.

Whittaker, R. H., 1969, New concepts of kingdoms of organisms, *Science* **163**:150–160.

Wiborg, O., Hyldig-Nielsen, J. J., Jensen, E. O., Paludan, K., and Marcker, K. A., 1982, The nucleotide sequences of two leghemoglobin genes from soybean, *Nucleic Acids Res.* **10**:3487–3494.

Wildeman, A. G., 1988, A putative ancestral actin gene present in a thermophilic eukaryote: Novel combination of intron positions, *Nucleic Acids Res.* **16**:2553–2564.

Williamson, R. E., and Toh, B. H., 1979, Motile models of plant cells and the immunofluorescent localization of actin in a motile *Chara* cell model, in: *Cell Motility: Molecules and Organization* (S. Hatano, H. Ishikawa, and H. Sato, eds.), pp. 339–346, University Park Press, Baltimore, Maryland.

Wilson, C., Cross, G. S., and Woodland, H. R., 1986, Tissue-specific expression of actin genes injected into *Xenopus* embryos, *Cell* **47**:589–599.

Wohlfarth-Botterman, K. E., 1986, Biological aspects of motility, in: *The Molecular Biology of Physarum polycephalum* (W. F. Dove, J. Dee, S. Hatano, F. B. Haugli, and K. E. Wohlfarth-Bottermann, eds.), pp. 151–164, Plenum Press, New York.

Xie, Y., and Wu, R., 1989, Rice alcohol dehydrogenase genes: Anaerobic induction, organ specific expression and characterization of cDNA clones, *Plant Mol. Biol.* **13**:53–68.

Youngblom, J., Schloss, J. A., and Silflow, C. D., 1984, The two β-tubulin genes of *Chlamydomonas reinhardtii* code for identical proteins, *Mol. Cell. Biol.* **4**:2686–2696.

2

Evolutionary and Population Perspectives of the Human HLA Complex

WILLIAM KLITZ, GLENYS THOMSON, NICOLAS BOROT, and ANNE CAMBON-THOMSEN

What are these creatures, pretentiously reared up on their hind legs, looking down on the rest of creation? They have emerged from the same biological matrix as all the others, and yet seem to imagine themselves distinct. As an appropriate rebuke, we suggest putting them *under the glass to thereby reveal evolutionary mechanisms operating in all of life.*

INTRODUCTION

Despite the contributions of iconoclasts such as Copernicus, Darwin, and more recently the cognitive psychologists (see, for example, Lakoff, 1988), it remains difficult to overcome the preconception that humans are somehow separate, special, and occupy the center of the universe. One consequence of this is that studies on human population biology and genetics do not seem to have an established place at the table of evolutionary biology—it appears that we can speak with greater confidence on the evolutionary forces governing the creation of powdery mildew and house sparrows. In this light we

WILLIAM KLITZ and GLENYS THOMSON • Department of Integrative Biology, University of California, Berkeley, California 94720. NICOLAS BOROT • CRPG/CNRS, UPR 8291, CHU Purpan, F-31300 Toulouse, Cedex, France. ANNE CAMBON-THOMSEN • CRPG/CNRS, UPR 8291, and INSERM U100, CHU Purpan, F-31052 Toulouse, Cedex, France.

Evolutionary Biology, Volume 26, edited by Max K. Hecht *et al.* Plenum Press, New York, 1992.

present an analysis of a large family-based survey of the people of France typed for the polymorphic loci which regulate the immune system.

The major histocompatibility complex (MHC), consisting of a group of several loci involved in various aspects of immune function, have several features which make it especially appealing for evolutionary genetics. The high polymorphism at each of several loci creates an almost unheard of richness of genic variation. Further, these loci are tightly linked. In the human MHC, termed the human leucocyte antigen (HLA) system, recombination fractions vary from virtually complete linkage to 3%. This range of recombination permits the elaboration and testing of a variety of population-genetics models (Klitz and Thomson, 1992). That is not all, for the MHC loci are not merely markers, supplying gratefully accepted grist for the machinery of population genetics: their gene products directly control a variety of functions concerned with the immune response. The *HLA* region is thus not only an extraordinary resource, but also an exemplary system for studies of population genetics.

The rejection reaction of vertebrates to transplanted foreign tissue is due to the extreme responsiveness of the immune system to nonself MHC alleles. It turns out that this histocompatibility response is but a reflection of the central role of the MHC in immune system function. Because of this, both human (*HLA*) and mouse (*H-2*) systems have been intensively scrutinized to clarify the physiological role of the various loci. Of particular interest here is a large body of work pursuing the functional significance of the allelic variation in the recognition and control of microbial and parasitic invasion of the vertebrate host (Klein, 1986; Crumpton, 1987; Srivastava, 1991). Despite its great complexity, the genetics of the HLA system has been steadily unraveled over the last 30 years, thanks in real measure to a series of collaborative international workshops [for the most recent see Dupont (1989)].

THE MHC AND *HLA*

The *HLA* loci are located on the short arm of chromosome 6, and are classified into three groups (Fig. 1). Class I loci, located at the telomeric end of the region, code for molecules expressed on the surface of almost all nucleated cells. Class I molecules bind to virus-derived peptides in the endoplasmic reticulum. The complex crosses the cell membrane, where (see Fig. 2) it is recognized by cytotoxic T cells, which then destroy the infected cell. Although many class I loci exist (Koller *et al.,* 1989), the three prominent in histocompatibility reactions are *A*, *C*, and *B*. The class II loci are at the centromeric end of the MHC, and consist of the *DR*, *DQ*, and *DP* regions,

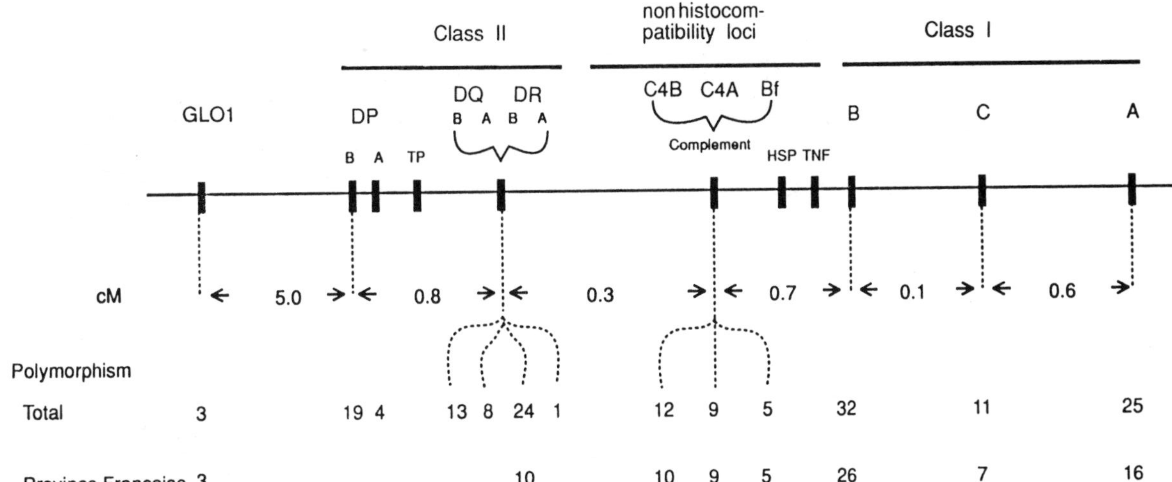

FIG. 1. Recombinational map of the *HLA* region indicating the known number of alleles at each locus, as well as the polymorphism accounted for in the Provinces Francaises survey. Total allelic polymorphism is from WHO Nomenclature Committee (1990). *TNF, HSP,* and *TP* are the loci for tumor necrosis factor, heat-shock protein, and transporter protein, respectively.

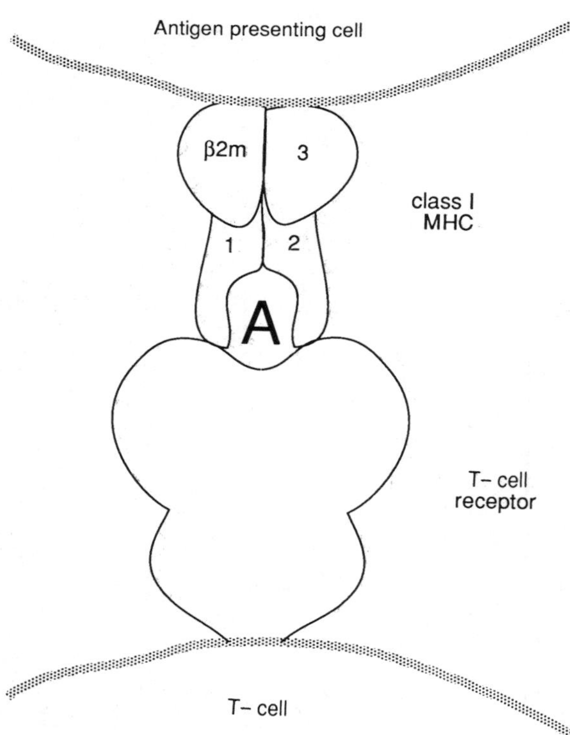

FIG. 2. The T-cell receptor interaction with an MHC molecule and antigen (A). This illustration shows a class I molecule with its three (labeled) external domains, and beta 2-microglobulin (chromosome 17). The antigen becomes associated with class I molecules intracellularly.

each of which codes for alpha and beta chains, which together comprise the functional molecule (Charron, 1990). Class II molecules, which are expressed on the surface of antigen-presenting cells such as macrophages, present antigenic peptides to T-helper cells, which then direct the activity of both the humoral (antibody) and cellular (effector T-cell) components of the immune system. Recently, an apparent member of the family of transmembrane transporter (TP) genes coding for a protein which moves peptides intracellularly to the endoplasmic reticulum for assembly with class I molecules has been mapped to the *HLA* class II region [reviewed by Parham (1990)]. An important aspect of both class I and class II molecules is their behavior as restriction elements: a particular T-cell clone requires for stimulation both a particular allelic form of the MHC molecule and a specific antigenic peptide fragment presented by the MHC molecule (Fig. 2). This specificity of three interacting components is at the heart of MHC function, and must be central to any interpretation of *HLA* polymorphism.

The class III region, located between the class I and class II regions, has become something of a catch-all category. It includes a tightly linked cluster of loci coding for components of the complement system, including C4B, C4A, and Bf (Fig. 1). Several additional loci are present in the class III region, including that for 21-hydroxylase, the tumor necrosis factor loci (*TNFA* and *TNFB*), and heat-shock protein loci (e.g., Spies *et al.*, 1989). The glyoxylase I locus (*GLOI*) is a nearby marker located some 5 cM centromeric of the class II region.

The polymorphism of the *HLA* loci is extraordinary (Fig. 1), ranging up to the 37 alleles currently described for *HLA-B*. Allelic variation at the class I loci and *DR* has been defined serologically, using one or more, usually polyclonal, antisera to distinguish allelic products. Molecular methods of typing employing allele-specific oligonucleotides are being developed, and may eventually supersede other techniques (e.g., Bugawan *et al.*, 1990; Gao *et al.*, 1990; Scharf *et al.*, 1991). Serological typing is still responsible for almost all of the available population data. Nucleotide sequence determination has allowed the subdivision of serologically defined alleles, so that knowledge of polymorphism is still increasing at these loci. Protein electrophoresis is used to define variation at the complement loci and *GLOI*.

The major histocompatibility loci are members of the immunoglobulin supergene family (Williams and Barclay, 1988) (see Fig. 3). The genes in this family produce cell-surface proteins involved in cell recognition. Several play central roles in immunity and often include the production of multiple distinct gene products, generated by germline polymorphism or somatic expansion. Three classes of proteins in this gene family are involved in the recognition, sequestering, or presentation of foreign peptides, which comprise the critical steps in the specific immune response of vertebrates. The immunoglobulin and T-cell receptor loci each generate diversity by somatic rearrangements occurring in lymphocytes, whereas MHC variability is manifest in the germ line. Many vertebrate loci involved in immune function and cell adhesion have been shown, through sequence affinity, to be members of this gene family. Several loci found in insects are also members of this gene lineage (Sun *et al.*, 1991).

Besides the flourishing presence of the class I and class II MHC in mammals, these loci are also present and expressed in birds, reptiles, and amphibians (Kaufman, 1990). Although immunoglobulin molecules trace back through to the agnatha (lamprey), fish have not been demonstrated to possess MHC loci.

The basic requirements of defense against microbial and parasitic invasion, including self–nonself recognition, communication, and effector mech-

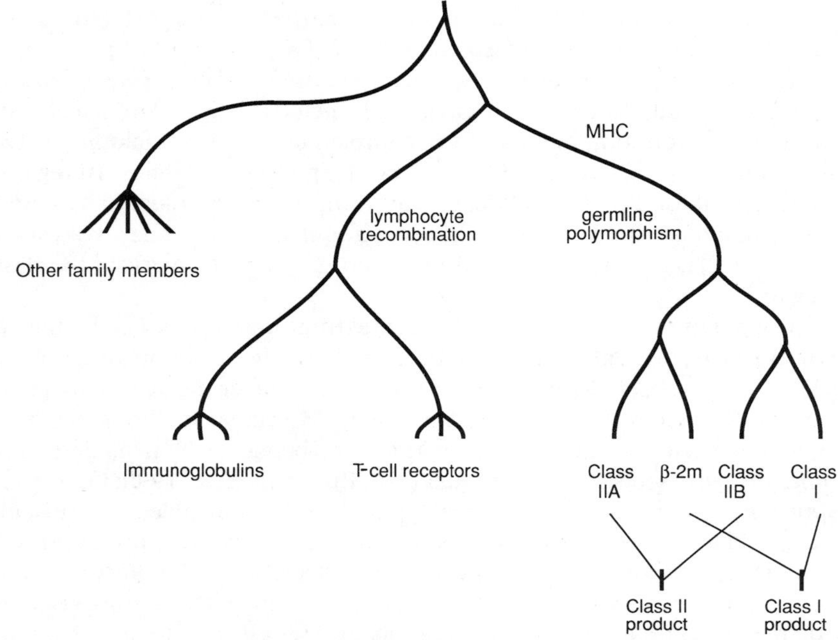

FIG. 3. The immunoglobulin supergene family. Three major subgroups of this family demonstrating expressed variation are the immunoglobulin, the T-cell receptor, and the MHC loci.

anisms, are general, but the particular adaptive solutions utilized by a taxon may not be identical with the human/mouse immune defense paradigm. The MHC of chickens (termed the *B* complex), for example, has expressed polymorphic class I and class II MHC loci with T-cell receptor restriction, but in addition they carry a third class of highly polymorphic cell-surface molecules (B-G), which are part of the *B* complex, and yet apparently are not members of the immunoglobulin supergene family (Guillemot *et al.,* 1989). In addition, the molecular structure of the *B* complex stands apart from the mammalian model in being highly compacted—the class I and class II genes are virtually adjacent in the *B* complex, with only 10–20 kb separation, compared to the several hundred kilobases separating the class I and class II regions in the HLA complex (Lawrance *et al.,* 1987). In addition, the intron–exon structure of the chicken class II B genes, while identical to that of mammals, possesses introns which are only 10% as long (Kroemer *et al.,* 1990).

The high polymorphism and heterozygosity familiar to the human and mouse MHC loci have been found in most other mammals examined, including diverse rodents, pigs, horses, sheep, cattle, various felids, and nonhuman primates [for references see Yuhki and O'Brien (1990) and Klein (1986)]. Some mammalian populations and species are virtually monomorphic at the MHC, however. Sometimes these exceptions to the typical pattern have been interpretable in terms of evolutionary sampling effects or lifestyle. African cheetahs are depauperate of genetic variation at all loci, including the MHC. This is dramatized by the ability of individual cheetahs to share skin grafts. A predictable vulnerability of these populations to infection is borne out by the occurrence of devastating viral epidemics in captive cheetah populations. Severe population bottlenecks in the recent evolutionary past have been proffered to explain the virtually monomorphic genetic state. As another example, Syrian hamsters have no detectable class I variation, but do have the typical rodent complement of polymorphism measured across the genome, including class II polymorphism. In this case a possible explanation may lie with the ecology and behavior of the species (Darden and Streilein, 1986). Syrian hamsters are at one end of the sociality scale in rodents (Lidicker and Patton, 1987), having a minimum of social contacts through the life cycle. Since the function of class I genes is to present viral antigens on the cell surface, and thus advertise an infected cell to the immune system, Syrian hamster class I monomorphism may reflect the sociality of the host which is necessary for viral transmission.

Natural selection has long been suspected to operate in the *HLA* region. This idea has been buttressed by inferences based on allele frequency distributions (Hedrick and Thomson, 1983; Klitz *et al.,* 1986), and patterns of association among alleles at different loci (Klitz and Thomson, 1987). These two approaches lie at the heart of this chapter, and are taken up below. Detailed analyses of nucleotide substitutions in coding regions for the antigen recognition site at both *HLA* and *H-2* histocompatibility loci show a higher rate of change for expressed, rather than silent, codon position substitutions (Hughes and Nei, 1988). This is in contrast to other loci, in which the rate of nucleotide substitution is generally higher at silent sites. In addition, greater heterozygosity is present at the functionally critical residues coding for the antigen recognition site, making a direct link between variability and function (Hedrick *et al.,* 1991*a*).

Several recent studies have shown that many MHC alleles have existed for millions of years (Figueroa *et al.,* 1988; Gyllensten and Erlich, 1989; Lawlor *et al.,* 1988; Mayer *et al.,* 1988). For example, many of the same allelic families and even specific alleles at the human *DQA* locus are present in other primates. This could only be possible with selection acting to maintain the variation over long periods of evolutionary time (Klein and Takahata, 1990).

Another line of evidence pertinent to selection governing MHC polymorphism comes from studies which correlate allelic variation with a particular disease. Such relationships have been documented in mice (Wakelin and Blackwell, 1988), domestic livestock (Ostergard *et al.,* 1989), and chickens (Scheiman and Collins, 1987). In the HLA system many dozens of associations with disease are known. Tiwari and Terasaki (1985) present an encyclopedic review of these data. Many of the diseases have a demonstrated or suspected autoimmune etiology. One very suggestive case of an *HLA* association with infectious disease is available from a study of the descendants of Dutch emigrants to Surinam in South America. Shortly after their arrival in the 1830s, the settler population of 300 individuals was devastated by epidemics of typhoid disease and yellow fever, which reduced their numbers by half. This "postselection" population was too large to result in the sampling effects on allele frequencies produced by population bottlenecks (Tajima, 1989). However, *HLA-DR* frequencies measured in the descendants of the emigrants show both deficiencies and excesses of particular alleles compared to contemporary Dutch populations (deVries, 1989). This can be interpreted as indications of both protective and susceptible effects determining survival according to class II allelic variation. Van Enden and co-workers (1982) review several cases of associations between *HLA* and infectious disease.

A second case of differential survival of a pathogen according to *HLA* type is observed in the action of *falciparum* malaria in West Africa (A. V. S. Hill *et al.,* 1991). Common West African alleles of both class I and class II loci were shown to reduce susceptibility to childhood mortality to malarial fever. This protective effect was suggested to be larger than that conferred by the sickle-cell hemoglobin variant in the Gambian population sampled.

These various observations all fit the notion that some form of balancing selection, operating through differential viability of individuals in response to infection, is the best candidate to explain MHC polymorphism (Takahata and Nei, 1990). Selection modes other than viability selection, however, may also be acting in this region. These include segregation distortion (Klitz *et al.,* 1987), maternal–fetal effects (Hedrick, 1988), and nonrandom mating (Potts *et al.,* 1991) [also see Hedrick *et al.* (1991*b*) for discussion].

A portion of the variation in allele frequencies from 26 polymorphic loci, including *HLA-A* and *HLA-B,* in populations sampled across contemporary Europe has been correlated with archeological dates for the first appearances of agriculture in its spread across Europe from an origin in the Near East 10,500 years ago (Sokal *et al.,* 1991). This suggests that the neolithic wave affecting the preagricultural inhabitants of Europe came not via clay tablets, nor by genocidal replacement, but by a process of "demic diffusion," in which the new lifestyle was spread by population admixture. This process seems to have resulted in the increasing dilution of Near Eastern genes coincident with the northward spread of agriculture. It is implied that the change in the allele frequencies of the six genetic systems, including *HLA-A* and *HLA-B*, which were distinctly correlated with the spread of agriculture can be explained by a simple function of admixture rates coincident with the dilution of neolithic peoples into the mesolithic European natives.

Neolithic behavior is, of course, not in the genes, but its practice must have genetic consequences for the practitioners. The agricultural revolution results in a sharp rise in population densities, which in turn supports a whole new suite of pathogens. Also accompanying the switch to farming is radical dietary change highlighted by greatly increased uniformity and low diversity of foods. Both of these aspects of the neolithic environment would be expected to create conditions for the differential survival of individuals based on their genetic makeup. The hemoglobinopathies and adult lactase expression are responses of human populations to these new environments. Evolution of *HLA* alleles might also be expected here. Could variation in *HLA* allele frequencies track both the movement of agriculture across Europe and balancing selection in response to pathogens and possibly other causes? The low correlation coefficients of the associations of *HLA* with the origin of agriculture (0.26 for *HLA-A* and 0.20 for *HLA-B*) make it clear that sufficient *HLA* variability is present (i.e., in the unexplained variation) to record selection pressures over the intervening 2–400 human generations during which agriculture has existed in Europe.

Efforts to sample the polymorphism of the *HLA* region in populations worldwide have been pursued since antisera capable of distinguishing *HLA* alleles first became available (Dausset and Colombani, 1973). The first alleles to be identified in the HLA system, used to establish the genetic basis of the observed variation, were obtained from Caucasian donors. The large size of the "blank" allelic class, i.e., alleles not recognized by the available sera, at the sampled loci (*A*, *C*, *B*, and *DR*) in non-Caucasian populations led to the belief that many population-specific alleles waited to be discovered. The 9th International Histocompatibility Workshop (Albert *et al.*, 1984) used family-based sampling, computer analysis of antisera reactivities, computer assignment of haplotypes, and the largest pool of antisera ever assembled, contributed by laboratories worldwide. The blank allelic class shrunk dramatically, not just in the Caucasian samples, but in Asian and Black populations as well. Instead of the discovery of many new alleles unique to one or a few populations, the great majority of allelic types are found worldwide. Studies comparing serological to DNA-based typing methods have affirmed the fact that the serological blank alleles consist of known, but unreactive, allelic products (Mytilineos *et al.*, 1990). Population differentiation due to differing allele frequencies and distinct haplotypes is, however, observed.

Some examples include the fact that allele *Bw46* is found exclusively in Asian populations, the virtual absence in Japanese of the common Caucasian allele *DR3*, the nearly exclusive occurrence of *Aw43* in African populations (du Toit *et al.*, 1990), and the fact that some populations reveal less overall polymorphism, e.g., South American Indians (Black and Salzano, 1981). Nonetheless, we want here to emphasize the great extent of shared *HLA* polymorphism among the world's populations. We propose that answers to the substantial universality of most *HLA* alleles are best sought in population level forces acting through time to maintain and preserve the polymorphism we observe today.

Can the broad strokes outlining the major themes of evolutionary history in the MHC be connected to the intragenerational events which must lead ultimately to an explanation of MHC polymorphism? Here we examine population level phenomena capable of leaving long-lasting signatures—heterozygosity and gametic disequilibrium—revealed through the distribution of alleles at a locus, and the patterns of association observed in two-locus haplotypes. Under neutrality conditions these measures require on the order of N generations to reach equilibrium (where N is the effective population size) and considerably less when selection is present (Kimura and Ohta, 1971). This time period is conceptually sandwiched between the millions of years over which phyletic evolution is observed, and the generation-to-generation periods in which mutation events, selection events, recombination, gametes, individuals, and families come and go. The tactic employed here is to analyze a generation of intensively sampled and HLA-typed *Homo sapiens.*

THE PROVINCES FRANCAISES SURVEY

Because of the growing interest in HLA matching in transplantation, a large genetic survey was initiated in the early 1980s to estimate the geographic distribution of genetic polymorphism in the population of France (Ohayon and Cambon-Thomsen, 1986). In order to enable the assignment of haplotypes, sampling was family based, requiring both parents and at least two children. To give some impression of the historical distribution of populations, a sampled family had to have been settled in an area for at least three generations. No more than five families were selected from a single village. A total of 1287 (mostly) nuclear families were sampled (resulting in 5148 sampled haplotypes), divided among 15 regions across France, including Corsica (Fig. 4). The total also includes 89 French-speaking families from Quebec.

FIG. 4. Distribution of regions sampled in the Provinces Francaises genetic survey.

Twenty-five polymorphic loci corresponding to 13 independent genetic systems were studied, including *HLA*. Here we examine variation in the *HLA*-region loci *A*, *C*, *B*, *Bf*, *C4A*, *C4B*, *DR*, and *GLOI*.

Allele frequencies, haplotype frequencies, and gametic disequilibrium values were estimated using the Family Analysis Program, FAP (Neugebauer *et al.,* 1984; Borot *et al.,* 1986). The program estimates these parameters in analyzing phenotypic pedigree data by likelihood maximization. The genetic model applied to the *HLA* loci was for codominant alleles, except for the blank allelic classes in each of the histocompatibility loci and in *C4A* and *C4B,* which were modeled as recessive. A common set of 192 preselected HLA antisera (48 anti-HLA-A, 85 anti-HLA-B, 20 anti-HLA-C, and 39 anti-HLA DR) was used by the 17 histocompatibility-typing laboratories which performed the typing. Analysis of the serological data (Sierp *et al.,* 1986) gave rise to computerized cell typing which was taken into account for the FAP analysis.

Although humans share the great majority of genetic variability across all populations and races (Lewontin, 1974), patterns of microdifferentiation can nonetheless be distinguished when sufficiently informative loci are analyzed. Multivariate analysis of *HLA* allele frequencies have uncovered subtle patterns of clinal variation across Europe (Menozzi *et al.,* 1978; Sokal *et al.,* 1991). These have been interpreted as revealing past population movements, including the admixture of neolithic farmers with the mesolithic peoples native to Europe at that time (Ammerman and Cavalli-Sforza, 1984). Analysis of the Provinces Francaises data has revealed trends of *HLA* variation among populations within France (Cambon-Thomsen *et al.,* 1989). In this study the *HLA* loci *B* and *DR* were considered most informative in defining interregional variability. Corsica and Bearn, which are both isolated by geography and time from other French populations, are most divergent genetically. Multivariate analyses utilizing all of the genetic data uncovered north–south, northeast–southwest, and east–west clines across France, presumably reflecting past population movements.

SINGLE-LOCUS ANALYSES

Single Locus—Genotype Frequency Distributions

Our intention at this point is to evaluate the overall quality of the data base, and form a basis for combining the 16 separate population samples, as is done in the analysis of gametic disequilibrium below. We examine the interpopulation differences by looking at the genotypic ratios in the combined sample for the four histocompatibility loci (*A*, *C*, *B*, and *DR*).

Hardy–Weinberg expectations for *HLA-A* and *HLA-B* were close to observed genotype frequencies in the total Provinces Francaises data. Departures from Hardy–Weinberg expectations were observed for both *HLA-C* and *HLA-DR* when tested in the combined sample (Table I). For *HLA-C* the results suggest that available typing produces assignments that contain a degree of inaccuracy, even with complete pedigree data. For the *C* locus all identified homozygotes (alleles *Cw1* to *Cw7*) were present in deficit, except for the blank homozygotes. HLA-C specificities are known to produce weak serological reactions, due to reduced mRNA production of the *HLA-C* locus (Davidson *et al.,* 1985). The frequency of the blank allelic class (all alleles for which the product is not detectable by known antisera) at the *C* locus, despite international efforts to develop specific antisera for all alleles of the class I loci, remains quite high (32% in the Provinces Francaises data). This sizeable fraction of blank alleles may largely represent individuals whose *HLA-C* expression is too low for serological typing, rather than unidentified alleles for which no antisera are known. We suspect that the high frequency of untypeable genes, as reflected in the incidence of blank alleles, may in some way be contributing to the consistent deficiency of the homozygote classes.

The results for the *DR* locus must be viewed against the established difficulties in serological DR typing based on comparison to DNA typing

TABLE I. Observed and Expected Numbers of *HLA-C* and *HLA-DR* Genotypes under the Assumption of Hardy–Weinberg Equilibrium for the 16 Provinces Francaises Regions Combined[a]

HLA-C			*HLA-DR*		
Genotype	Observed	Expected	Genotype	Observed	Expected
* X/X	15	19.3	* 1/1	44	30.5
X/w3	52	43.6	1/2	87	78.7
X/w4	57	53.6	1/3	86	94.1
X/w5	49	38.3	1/4	95	80.4
X/w6	44	37.0	1/7	70	88.7
X/w7	89	94.3	1/5	62	76.8
X/0	131	146.7	1/6	51	50.6
* w3/w3	16	24.9	1/Y	30	38.8
w3/w4	57	61.7	* 2/2	57	50.8
w3/w5	38	43.7	2/3	107	121.3
w3/w6	52	42.7	2/4	108	103.7
w3/w7	118	108.4	2/7	106	114.4
w3/0	168	167.3	2/5	77	99.1
* w4/w4	21	38.6	2/6	90	65.2
w4/w5	54	54.2	2/Y	45	50.1
w4/w6	59	53.5	* 3/3	90	72.6
w4/w7	151	135.2	3/4	107	123.9
4/0	223	207.5	3/7	127	136.9
* w5/w5	8	19.2	3/5	146	118.5
w5/w6	39	37.5	3/6	67	78.1
w5/w7	123	95.1	3/Y	58	59.8
w5/0	135	146.9	* 4/4	68	52.9
* w6/w6	9	18.5	4/7	100	116.9
w6/w7	111	93.6	4/5	103	101.3
w6/0	122	143.6	4/6	49	66.7
* w7/w7	80	118.5	4/Y	52	51.2
w7/0	376	364.3	* 7/7	84	64.6
0/0	292	281.4	7/5	119	111.8
			7/6	81	73.6
			7/Y	57	56.4
			* 5/5	51	48.4
			5/6	55	63.7
			5/Y	53	48.9
			* 6/6	23	20.9
			6/Y	33	32.2
			* Y/Y	17	12.3

[a] To reduce the influence of small sample sizes, the alleles *Cw1* and *Cw2* are combined into the single class X. Likewise, the *DR* alleles 8, 9, 10, and 0 (blank) are combined into the class Y. Homozygotes are indicated by an asterisk.

(Mytilineos *et al.*, 1990). In contrast to the pattern of genotypic ratios in *HLA-C*, the *HLA-DR* locus had an excess of homozygotes in each of the eight possible homozygote classes (Table I). Homozygote excess is of course one possible indicator of the Wahlund effect, resulting from the admixture of populations with differing allele frequencies. That *HLA-A* and *HLA-B* showed no deviations from Hardy–Weinberg ratios suggests that the *DR* pattern requires another explanation. One interpretation of this observation is that the allelic products of *DR* from each chromosome are expressed in uneven amounts. When overall expression is weak, an individual might be typed as homozygous for the DR antigen most common on the cell surface. Support for this possibility comes from evidence that the regulatory regions of class II loci differ from allele to allele (Andersen *et al.*, 1991), which might result in consistently different amounts of expressed protein.

Contingency tests of allele frequency versus sex suggest possible heterogeneity at *HLA-C* ($P < 0.05$). One allele, *Cw1*, was the primary contributor to this effect, with a frequency of 0.042 in males and 0.030 in females. The *HLA-C* result might be due to sampling error, because *HLA-A, HLA-B,* and *HLA-DR* allele frequencies did not differ between the sexes.

Single Locus—Allele Frequency Distributions

Combatants in the selection–neutrality debate argue over the ubiquity and intensity of selection at any given locus across the genome. Neither side takes the position that outstanding examples of their opposition's claims might not exist at some loci. Given this standoff, the inquiry then shifts to discovering the frequency, overall incidence, and intensity of selection across the genome. With the examination of single-locus allele frequency distributions it is possible to get an initial impression of the selection history of a locus.

The expected frequency distribution of alleles at a locus under neutrality conditions at equilibrium has been described by Ewens (1972). Watterson (1978) subsequently presented the homozygosity F statistic as a single index, conditional on the number of alleles and sample size, for testing the frequency distribution of alleles at a locus against neutrality expectations:

$$F = \sum p_i^2$$

where p_i is the frequency of allele i at a locus. (Note that the test is of the allele frequency distribution and not for deviations from Hardy–Weinberg expectation.) In examining natural populations of *Drosophila pseudoobscura* for allelic variation at the esterase-5 locus, Kieth (1983) found evidence for purifying selection, due to homozygosity F values being higher than that expected under neutrality.

Considering all points for the 16 populations and seven *HLA*-region loci in the Provinces Francaises data, the number of alleles per locus varies from 4 to 28, while the homozygosity F statistic ranges from 0.07 to 0.68 (Fig. 5). The minimum homozygosity possible for a given number of alleles at a locus occurs when all of the alleles have the same frequency, and creates a lower bound. The F values expected under neutrality supply a useful point of reference for interpreting the evolutionary history of a locus. The expected F values under neutrality for a given number of alleles in Fig. 5 are for a sample size of 400 genes, which approximates the sample size obtained from most of the individual regions in the Provinces Francaises study. The F values for each locus occupy a discrete space defined by the 16 populations. Only the loci *C4A* and *C4B* overlap in their range.

Although the loci are distinguishable from each other on this basis, what of the individual populations within each locus? Does any particular population have a consistent tendency to be closer or farther from the neutrality expectation? This can be clarified by examining the F values of the two populations expected to be most divergent from the rest of the Provinces Francaises samples, i.e., Bearn (in the Pyrenees) and Corsica (isolated from the French mainland) (Cambon-Thomsen and Ohayon, 1986; Cambon-Thomsen *et al.*, 1989). The apparent random occurrence of these two potential outliers can be seen from their locations within the range of values for the 16 populations defining each locus (Fig. 5). This can be more formally estab-

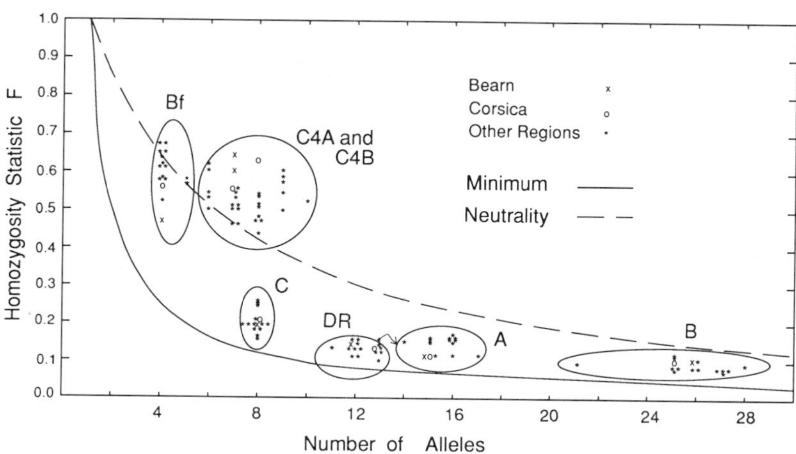

FIG. 5. Homozygosity F values plotted against polymorphism (number of alleles) in the Provinces Francaises population samples for each of seven *HLA*-region loci.

lished by a two-way ANOVA test of locus × population, using the variable $F_{obs} - F_{exp}$. The ANOVA was not significant for any of the following combinations tested: all seven loci, the three class III loci alone, or the four histocompatibility loci alone. Thus, we find no evidence for distinct selection between the populations.

To evaluate results from the allele frequency distributions, we need to ask what assumptions or perturbations might impact the homozygosity statistic. Two pertinent considerations for these data are allelic misclassification and undetected allelic heterogeneity or allelic splits. If we assume some fraction of the typings are actually randomly assigned (due, for example, to transcribing errors), then this will tend to even the allele frequencies, and thus lower the observed *F*. The magnitude of this effect depends on both the distribution of the accurately typed alleles and on the fraction of random typings. The other category of misclassification occurs if typing errors are biased, which might be toward either common or rare alleles. The former case raises, and the latter case lowers, *F*. Based on the examination of genotypic ratios discussed above, only the typing of HLA-C suffers from the flagrant commitment of both of these errors, while for HLA-A and -B the serological typing seems to be of consistent quality. For the class II loci (here *DR*), typing using sequence-specific oligonucleotide probes of the primary polymorphic (first) domain has been used to achieve virtually complete allelic identification in Caucasian population samples (Scharf *et al.*, 1991), and the *F* values did not differ from the results gained from serological DR typing (Begovich *et al.*, 1992). Results from RFLP typing (Yu *et al.*, 1986) speak against the possibility of consistent mistyping of C4 alleles into the common allelic class at each locus, which would artificially inflate *F*. In conclusion, we are confident of typing quality, except in the case of HLA-C.

The histocompatibility loci *A*, *B*, *C*, and *DR* are all uniformly lower in the homozygosity statistic *F* conditional on the number of alleles at the locus than the neutrality expectation, indicating a more even allele frequency distribution than expected under neutrality. The three complement loci (Fig. 5), in contrast, show different patterns. The *F* values for the *Bf* populations seem to cluster around the neutrality expectations, while the two loci *C4A* and *C4B* have *F* values which tend to exceed the neutrality expectation.

Allele frequencies at the histocompatibility loci (*A*, *C*, *B*, and *DR*) are all more even than expected under neutrality, suggesting the past operation of some form of balancing selection. Selection models of both overdominance and frequency-dependent selection can be entertained. It must, however, be emphasized that the form of the selection is unknown, and potential selective mechanisms should be examined in light of classical population-genetics theory, which shows that it is very difficult to maintain so many alleles even with strong balancing selection. The complement loci, although located in the 1-cM gap between the class II *DR* locus and the class I *B* locus (see Fig. 1), seem to be under quite different selection regimes. Evolutionarily recent directional selection in favor of the common *C4A* and *C4B* alleles and biased gene conversion are interpretations concordant with these observations.

Examination of Chinese, Japanese, and other Caucasian samples at this same set of class III loci revealed similar patterns (Klitz *et al.*, 1986). Not all human populations typed at the histocompatibility loci suggest a history of balancing selection according to the homozygosity statistic. The results for populations from Papua New Guinea, sampled for the *HLA* loci *A*, *B*, and *C*, seem to be indistinguishable from neutrality expectations (Bahtia *et al.*, 1989). Anthropological and genetic data suggest that these populations may have originated from a small founder population some 40,000 years ago (Serjeantson, 1989). Eventually a complete understanding of *HLA* selection must incorporate these divergent results.

MULTILOCUS ANALYSES

Linked polymorphic loci create the opportunity to extend the evolutionary analysis beyond single-locus theory. Unlike genotypic ratios, which return to Hardy–Weinberg frequencies in a single generation under random mating, gametic (or linkage) disequilibrium between two loci decays as a function of the recombination rate, and so may persist for many generations after the occurrence of forces responsible for the generation of the disequilibrium in the first place. Given the availability of both physical and genetic maps for the *HLA* region, the primary forces (population size, mutation rate, migration, selection, and recombination) present themselves for our dissection.

Recently it has become feasible to analyze experimentally mutation and recombination in the MHC. It is now possible to type single sperm for analysis of recombination between desired loci using informative males (Cui *et al.*, 1989), and to detect spontaneous genetic events in *H-2* loci by using a graft rejection assay to reveal changes in class I loci (Nathenson *et al.*, 1986; Melvold and Kohn, 1990). An experimental system has been created for observing class I gene conversion events by integrating class I sequences into yeast (Wheeler *et al.*, 1990). With the availability of increasingly detailed knowledge of both mutational and recombinational phenomena, it is becoming clear that in neither of these cases will a single summary statistic be sufficient. Recombination does not occur uniformly across the genome, nor even necessarily at a uniform rate among different haplotypes. In the mouse, for example, MHC haplotypes have been found which promote recombination in a particular region (Uematsu *et al.*, 1988; Shiroishi *et al.*, 1990). Mutational mechanisms have also demonstrated a variety of novel influential effects that force us to replace past views that the primary mover of genetic change is simple point mutations. Molecular analysis has shown that both segmental exchange (Kuhner *et al.*, 1990) and recombination (Holmes *et al.*, 1987) events have contributed to the genetic landscape observed in the vertebrate MHC. In addition, discovery of retroviral-like sequences in the introns of histocompatibility genes (Bosch *et al.*, 1989) and variation in haplotype sizes of up to 100 kb among HLA haplotypes (Dunham *et al.*, 1989) suggest the possible operation of transposon-mediated insertion–deletion events. Despite these complexities, we assume here that recombination estimates based on population samples might still be richly informative as to the overall evolutionary forces which have produced HLA disequilibrium.

Multilocus Analyses—Measuring Gametic Disequilibrium

The analysis of pairwise disequilibrium among multiallelic loci presents special challenges requiring both the modification of statistics designed for the analysis of diallelic loci and the development of entirely new measures and approaches.

The disequilibrium for a single haplotype is defined as

$$D_{ij} = x_{ij} - p_i q_j$$

where x_{ij} is the observed frequency of haplotype A_iB_j at loci A and B. The expected frequency of gamete A_iB_j is then p_iq_j, assuming no allelic association. A haplotype may be either more or less frequent than expected, resulting in D_{ij} values that are either positive or negative, respectively. The range of this measure is a function of allele frequencies. Although it is clear that no ideal measure of allelic association exists (Hedrick, 1987; Lewontin, 1988), it is often useful to employ D'_{ij}, which is the value of D_{ij} normalized by the maximum value it could have for this combination of allele frequencies and sign (Lewontin, 1964). D'_{ij} ranges from -1 to $+1$. The value -1 occurs when no samples of haplotype A_iB_j are present, while $+1$ occurs when the rarer allele A_i or B_j is found only in haplotype A_iB_j. The most common HLA haplotypes in the Provinces Francaises are displayed along with their two-locus D' values in Table II.

As a statistical test to establish the presence of disequilibrium between two multiallelic loci, we perform a row by column (R × C) test of heterogeneity (Sokal and Rohlf, 1981), using the log likelihood (G) or Pearson (χ^2) test

TABLE II. The Most Common Seven-Locus Haplotypes in the Provinces Francaises Survey

Locus:	Haplotypes							Estimated frequency	D′	
	A	*C*	*B*	*Bf*	*C4A*	*C4B*	*DR*		*A-B*	*B-DR*
	1	7	8	S	Q0	1	3	0.0342	0.64	0.64
	29	0	44	F	3	1	7	0.0212	0.55	0.27
	3	7	7	F	3	1	2	0.0169	0.34	0.46
	2	5	44	S	3	Q0	4	0.0121	0.11	0.09
		5	18	F1	3	Q0	3	0.0108	—	0.31
		4	35	S	3	1	11 (5)	0.0090	—	0.06
	2	3	62	S	—	—	4	0.0089	0.24	0.13
	1	6	51 (17)	S	6	1	7	0.0081	0.31	0.43
		6	13	S	3	1	7	0.0081	—	0.57
		3	60	S	3	1	4	0.0067	—	0.21
	3	4	35	F	—	Q0	1	0.0060	0.12	0.18
		6	50	S07	2	1	7	0.0036	—	0.41
	2	7	18	S	3	1	11	0.0032	0.02	0.04
	2	0	51	S	3	1	2	0.0025	0.20	0.13

statistic. The problem of small expected numbers in many cells, which is inevitably present when examining sets of highly polymorphic loci, was reduced by combining rare alleles to give a minimum total of 10 (for a given row or column). [Larntz (1978) has shown that expected values as low as one per cell result in reliable R × C tests.]

We can now check for the presence or absence of disequilibrium between any pair of loci, testing the null hypothesis: Disequilibrium = 0. All pairwise combinations of the seven *HLA*-region loci proved significant for nonrandom association. The most distant combination of *HLA* loci sampled in Provinces Francaise, i.e., *A* and *DR* (1.7% recombination), having relatively low overall allelic association (as predicted), were still very significantly associated. The R × C test on the *A-DR* haplotypes has a *G* value of 1003 and a χ^2 of 1174, and so with 150 degrees of freedom is highly significant. In contrast, disequilibrium values of these same seven loci with *GLOI*, the highly heterozygous diallelic locus located some 5 cM centromeric to *DR*, were all nonsignificant.

In the presence of significant nonrandom association among the alleles of different loci on a chromosome, we can next examine the effect of recombination on the observed disequilibrium. Under neutrality a finite population at equilibrium has expected nonzero gametic disequilibrium—the combined effects of genetic drift and mutation result in interlocus association, which is highest when recombination is limited (Ohta and Kimura, 1969; W. G. Hill, 1975).

For examining the degree of disequilibrium among pairs of multiallelic loci, we adapt from Cohen (1988), pp. 215–271, the following statistic:

$$W = \sum \sum D_{ij}^2/p_i q_j$$

W expresses the total absolute deviation of observed from expected values in a contingency table. When W is squared and multiplied by N, the total sample size, it becomes the standard Pearson χ^2 statistic.

Values of W for all pairwise combinations of the seven HLA loci have been plotted against recombination distance in Fig. 6. The two-point haplotypes of the four histocompatibility loci *A*, *C*, *B*, and *DR* form a trend of increasing disequilibrium with decreasing map distance. The addition of the pairwise disequilibrium values involving the complement loci (*Bf*, *C4A*, and *C4B*) leaves two impressions. First, there seems to be more variability in W, and second, the relationship of W with map distance is less than that for the other sets of loci, especially as the recombination distance approaches zero. Overall disequilibrium calculated for the *C4A-C4B* haplotypes is higher than for *C4A-Bf* and *C4B-Bf*. The tight linkage of the *C4* loci, separated by just 4 kb of DNA (Campbell *et al.*, 1988), would seem to fit this observation. However, the *Bf* locus is also within 100 kb of the *C4* loci, and the pairwise disequilibrium values for *Bf-C4A* and *Bf-C4B* seem anomalously low. The relatively lower polymorphism and heterozygosity present in the complement loci may contribute to these observations. We have also included in Fig. 6 two-point disequilibrium values among the *DRB1*, *DQA1*, and *DQB1* loci taken from a separate study of Caucasian haplotypes (Begovich *et al.*, 1992), which are separated by some 100 and 20 kb, respectively (Trowsdale and Campbell, 1988). The virtually complete disequilibrium among these three loci results in high values of W.

Our results here suggest that for *HLA* loci, at least, random association among alleles at different loci declines to zero in the range of 2–5% recombination. Because virtually no other data are available, we cannot gain a wider perspective utilizing comparable data from other parts of the genome or from other organisms. The analysis hints that genic variation at the two C4 complement factor loci is out of step with the pattern suggested by other HLA loci. This distinct characteristic of *C4A* and *C4B* adds to the observations made on the *C4* loci in the single-locus analyses.

Multilocus Analyses—Disequilibrium Patterns

Recombination among tightly linked loci not only determines the overall disequilibrium, but also delineates the kind of information that can be

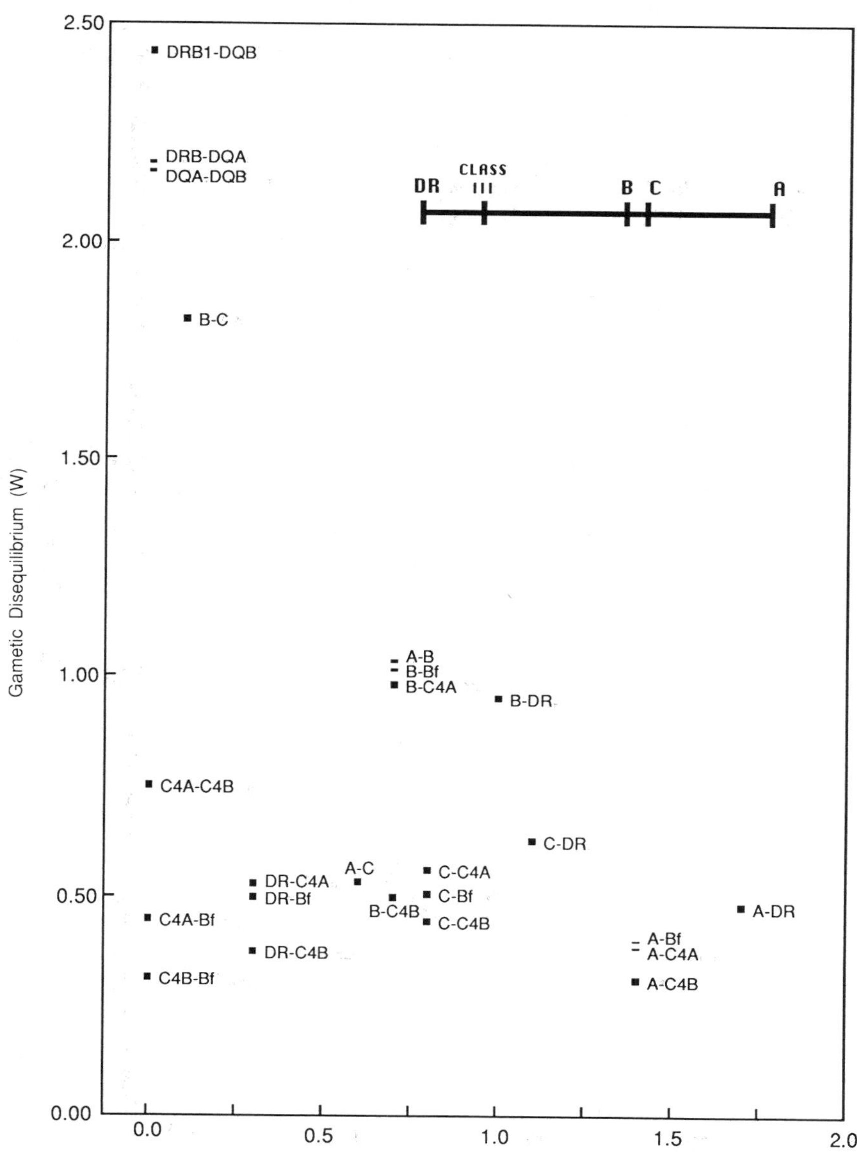

FIG. 6. The relationship of two-locus disequilibrium to recombination map distance in the *HLA* region. All of the two-locus disequilibrium values are significantly different from $D = 0$. All data are from the Provinces Francaises survey except for pairwise disequilibrium among *DRB1*, *DQA*, and *DQB*, which is taken from Begovich *et al.* (1992).

extracted. When linkage is exceptionally tight so as to put the recombination rate on an evolutionary time scale, then the observed associations among the alleles on a chromosome may be a record of the relative ages of alleles and haplotypes (Begovich *et al.*, 1992). Recombination, when present at measurable levels, becomes a population-wide force, reassorting allelic combinations and reducing any associations. By examining the disequilibrium of each haplotype and the patterns of haplotypes in the entire population, it is possible to make inferences about admixture and selection, and to identify those haplotypes which have been distinctly affected by selection within the evolutionarily recent past (Thomson and Klitz, 1987).

A useful starting point for this detailed analysis of haplotype frequencies is examination of the overall distribution of haplotypes according to the normalized disequilibrium D' against the expectations under conditions of selective neutrality. Using a Fisher–Wright neutral allele model, Klitz and Thomson (1987) simulated a population with two loci having a recombination fraction of 0.8% (assuming an effective population size of 5000), and a mutation rate sufficient to generate 10–14 alleles at one locus and 17–23 at a second locus. These simulated results, based on 1010 replicate runs, form the basis for comparison to the haplotypes *A-B* and *B-DR*. The loci *A* and *DR* have similar degrees of polymorphism and similar recombination distances from *HLA-B*. *DR* is a class II locus, while *A* and *B* share the function and structures common to class I loci.

Both *A-B* and *B-DR* indicate clear departures from that expected under strict neutrality (Fig. 7). This is most noticeable in the D' interval 0.0–0.1 and at -1.0, where the Provinces Francaises data fall above and below the neutrality expectation, respectively. The -1 class are haplotypes missing from the population. The large fraction of haplotypes observed in this class under the neutrality case possess a new allele which has just appeared in the population. The one or very few haplotypes on which this new allele exists have not undergone recombination. Most of these new haplotypes disappear from the population before much recombination occurs. An additional contrast with neutrality is the tendency for the Provinces Francaises haplotypes in the space from 0.0 to -0.7 to be more common than under neutrality. Overall the *A-B* and *B-DR* haplotype frequency distributions are similar, although some differences appear to exist in the -1.0 class, and in the interval from 0.0 to -0.7. Having introduced the disequilibrium space indicating the location of all haplotypes in a two-locus system, and demonstrated the overall distribution of haplotypes in that space, we are now prepared to identify those few haplotypes which are largely responsible for the disequilibrium observed in the entire population.

Disequilibrium pattern analysis is a method of examining the evolutionary dynamics of a multiallelic two-locus system (Thomson and Klitz,

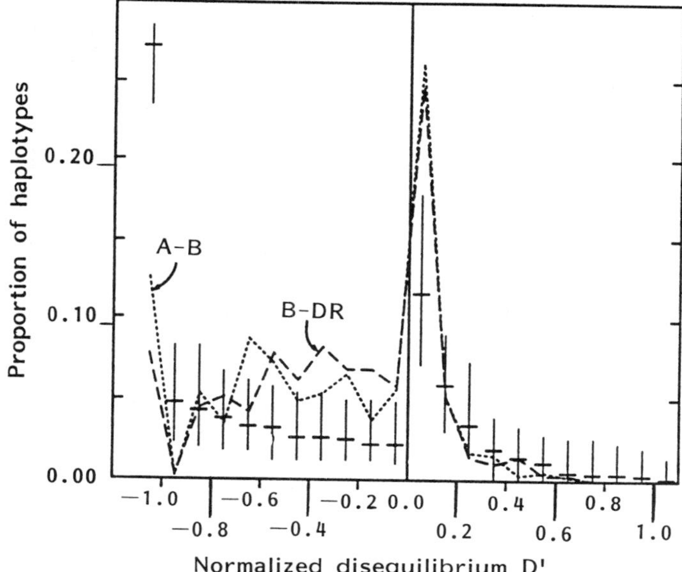

FIG. 7. The frequency distribution of the normalized disequilibrium D' of all haplotypes for *HLA-A-B, HLA-B-DR,* and the simulated neutrality expectations. For each interval of the simulated values, the mean and 95% interval are indicated.

1987). The disequilibrium values of the haplotypes in a two-locus system can be usefully displayed in the disequilibrium space defined by the expected haplotype frequency $p_i q_j$ as the ordinate and the disequilibrium (expressed as either D or D') as the abscissa (Fig. 8). For simplicity, we imagine a population which is initially in linkage equilibrium ($D_{ij} = D'_{ij} = 0$) and in which the alleles at each of the loci have the same frequency. If selection occurs favoring one haplotype at the expense of all other haplotypes, that haplotype increases in frequency and moves into the positive portion of the disequilibrium space. Each of the two classes of "related" haplotypes (those sharing one allele with the selected haplotype) move equidistantly into the negative space. This action requires that all other haplotypes—the "unrelated" (those haplotypes sharing no alleles with the selected haplotype)—move into the positive space. The extent of movement of the related haplotypes is determined by the magnitude of selection, while the extent of unrelated movement is due to both to the degree of selection and the total number of alleles present at both loci. The algebraic necessity for this haplotype movement can be seen by examining the system as a 2 × 2 table, where the system is analyzed as two diallelic loci (Fig. 9). The convention $\overline{A1}$ and $\overline{B1}$ denotes all

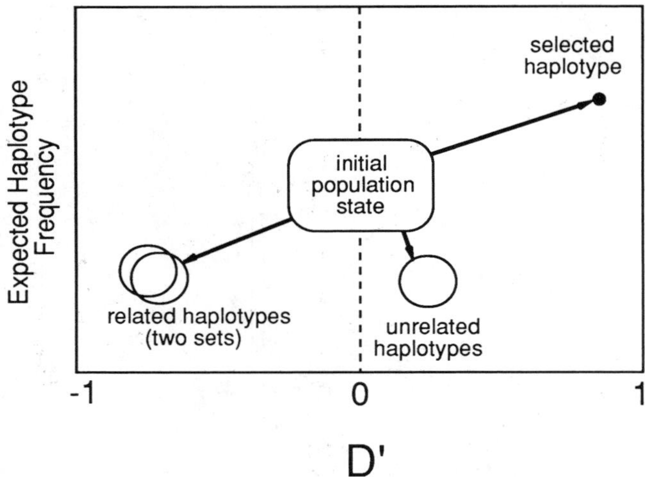

FIG. 8. The response of a system of two-locus multiallelic haplotypes in linkage equilibrium to selection favoring one haplotype.

alleles except *A1* or *B1*. By definition, the sum of the disequilibrium values for any row or column must be zero, so that the single positive *D* value present on the selected haplotype forces all other haplotypes to possess either positive or negative disequilibrium values. Selection favoring one haplotype still pushes all haplotypes in the population in the directions indicated in Fig. 8, regardless of initial haplotype frequencies. The force of recombination acts continuously to push all haplotypes toward $D = 0$. Under conditions of selection favoring a haplotype, the related haplotypes are uniformly held in the negative space by the counterbalancing forces of selection and recombi-

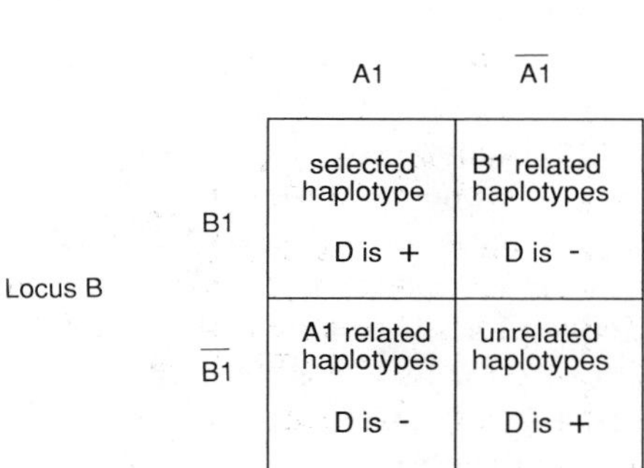

FIG. 9. The dynamics of selection favoring the haplotype *A1-B1* in a two-locus multiallelic system depicted as a 2 × 2 table. The sum of the disequilibrium values *D* for any row or column must equal zero.

nation, and have a specific pattern of expectation under selection, namely, the disequilibrium is proportional to the frequency of the unshared allele and have a single expected value of the normalized disequilibrium. The admixture of two populations with differing allele frequencies can generate disequilibrium, but the selection pattern of haplotype distribution is mimicked by admixture only under very stringent conditions. The only feasible case occurs when a single haplotype is fixed in one of the parental populations, which is unlikely in the highly polymorphic HLA system.

The array of disequilibrium values at all two-point haplotypes for the loci *A*, *B*, and *DR* were examined in the Provinces Francaises data for patterns suggestive of selection. The complement proteins coded by *Bf*, *C4A*, and *C4B* are not suitable for disequilibrium pattern analysis because of their lower polymorphism and lower heterozygosity. Our description of the disequilibrium patterns found in Provinces Francaises data begins with the haplotype *A1-B8*, long renowned for its high disequilibrium. Figure 10 displays all *B8* haplotypes defined by the 16 *A* alleles in the normalized disequilibrium space. Disequilibrium pattern analysis of these haplotypes was first described in a sample of Danes (Klitz and Thomson, 1987). Here we find the haplotype *A1-B8* having an observed frequency of 0.068, an expected frequency of 0.0128 under random association, and a D' of +0.64. The only other *B8* haplotype having positive disequilibrium is the rare *A10n* (for "10 new," which includes the non-*A25* and non-*A26* *HLA-A10* variants). The rest of the *B8* haplotypes occupy the negative space, and have a D' value close to −0.64, as expected under selection. Even with the large sample size available from the Provinces Francaise study, sampling error becomes evident in the rarer haplotypes. The other two-point haplotypes (*A* with *DR*, and *B* with *DR*) comprising the *A1-B8-DR3* haplotype show similar, if less striking patterns to that found in the *B8-A* haplotypes (data not shown). These three alleles are part of the most frequent extended HLA haplotype found in Northern Europe and in the Provinces Francaises data: *A1-Cw7-B8-C2C-BfS-C4AQ0-C4B1-DR3-DQw2*.

Rare alleles exhibit selection patterns as well. The allele *B13*, with a frequency of only 0.016, is a case in point (Fig. 11). The haplotype *B13-DR7* is alone in the positive space, with the other *DR* haplotypes clustered around $D' = -0.5$. Sequence analysis has shown that the rare allele *B47*, which is also in high linkage disequilibrium with *DR7*, arose via complex mutational events from *B13*, *B27*, and *B44* (Zemmour *et al.*, 1988). At the least, this suggests that the selectively favorable characteristic of the *B13-DR7* haplotype is not due to the recent appearance of the *B13* allele as such, because *B13* has been around long enough to give rise to another allele.

For contrast to cases indicating apparent selection, we include the illustration of the *A2-B* haplotypes (Fig. 12) as an example illustrative of the great

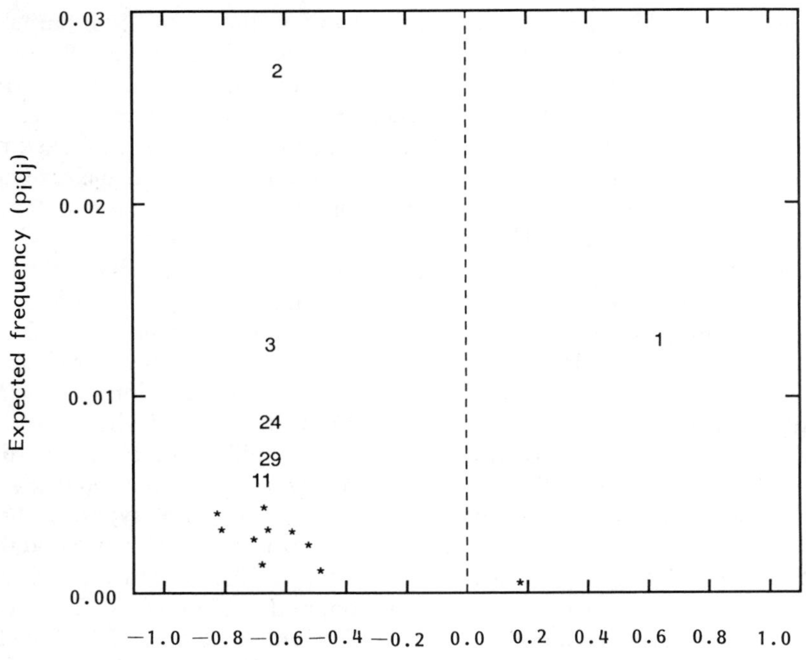

FIG. 10. All *A-B* haplotypes containing the allele *B8,* where numbers indicate the allele designation at the *A* locus in the disequilibrium space, defined by the normalized disequilibrium measure *D'* and by the expected haplotype frequency under random association (the product of each of the constituent alleles). This pattern of haplotype, with one (or a few) frequent haplotypes in the positive space (here *A1-B8*), and all other haplotypes clustered around a single negative *D'* value is indicative of selection. Rare haplotypes are indicated with an asterisk.

majority of cases when the distinctive pattern indicative of recent selection is not present. As is apparent, the disequilibrium values are scattered throughout the positive and negative space.

Altogether six haplotypes from the Provinces Francaises data demonstrated disequilibrium patterns strongly suggestive of selection: *A1-B8-DR3, A3-B7-DR2, A29-B44-DR7, A1-B17-DR7, A3-B35-DR1,* and *B13-DR7.* At this point, ascribing selection to a haplotype is based on the presence of a particular pattern of haplotypes in the disequilibrium, not on formal statistical testing. It seems likely that only strong or recent cases of selection can be uncovered in this way. Older selection events, diluted by recombination, and covered by subsequent selection, would be difficult to detect. Another contribution to the inability to detect events occurs when the defined allele does not identify the entity under selection.

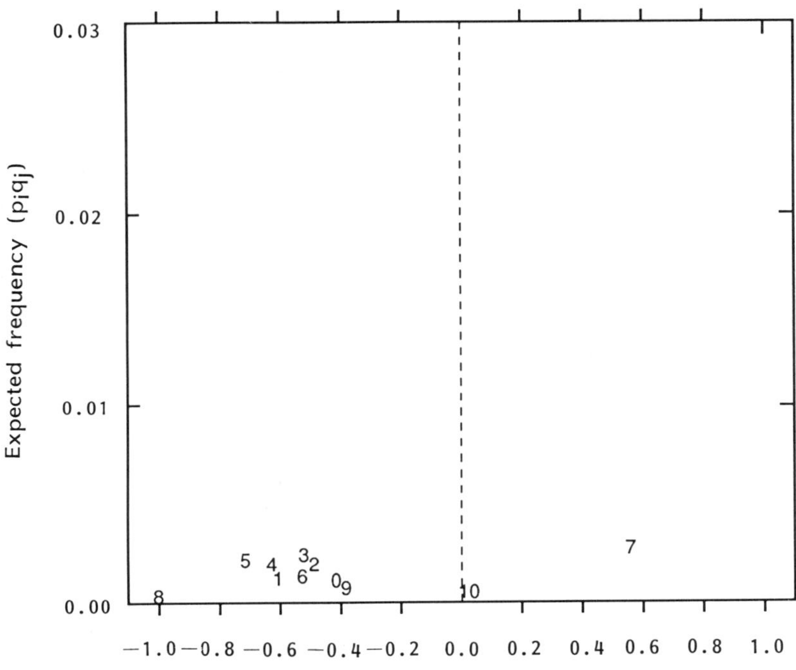

FIG. 11. The *B-DR* haplotypes containing the allele *B13* in the disequilibrium space. Despite its rareness, *B13-DR7* appears to have been selected.

Increased discrimination of allelic specificities might allow better identification of functionally distinct haplotypes. Allelic subdivision can produce a striking impact on the interpretation of the disequilibrium patterns. DR4 is a common serologically defined specificity coded by the *DRB1* locus with a frequency of 0.13 in the combined Provinces Francaises data. It also has been shown to confer susceptibility to several autoimmune diseases, for example, rheumatoid arthritis and juvenile diabetes (Tiwari and Terasaki, 1985). The disequilibrium pattern of the *DR4-B* haplotypes is shown in Fig. 13. The four common DR4 haplotypes in positive disequilibrium are defined by the *B* alleles *44, w62, 27,* and *w60.* The common DR4 haplotypes in the negative space are *B7, B8,* and *B35.* No selection pattern is apparent.

Sequencing of the hypervariable second exon of the *DRB* locus of DR4-typed haplotypes has split *DR4* into eight subtypes (WHO Nomenclature Committee, 1990). Two of these subtypes (designated *DR0401* and *DR0404*) are common in Caucasians. The disequilibrium space locations of the most common *DR4-B* haplotypes have been determined (Begovich *et al.*,

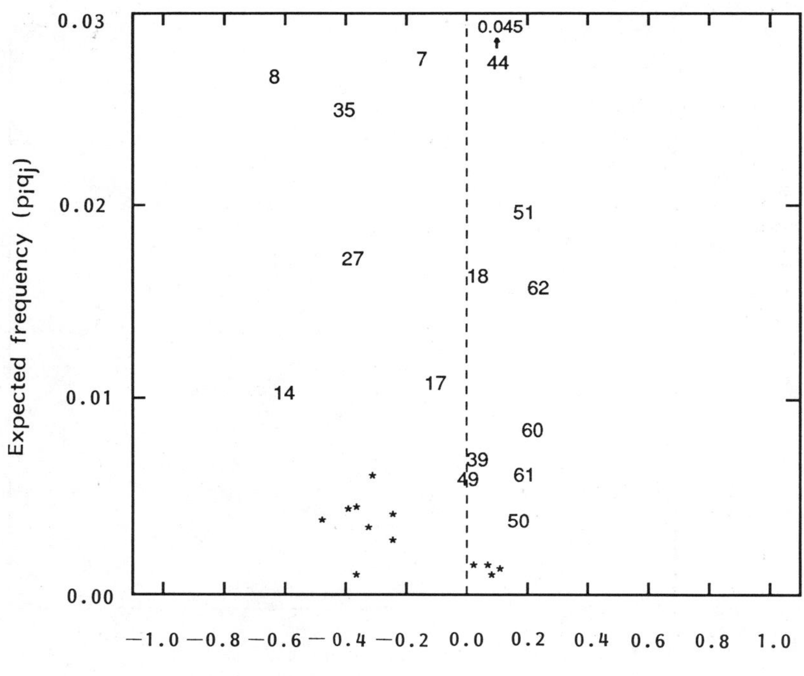

FIG. 12. The *A-B* haplotypes containing the allele *A2* in the disequilibrium space. Numerous haplotypes in both the positive and negative space suggest the absence of recent selection favoring *A2.*

1992). For *DR0401,* the *B44* and *B62* haplotypes both increase in D', while all the other common haplotypes move to the extreme negative side of the D' space. For *DR0404,* haplotypes *B27* and *B60* increase in D', with *B7* moving into the positive space, while all other haplotypes are pushed to more negative D' values. This picture suggests that splits of *DR4* more accurately reveal selection events and come closer to characterizing the actual haplotype under selection. Supertypic allelic classes may be covering up as yet unidentified alleles which may define selected haplotypes.

Multilocus Analyses—Constrained Disequilibrium Values

Given a recombination rate in the vicinity of $c = 0.01$ and highly polymorphic loci, disequilibrium pattern analysis supplies a method for identify-

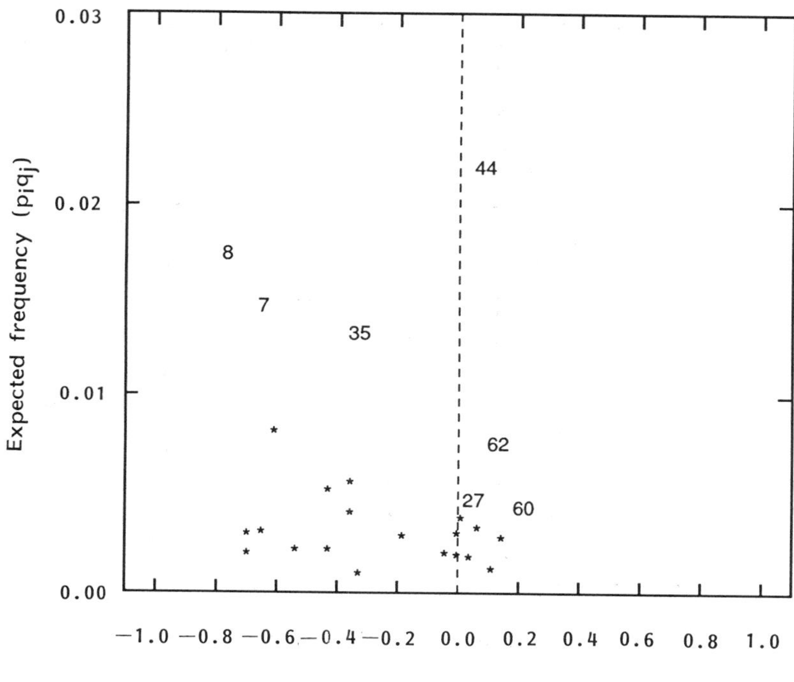

FIG. 13. The *DR-B* haplotypes containing the allele *DR4* in the disequilibrium space. When *DR4* is subdivided into additional alleles, haplotypes change position to assume a selected pattern.

ing haplotypes which have been under positive selection during the previous few hundred generations. Traces of that selection may be left in the population records of pairwise disequilibrium values. If *c* is much greater that 0.01, then evidence of all but very strong or recent selection disappears rapidly, while if *c* is too small, then allelic associations on haplotypes can be created by stochastic forces alone, thus making selective interpretations of disequilibrium more difficult.

Could the disequilibrium values in linked systems also contain data directing us to just which locus is under selection or in the subregion under selection, and therefore worthy of more detailed analysis? A novel approach to the study of disequilibrium in systems having three or more linked loci utilizes the additional constraints on two-way disequilibrium values imposed by a three-locus system to localize the chromosomal region under selection (Robinson *et al.,* 1991*a*).

The behavior of the normalized pairwise disequilibrium measures, using two-locus (D') and three-locus (D'') constraints in a three-locus system, when only one of the three loci is under positive selection, has been examined (Robinson *et al.*, 1991*a*). It turns out that the difference between these pairwise disequilibrium measures, $\delta = D' - D''$, has a distribution for the two unselected loci differing from that for the selected locus with an unselected locus. The hallmark of selection is a high positive value of δ for the two unselected loci. Before analyzing the Provinces Francaises data in this fashion, we should point out that Robinson *et al.* (1991*b*) have also examined the effects of genetic drift on δ, which indicated that positive δ values are unlikely to be found in the absence of selection when recombination occurs at a rate greater than 0.001. These guidelines, then, suggest a fruitful examination of the pairwise disequilibrium among all the typed HLA loci except for the complement group *Bf, C4A,* and *C4B,* which are too closely linked. The 14 most common haplotypes in the total Provinces Francaises sample (having frequencies greater than 0.25%) were chosen for study.

Delta values indicating past selection were present on nine distinct haplotypes, including the six previously detected as selected by disequilibrium pattern analysis, and two indicating selection on rare *Bf* alleles. The results of the constrained disequilibrium analysis for the *A1-B8-DR3* haplotype are shown in Table III. The alleles at three loci of the given haplotype are examined at one time. Pairwise δ values are reported under the "constraining" locus, i.e., the locus that imposes additional constraints on the pairwise dis-

TABLE III. Delta Values for Three-Way Combinations of Alleles at the Complete *A1–B8–DR3* Haplotype[a]

A1	*C7*	*B8*	*BfS*	*C4AQ0*	*C4B1*	*DR3*
−0.024	−0.079	0.125	—	—	—	—
−0.184	—	0.133	—	−0.0184	—	—
−0.124	—	0.159	—	—	—	−0.141
—	−0.122	0.101	—	−0.015	—	—
—	−0.064	0.096	—	—	—	−0.026
—	—	0.013	—	−0.069	—	0.0
—	—	0.051	0.0	—	—	0.0
—	—	0.109	—	0.0	—	−0.060

[a] The alleles at three loci are examined at one time. Pairwise δ values are reported under the "constraining" locus, the locus that imposes additional constraints on the pairwise disequilibrium value between the two other loci. For example, when considering the trio *A1-B8-DR3,* the δ value for *A1-DR3* (0.159) is reported under *B8.* When a large positive δ value between two of three loci is observed, it usually indicates that the constraining locus is being selected, e.g., *B8.* As there are 35 three-way combinations for each locus, only combinations yielding a δ value of magnitude 0.05 or greater are reported. [Adapted from Robinson *et al.* (1991*b*).]

equilibrium value between the two other loci. The positive δ values occurred exclusively at the *HLA-B* locus (*B8*), indicating selection at, or near, this locus. The high disequilibrium and relatively high frequency of the haplotype *A1-B8* caught the attention of HLA investigators 20 years ago, when it was first realized that the originally identified histocompatibility alleles belonged to a system consisting of multiple linked loci. This haplotype has long served as the old HLA war horse, forever doing battle as a demonstration of both HLA disequilibrium and presumed selection. With disequilibrium pattern analysis it has been possible to document more explicitly the positive selective factors conferred on this haplotype. With the method of constrained disequilibrium values we can now determine that this evolutionary event has been due to selection on or near the *B8* allele.

All six of the haplotypes identified by disequilibrium pattern analysis as demonstrating a pattern indicative of selection were also identified by the constrained disequilibrium values method as selected haplotypes, thus, for this case, validating these approaches. In each of these cases the *B*-locus allele was that identified as the selected region of the haplotype. The constrained disequilibrium values method further identified evidence for selection on two rare *Bf*-locus alleles (*F1* and *S07*): haplotypes *DR3-BfF1-B18-A30* and *DR7-BfS07-B50-Cw6*. Additionally, weaker evidence for *DR*-locus selection was found on the haplotype *DR4-B44-A2*.

Multilocus Analyses—Very Tight Linkage

When a new allele is created at a locus, it will be associated with the particular alleles present at other loci on the haplotype where it appears. Initially the association is complete in the founding haplotype, declining thereafter as a function of recombination. When the recombination rate between two loci is low enough to approach the rate of appearance of new alleles, the haplotype on which a new allele appears will be maintained and evident through strong positive disequilibrium. Allelic diversification may of course proceed at different rates and times in two linked loci. This method of reconstructing ancestral combinations of alleles on a haplotype has been employed by Begovich *et al.* (1992) to analyze the haplotypes of the tightly linked *DRB-DQ* loci. Here we examine the *B-C* haplotypes from the Provinces Francaises survey in this way.

The class I loci *HLA-B* and *HLA-C* are tightly linked, separated by a recombination fraction of 0.001 (Fig. 1). According to the simulations of Robinson *et al.* (1991*b*) described above, this fraction marks the border between the region in which disequilibrium due to haplotype creation predominates and the region in which recombination, in the absence of population forces (admixture and selection), operates to actively randomize interlocus associations. For the *B-C* haplotypes, two suggestive instances of haplotype origin stand out. The alleles *B7* and *B8* are each in positive disequilibrium with only *Cw7* (D' = 0.66 and 0.59, respectively). As discussed earlier, these two haplotypes are candidates for haplotypes which have been under selection. Since other *B* alleles are in positive disequilibrium with *Cw6* and *Cw7*, we might attribute these instances to chance. However, the nucleotide similarities of the pairs *B7* and *B8* is very high, indicating common ancestry (Kuhner *et al.*, 1991). We reconstruct this history, then, as an ancestral haplotype, perhaps *B7-Cw7* giving rise to *B8-Cw7*. Bolstering this line of thought is the allele *Bw42*, another member of the *B7* and *B8* group, which is present in Asian populations, but absent from Caucasians. In the Ninth International Histocompatibility Workshop (Albert *et al.*, 1984), *Bw42* in Asians was associated only with *Cw7*.

SUMMARY

The major histocompatibility complex found in vertebrates is a richly polymorphic set of linked loci, involved in regulation of the immune response. This study of variation and evolution in the human MHC is based on the sampling of 16 French populations—the Provinces Francaises survey—comprising a total of 5148 haplotypes typed for seven HLA-complex loci. The analysis of allele frequency distributions at each of the seven sampled loci strongly supports an evolutionary past of stabilizing selection for each of the four histocompatibility loci. In contrast to this, variation in loci coding for the complement proteins was close to, or exceeded in a homozygosity measure, the allelic distribution expected under random genetic drift alone.

The association between alleles at two linked loci is influenced by a variety of forces. Separating out the role of recombination as a primary determinant of disequilibrium can help elucidate the complex disequilibrium patterns present in linked multiallelic systems. As found in the Provinces Francaises data, disequilibrium among the *HLA* loci declines with increasing recombination fraction, with measures of overall disequilibrium being not significantly different from zero when recombination exceeds 2%. Genetic variation in the complement loci *C4A* and *C4B* demonstrated their discordant nature again by having lower two-locus disequilibrium values than that observed among the histocompatibility loci. As might be expected from analysis of the histocompatibility locus allele frequencies, the frequency distribution of two-locus disequilibrium values departs dramatically from that expected under neutrality conditions. The much greater proportion of missing haplotypes found in the (simulated) neutrality case compared to that found in the locus pairs *A-B* and *B-DR* demonstrates the much greater survival time of allelic variants in the histocompatibility loci. Further dissection of disequilibrium in these same locus pairs identified a handful of haplotypes apparently under selection during the past few hundred generations. This group of selected haplotypes is responsible for the great majority of disequilibrium observed between the loci. A method based on two-locus disequilibrium values constrained by the disequilibrium present at a third linked locus was able to identify which allele of a selected haplotype is most closely associated with the selection favoring the haplotype. Our demonstration of selection on specific *HLA* alleles in a haplotypic combination does not, however, elucidate the mechanism maintaining MHC polymorphism over long evolutionary periods. Although balancing selection is a strong candidate for the maintenance of MHC variation, it seems unlikely that one selective mechanism is operating in this complex multigene family, and it would not be surprising if novel mechanisms are eventually uncovered.

When the recombination fraction is very low (i.e., 0.001 or less) the observed haplotypes may represent primordial haplotypes which appear when a mutation first produces a new haplotype and have remained unbroken by recombination up to the present. Candidates for such haplotypes based on sequence similarity in the locus pair *B-C*, separated by a recombination fraction of 0.001, are identified in the Provinces Francaises data.

The difficulty of detecting selection events has long been recognized, and is reflected in the literature by the paucity of examples of selection in natural populations. Here we have examined data in several complementary ways to detect the footprints of selection in an exceptional part of the vertebrate genome.

Acknowledgments

We thank Hanne Ostergard and Mary Kuhner for many useful comments on the manuscript. This work was supported by National Institutes of Health Grants GM35326 and HD12731, by INSERM, CNRS, and by the Association Claude Bernard, France. The authors wish to acknowledge the help of all of the scientific participants in the Provinces Francaises survey in France and Quebec, and the collaboration of the institutes of Prof. Max Baur (Bonn) and Prof. Ekkehard Albert (Munich). We also acknowledge the kind participation of all the Provinces Francaises survey families.

REFERENCES

Albert, E. D., Baur, M. P., and Mayr, W. (eds.), 1984, *Histocompatibility Testing 1984,* Springer-Verlag, New York.

Ammerman, A. J., and Cavalli-Sforza, L. L., 1984, *The Neolithic Transition and the Genetics of Populations in Europe,* Princeton University Press, Princeton, New Jersey.

Andersen, L. C., Beaty, J. S., Nettles, J. W., Seyfried, C. E., Nepom, G. T., and Nepom, B. S., 1991, Allelic polymorphism in transcriptional regulatory regions of HLA-DQB genes, *J. Exp. Med.* **173:**181–192.

Bahtia, K., Jenkins, C., Prasad, M., Koki, G., and Lombange, J., 1989, Immunogenetic studies of two recently contacted populations from Papua New Guinea, *Hum. Biol.* **61:**45–64.

Begovich, A., McClure, G. R., Suraj, V. C., Helmuth, R. C., Fildes, N., Bugawan, T. L., Erlich, H. A., and Klitz, W., 1992, Polymorphism, recombination, and linkage disequilibrium within the HLA class II region, *J. Immunol.* **148:** 249–258.

Black, F., and Salzano, H., 1981, Evidence of heterosis in the HLA system, *Am. J. Hum. Genet.* **33:**894–899.

Borot, N., Neugebauer, M., Sevin, A., Cambon-Thomsen, A., and Hillat, L., 1986, Methods for the genetic analysis of family data in the survey 'French Provinces', in: *Human Population Genetics* (E. Ohayon and A. Cambon-Thomsen, eds.), Vol. 142, pp. 169–175, Colloque INSERM.

Bosch, M. L., Termijtelen, A., van Rood, J. J., and Giphart, M. J., 1989, Evidence for a hot spot for generalized gene conversion in the second exon of MHC genes, in: *Immunobiology of HLA* (B. Dupont, ed.), Vol. II, pp. 266–268, Springer-Verlag, New York.

Bugawan, T. L., Begovich, A. B., and Erlich, H. A., 1990, Rapid HLA DPB typing using enzymatically amplified DNA and nonradioactive sequence-specific oligonucleotide probes, *Immunogenetics* **32:**231–241.

Cambon-Thomsen, A., and Ohayon, E., 1986, Analyse des données génétiques sur l'échantillon global des Provinces Francaises, in: *Human Population Genetics* (E. Ohayon and A. Cambon-Thomsen, eds.), Vol. 142, pp. 297–322 Colloque INSERM.

Cambon-Thomsen, A., Borot, N., Neugebauer, M., Sevin, A., and Ohayon, E., 1989, Interregional variability between 15 French provinces and Quebec, *Collegium Antropologicum* **13:**24–41.

Campbell, R. D., Law, S. K., Reid, K. B., and Sim, R. B., 1988, Structure, organization and regulation of the complement genes, *Annu. Rev. Immunol.* **6:**161–195.

Charron, D., 1990, Molecular basis of human leucocyte antigen class II disease associations, *Adv. Immunol.* **48:**107–159.

Cohen, J., 1988, *Statistical Power Analysis for the Behavioral Sciences,* 2nd ed., L. Erlbaum Associates, Hillsdale, New Jersey.

Crumpton, M. L. (ed.), 1987, HLA in medicine, *Br. Med. Bull.* **43(1):**1–240.

Cui, X., Li, H., Goradia, T. M., Lange, K., Kazazian, H. H., Galas, D., and Arnheim, N., 1989, Single sperm typing: Determination of genetic distance between the G-gamma globin and parathyroid hormone loci by using the polymerase chain reaction and allele-specific oligomers, *Proc. Natl. Acad. Sci. USA* **86:**9389–9393.

Darden, A. G., and Streilein, J. W., 1986. Can a mammalian species with monomorphic class I MHC molecules succeed?, in: *Paradoxes in Immunology* (G. W. Hoffmann, J. Levy, and G. T. Nepom, eds.), pp. 9–26, CRC Press, Boca Raton, Florida.

Dausset, J., and Colombani, P. J. (eds.), 1973, *Histocompatibility Testing 1972,* Munksgaard, Copenhagen.

Davidson, W. F., Kress, M., Koury, G., and Jay, G., 1985, Comparison of HLA class I gene sequences: Derivation of locus-specific oligonucleotide probes specific for HLA-A, HLA-B and HLA-C genes, *J. Biol. Chem.* **260:**13414–13423.

DeVries, R. R. P., Geziena, M., Schreuder, Th., Naipal, H., D'Amaro, J., and van Rood, J. J., 1989, Selection by typhoid and yellow fever epidemics witnessed by the HLA DR locus, in *Immunobiology of HLA* (B. Dupont, ed.), Vol. II, pp. 461–462, Springer-Verlag, New York.

Dunham, I., Sargent, C. A., Dawkins, R. L., and Campbell, R. D., 1989, An analysis of variation in the long-range genomic organization of the human major histocompatibility complex class II region by pulsed-field gel electrophoresis, *Genomics* **5:**787–796.

Dupont, B. (ed.), 1989, *Immunobiology of HLA,* Springer-Verlag, New York.

Du Toit, E. D., Taljaard, D. G., Marshall, J., Ritchie, C., and Oudshoorn, M., 1990, The HLA genetic constitution of the Bushmen (San), *Hum. Immunol.* **28:**406–415.

Ewens, W. J., 1972, The sampling theory of selectively neutral alleles, *Theor. Popul. Biol.* **3:** 87–112.

Figueroa, F., Gunther, E., and Klein, J., 1988, MHC polymorphism predating speciation, *Nature* **335:**265–267.

Gao, X., Fernandez-Vina, M., Shumway, W., and Stastny, P., 1990, DNA typing for class II HLA antigens with allele-specific or group-specific amplification. I. Typing for subsets of HLA-DR4, *Hum. Immunol.* **27:**40–50.

Guillemot, F., Kaufman, J. F., Skjoedt, K., and Auffray, C., 1989, The major histocompatibility complex in the chicken, *Immunol. Rev.* **5:**300–304.

Gyllensten, U., and Erlich, H. A. 1989, Ancient origin for the polymorphism at the DQA locus in primates, *Proc. Natl. Acad. Sci. USA* **86:**9986–9990.

Hedrick, P. W., 1987, Gametic disequilibrium measures: Proceed with caution, *Genetics* **117:**331–341.

Hedrick, P. W., 1988, HLA-sharing, recurrent abortion and the genetic hypothesis, *Genetics* **119:**199–204.

Hedrick, P. W., and Thomson, G., 1983, Evidence for balancing selection at HLA, *Genetics* **104:**449–456.

Hedrick, P. W., Whittam, T. S., and Parham, P., 1991*a,* Heterozygosity at individual amino acid sites: Extremely high levels for the HLA-A and HLA-B genes, *Proc. Natl. Acad. Sci. USA* **88:**5897–5901.

Hedrick, P. W., Klitz, W., Robinson, W. P., Kuhner, M. K., and Thomson, G., 1991*b,* Evolutionary genetics of HLA, in: *Evolution at the Molecular Level* (R. K. Selander, A. G. Clark, and T. S. Whittam, eds.), pp. 248–271, Sinauer Associates, Sunderland, Massachusetts.

Hill, A. V. S., Allsopp, C. E. M., Kwiatkwoski, D., *et al.,* 1991, Common West African HLA antigens are associated with protection from severe malaria, *Nature* **352:**595–600.

Hill, W. G., 1975, Linkage disequilibrium among multiple neutral alleles produced by mutation in finite populations, *Theor. Popul. Biol.* **8:**117–126.

Holmes, N., Ennis, P., Wan, A. M., Denny, D. W., and Parham, P., 1987, Multiple genetic mechanisms have contributed to the generation of the HLA A2/A28 family of class I MHC molecules, *J. Immunol.* **139:**936–941.

Hughes, A., and Nei, M., 1988, Pattern of nucleotide substitution at major histocompatibility complex loci reveals overdominant selection, *Nature* **335:**167–170.

Kaufman, J., 1990, Evolution of the MHC: Lessons from nonmammalian vertebrates, *Immunol. Rec.* **9:**123–134.

Kieth, T. P., 1983, Frequency distributions of esterase-5 alleles in two populations of *Drosophila pseudoobscura, Genetics* **105:**153–155.

Kimura, M., and Ohta, T., 1971, *Theoretical Aspects of Population Genetics,* Princeton University Press, Princeton, New Jersey.

Klein, J., 1986, *Natural History of the Major Histocompatibility Complex,* Wiley, New York.

Klein, J., and Takahata, N., 1990, The major histocompatibility complex and the quest for origins, *Immunol. Rev.* **113:**27–46.

Klitz, W., and Thomson, G., 1987, Disequilibrium pattern analysis II. Application to Danish HLA A and B locus data. 1987, *Genetics* **116:**633–643.

Klitz, W., and Thomson, G., 1991, Interpreting MHC disequilibrium, in: *Molecular Evolution of the Major Histocompatibility Complex* (J. Klein and D. Klein, eds.), pp. 257–260, Springer-Verlag, Heidelberg.

Klitz, W., Thomson, G., and Baur, M. P., 1986, Contrasting selection histories among tightly linked HLA loci, *Am. J. Hum. Genet.* **39:**340–349.

Klitz, W., Lo, S. K., Neugebauer, M. P., Albert, E. D., and Thomson, G., 1987, A comprehensive search for segregation distortion in HLA, *Hum. Immunol.* **18:**163–180.

Koller, B. H., Geraghty, D. E., DeMars, R., Davick, L., Rich, S. S., and Orr, H. T., 1989, Chromosomal organization of the human major histocompatibility complex class I gene family, *J. Exp. Med.* **169:**469–480.

Kroemer, G., Guillemot, F., and Auffray, C. 1990, Genetic organization of the chicken MHC, *Immunol. Res.* **9:**8–19.

Kuhner, M. K., Watts, S., Klitz, W., Thomson, G., and Goodenow, R. S., 1990, Gene conversion in the evolution of both the H-2 and Qa class I genes of the murine major histocompatibility complex, *Genetics* **126:**1115–1126.

Kuhner, M. K., Lawlor, D. A., Ennis, P., and Parham, P., 1991, Gene conversion in the evolution of the human and chimpanzee MHC class I loci, *Tissue Antigens* **38:**152–164.

Lakoff, G., 1988, *Women, Fire and Dangerous Things: What Categories Reveal about the Mind,* University of Chicago Press, Chicago, Illinois.

Larntz, K., 1978, Small sample comparisons of exact levels for chi-square goodness-of-fit statistics, *J. Am. Statist. Assoc.* **71:**455–461.

Lawlor, D. A., Ward, F. E., Ennis, P. D., Jackson, A. P., and Parham, P., 1988, HLA-A and B polymorphism predates the divergence of humans and chimpanzees, *Nature* **335:**268–271.

Lawrance, S. K., Smith, C. L., Srivastava, R., Cantor, C. R., and Weissman, S. M., 1987, Megabase-scale mapping of the HLA gene complex by pulsed field gel electrophoresis, *Science* **235:**1387–1390.

Lewontin, R. C., 1964, The interaction of selection and linkage. I. General considerations: Heterotic models, *Genetics* **49:**49–67.

Lewontin, R. C., 1974, *The Genetic Basis of Evolutionary Change,* Columbia University Press, New York.

Lewontin, R. C., 1988, On measures of gametic disequilibrium, *Genetics* **120:**849–852.

Lidicker, W. Z., and Patton, J. L., 1987, Patterns of dispersal and genetic structure in populations of small rodents, in: *Mammalian Dispersal Patterns* (B. D. Chepko-Sade and Z. T. Halpin, eds.), pp. 144–160, University of Chicago Press, Chicago, Illinois.

Mayer, W. E., Jonker, M., Klein, D., Ivanyi, P., Van Seventer, G., and Klein, J., 1988, Nucleotide sequences of chimpanzee MHC class I alleles: Evidence for trans-species mode of evolution, *EMBO J.* **7:**2765–2774.

Melvold, R. W., and Kohn, H. I., 1990, Spontaneous frequency of H-2 mutations, in: *Transgenic Mice and Mutants in MHC Research* (I. K. Egorov and C. S. David, eds.), pp. 3–13, Springer-Verlag, New York.

Menozzi, P., Piazza, A., and Cavalli-Sforza, L., 1978, Synthetic maps of human gene frequencies in Europeans, *Science* **201:**786–792.

Mytilineos, J., Scherer, S., and Opelz, G., 1990, Comparison of RFLP-DR beta and serological HLA-DR typing in 1500 individuals, *Transplantation* **50:**870–873.

Nathenson, S. G., Geliebter, G. M., Pfaffenbach, M., and Zeff, R. A., 1986, Murine major histocompatibility complex class I mutants: Molecula- analysis and structure–function implications, *Annu. Rev. Immunol.* **4:**471–502.

Neugebauer, M., Willems, J., and Baur, M. P., 1984, Analysis of pedigree data by computer, in: *Histocompatibility Testing 1984* (E. D. Albert, M. P. Baur, and W. R. Mayr, eds.), pp. 52–57, Springer-Verlag, New York.

Ohayon, E., and Cambon-Thomsen, A. (eds.), 1986, *Human Population Genetics,* Colloque INSERM, Vol. 142.

Ohta, T., and Kimura, M., 1969, Linkage disequilibrium due to random genetic drift, *Genet. Res.* **13:**47–55.

Ostergard, H., Kristensen, B., and Andersen, S., 1989, Investigations in farm animals of associations between the MHC system and disease resistance and fertility, *Livest. Prod. Sci.* **22:**49–67.

Parham, P., 1990, Transporters of delight, *Nature* **348:**647–648.

Potts, W. K., Manning, C. J., and Wakeland, E. K., 1991, Mating patterns in seminatural populations of mice influenced by MHC genotype, *Nature* **352:**619–621.

Robinson, W. P., Asmussen, M. A., and Thomson, G., 1991*a,* Three-locus systems impose additional constraints on pairwise disequilibria, *Genetics* **129:**925–930.

Robinson, W. P., Cambon-Thomsen, A., Borot, N., Klitz, W., and Thomson, G., 1991*b,* Selection, hitchhiking and disequilibrium analysis at three linked loci with application to HLA data, *Genetics* **129:**931–948.

Scharf, S. J., Griffith, R. L., and Erlich, H. A., 1991, Rapid typing of DNA sequence polymorphism at the HLA-DRB1 locus using the polymerase chain reaction and nonradioactive oligonucleotide probes, *Hum. Immunol.* **30:**190–194.

Scheiman, L. W., and Collins, W. M., 1987, Influence of the major histocompatibility complex on tumor regression and immunity in chickens, *Poultry Sci.* **66:**812–818.

Serjeantson, S. W., 1989, HLA genes and antigens, in: *The Colonization of the Pacific—A Genetic Trail* (A. V. S. Hill and S. W. Serjeantson, eds.), pp. 120–173, Oxford University Press, Oxford.

Shiroishi, T., Hanzawa, N., Sagai, T., Ishiura, M., Gojobori, T., Steinmetz, M., and Moriwake, K., 1990, Recombinational hotspot specific to female meiosis in the mouse major histocompatibility complex, *Immunogenetics* **31:**79–88.

Sierp, G., Albert, E., Cambon-Thomsen, A., and Ohayon, E., 1986, Data analysis of HLA typing for 'Provinces Francaises', in: *Human Population Genetics* (E. Ohayon and A. Cambon-Thomsen, eds.), Vol. 142, pp. 177–195, Colloque INSERM.

Sokal, R. R., and Rohlf, F. J., 1981, *Biometry,* 2nd ed., Freeman, San Francisco.

Sokal, R. R., Oden, N. L., and Wilson, C., 1991, Genetic evidence for the spread of agriculture in Europe by demic diffusion, *Nature* **351:**143–145.

Spies, T., Bresnahan, M., and Strominger, J. L., 1989, Human major histocompatibility complex contains a minimum of 19 genes between the complement cluster and HLA-B, *Proc. Natl. Acad. Sci. USA* **86:**8955–8958.

Srivastava, R. (ed.), 1991, *Immunogenetics of the Major Histocompatibility Complex,* VCH, New York.

Sun, S.-C., Lindstrom, I., Boman, H. G., Faye, I., and Schmidt, O., 1991, Hemolin: An insect-immune protein belonging to the immunoglobulin superfamily, *Science* **250:**1729–1731.

Tajima, F., 1989, The effect of change in population size on DNA polymorphism, *Genetics* **123:**597–601.

Takahata, N., and Nei, M., 1990, Allelic geneology under overdominant and frequency-dependent selection and polymorphism of major histocompatibility complex loci, *Genetics* **124:**967–978.

Thomson, G., and Klitz, W., 1987, Disequilibrium pattern analysis. I. Theory, *Genetics* **116**:623–632.

Tiwari, J. L., and Terasaki, P. I., 1985, *HLA and Disease Associations*, Springer-Verlag, New York.

Trowsdale, J., and Campbell, R. D., 1988, Physical map of the HLA region, *Immunol. Today* **9**:34–35.

Uematsu, Y., Fischer Lindahl, K., and Steinmetz, M., 1988, The same MHC recombinational hotspots are active in crossing over between wild/wild and wild/inbred mouse chromosomes, *Immunogenetics* **27**:96–101.

Van Enden, W., deVries, R. R. P., and van Rood, J., 1982, HLA and infectious diseases, in: *Human Genetics*, Part B: *Medical Aspects*, pp. 37–54, Liss, New York.

Wakelin, D., and Blackwell, J. M. (eds.), 1988, *Genetics of Resistance to Infection: Bacterial and Parasitic Infections*, Taylor and Francis, London.

Watterson, G. A., 1978, The homozygosity test of neutrality, *Genetics* **88**:405–417.

Wheeler, C. J., Maloney, D., Fogel, S., and Goodenow, R. S., 1990, Microconversion between murine H-2 genes integrated into yeast, *Nature* **347**:192–194.

WHO Nomenclature Committee, 1990, Nomenclature for factors of the HLA system, 1989, *Immunogenetics* **31**:131–140.

Williams, A. F., and Barclay, G., 1988, The immunoglobulin gene superfamily, *Annu. Rev. Immunol.* **6**:381–398.

Yu, C. Y., Belt, K. T., Giles, C. M., Campbell, R. D., and Porter, R. R., 1986, Structural basis of the polymorphism of human complement components C4A and C4B: Gene size, reactivity and antigenicity, *EMBO J.* **5**:2873–2881.

Yuhki, N., and O'Brien, S. J., 1990, DNA variation of the mammalian major histocompatibility complex reflects genomic diversity and population history, *Proc. Natl. Acad. Sci. USA* **87**:836–840.

Zemmour, J., Ennis, P. D., Parham, P., and Dupont, B., 1988, *Immunogenetics* **27**:281–287.

3

Protein Heterozygosity, Protein Structure, and Taxonomic Differentiation

ROBERT D. WARD, DAVID O. F. SKIBINSKI, and MATHEW WOODWARK

INTRODUCTION

Soon after the introduction of electrophoretic techniques to study the genetic structure of natural populations of animals and plants came the realization that enzymes differed in their levels of variability. Some enzymes, such as esterases and phosphoglucomutases, were frequently polymorphic, whereas others, such as superoxide dismutase and glutamate dehydrogenase, were so only rarely. Initial explanations focused on physiological considerations, such as the degree of substrate specificity (Gillespie and Kojima, 1968; Gillespie and Langley, 1974) and the regulatory nature of the enzyme (Johnson, 1974). Correlations were observed in some groups of organisms but not others, and there were some conflicts of opinion concerning the proper allocation of enzymes to particular categories (Selander, 1976).

Subsequent hypotheses examined relationships between heterozygosity and structure. Monomeric enzymes (those made up of single polypeptide chains) were, on average, found to be more heterozygous than multimeric

---

ROBERT D. WARD • Department of Human Sciences, Loughborough University of Technology, Loughborough, Leicestershire, LE11 3TU, United Kingdom; *present address:* CSIRO Division of Fisheries, GPO Box 1538, Hobart, Tasmania 7001, Australia. DAVID O. F. SKIBINSKI and MATHEW WOODWARK • School of Biological Sciences, University College of Swansea, Swansea, SA2 8PP, United Kingdom.

Evolutionary Biology, Volume 26, edited by Max K. Hecht *et al.* Plenum Press, New York, 1992.

enzymes (Zouros, 1976; Ward, 1977), and there was a tendency for dimeric enzymes to be more variable than tetrameric enzymes (Ward, 1977; Harris *et al.*, 1977). Subunit molecular weight was found to be positively correlated with heterozygosity (Koehn and Eanes, 1977; Ward, 1978; Nei *et al.*, 1978). These findings were generally associated with reduced functional constraints in monomeric and larger subunit enzymes, although Harris *et al.* (1977) suggested that the reason multimeric enzymes were less variable than monomeric enzymes was because multimers forming interlocus hybrids were substantially less variable than those not forming such hybrids, the latter category not differing from monomers in heterozygosity. It was speculated that this may be attributable to a greater degree of metabolic flexibility in those enzymes forming interlocus hybrids, and hence reduced selection pressures for heterozygosity at such loci (Harris *et al.*, 1977).

Finally, subcellular location has an effect on heterozygosity. Mitochondrial isozymes have lower average heterozygosity than their cytoplasmic counterparts (Ward and Skibinski, 1988). Again, this may reflect a reduced degree of functional constraint in cytoplasmic enzymes, which do not have to target particular sites in the mitochondrial membrane.

The present chapter presents an analysis of genetic variability gathered from many enzymes and hundreds of animal species. The genetics literature already contains some examples of the analyses of population or species data bases, perhaps the best known of which is that established by Nevo and his collaborators. This has been used to test for associations between species heterozygosity and various life history traits (Nevo *et al.*, 1984). The data base we have established is, however, more detailed, as it is structured in terms of allele frequencies locus by locus rather than by average species or population heterozygosity. So not only can we examine variation in species heterozygosity (Nevo *et al.*, 1984) among taxa, and compare the extent of subpopulation differentiation in different taxa, we can also examine how different enzymes vary across species and explore the reasons for this variation. Enzyme-specific relationships among heterozygosity and genetic distance have previously been explored using this data base (Skibinski and Ward, 1981; Ward and Skibinski, 1985).

The average heterozygosities of 47 proteins in 648 vertebrate and 370 invertebrate species are given, and the data further partitioned into birds, reptiles, amphibia, mammals, and fishes, and into mollusks, insects, crustacea, and "others." The tabulated variation in heterozygosity among enzymes allows mean heterozygosity estimates of species to be adjusted for the particular suite of enzymes utilized, thus permitting such estimates for different species electrophoresed for different enzymes to be more properly compared. An appropriate adjustment method is suggested.

Subunit numbers and subunit molecular weights of those proteins routinely screened in surveys of genetic variation are given, and their relative contributions to heterozygosity assessed statistically. The need for these two factors to be considered jointly was stressed in a review article by Koehn and Eanes (1979), but has not hitherto been analyzed. We also assess the relative contributions of enzyme function and taxon to heterozygosity variation, and examine the data to see whether they lend support to the idea that multimeric enzymes forming interlocus hybrids are less variable than multimeric enzymes not forming such hybrids.

BACKGROUND INFORMATION

The Data Base and Heterozygosity Estimations

Computer data bases were established of published and, to a much lesser extent, unpublished values of allele frequency variation in electrophoretic surveys of natural populations of sexually reproducing animals. Two data bases were compiled, ENTIRE (comprising allele frequency data on all species, whether screened for one or many subpopulations) and WITHIN (comprising allele frequency data on species assayed for two or more subpopulations). Use of these two overlapping data bases permitted a variety of different analyses as described below. Minimum criteria for inclusion in either data base were that at least 15 individuals had been screened per subpopulation and that at least 15 loci had been screened.

Allele frequency data were input into the WITHIN data base locus by locus and subpopulation by subpopulation, so that the resulting matrix for each species was of the following size: total number of alleles by number of subpopulations. The ENTIRE data base was structured in a slightly different way. Here, where a species had been surveyed for more than a single subpopulation, allele frequencies across subpopulations were averaged in an unweighted manner. Thus, in the ENTIRE data base, the resulting matrix for each species was of the following size: total number of alleles by one. Care was taken to ensure that no species was represented in either data base more than once.

The data bases did not include estimates of observed heterozygosity (which were frequently not included in published data), so all heterozygosity estimates per locus per species are calculated from the Hardy–Weinberg equilibrium expectations, $H_j = 1 - \Sigma\, p_i^2$, where p_i is the frequency of the ith allele at the jth locus in that species. We have restricted our attention here to heterozygosity estimates and have not considered parameters such as number of alleles (which is heavily sample-size dependent) or degree of polymorphism (which is a crude figure, being either 0 or 1 for any locus, and definition dependent).

Note that in the ENTIRE data base, only a single value of H_j can be calculated for each locus in each species. This value of H_j is, for those species assayed in several subpopulations, equivalent to the parameter H_T of Nei (1973), the gene diversity in the total population. However, for the WITHIN data base, H_j is the estimate of heterozygosity at that locus in that subpopulation, and, averaged over subpopulations, gives, in Nei's (1973) terms, H_S (the average gene diversity within subpopulations). Alternatively, heterozygosity can be estimated for any locus in the WITHIN data base from the mean unweighted allele frequency across subpopulations, giving H_T. Variation among and within subpopulations of any one species in the WITHIN data base can then be partitioned following Nei (1973). The coefficient of gene differentiation G_{ST} is estimated as $(\bar{H}_T - \bar{H}_S)/\bar{H}_T$, where H_T and H_S are averaged over the k loci screened in that species. For that species, the average minimum genetic distance between subpopulations D_m may be estimated as $s(\bar{H}_T - \bar{H}_S)/(s - 1)$, where s is the number of subpopulations.

Other heterozygosity estimates used in the following analyses are as follows:

H = mean heterozygosity per species = $\sum H_j/k$, where k is the number of loci scored in that species

$\bar{H}$ = mean heterozygosity per taxon = $\sum H/t$, where t is the number of species scored in that taxon

H_E = mean heterozygosity per enzyme or protein = $\sum H_j/l$, where l is the number of loci scored for that enzyme or protein

Except where otherwise stated, all ensuing analyses are of the ENTIRE data base alone. All analyses of variance of heterozygosity values were carried out on arcsine-transformed values. Most of the statistical analyses were performed using SYSTAT 5.0, but the final multiple regression analysis used SPSS-X.

Determination of Subunit Number and Subunit Molecular Weight

Both subunit number (quaternary structure) and subunit molecular weight are highly conserved aspects of protein structure, yet difficulties sometimes arise in assuming homologies between proteins which are not truly homologous, and which may have differing structures. Such problems are more likely to arise in relatively nonspecific enzymes such as phosphatases and esterases than in the highly specific enzymes catalyzing, for example, the glycolytic pathway and the citric acid cycle.

Subunit numbers and subunit molecular weights of the enzymes used in the analyses of this chapter are given in Table I. The bulk of the molecular weight determinations for invertebrates stem from work on insects (primarily *Drosophila*) and crustaceans, and the majority of the vertebrate determinations come from research on enzymes in humans and other mammals. Where studies have been carried out both on mammals and on representatives of other vertebrate and invertebrate groups, subunit numbers (with the possible exception of G6PDH; see below) and subunit sizes are highly concordant within homologous enzymes. Ruth and Wold (1976) demonstrate this for a number of glycolytic enzymes, but the relationship is not restricted to such enzymes. Some additional examples of such conservation are included in Table II. Nadler (1986) has emphasised how enzymes which are monomers, dimers, or tetramers in vertebrates have identical subunit structures, where deduced, in ascaridoid nematodes. Nonenzymatic proteins such as albumin, transferrin, and hemoglobin are also highly conserved in terms of subunit size across different vertebrate groups (Peters, 1975; Putnam, 1975; Dickerson and Geis, 1983).

Where multiple loci code for enzymes with the same or very similar enzymic properties, it is generally the case that the different isozymes have the same subunit number and very similar subunit sizes (Hopkinson *et al.*, 1976). Many of the enzymes covered in this survey have both cytoplasmic and mitochondrial forms (MDH, ME, IDH, SOD, AAT, GDH, ACO), encoded by separate nuclear genes. While the amino acid sequences of the cytoplasmic and mitochondrial forms may be quite distinct [although with extensive regions of strong homology; see, for example, Barra *et al.* (1980)], the subunit molecular weights for a particular pair of enzymes are very similar and, with one exception (SOD), the numbers of subunits in the active enzyme are identical. Thus, the fact that the bulk of electrophoretic surveys do not attempt to discriminate among cytoplasmic and mitochondrial isozymes does not affect our analysis of the relationship of enzyme variability to subunit size and number.

Subunit numbers given in Table I are generally the highest-order complexes routinely observed, but dissociation of these complexes into lower-order structures may happen under specific experimental conditions without great loss of activity (Dixon and Webb, 1979; Gonzalez-Villasenor and Powers, 1985). While the great majority of electrophoretic surveys yield information about quaternary structures that is consistent with those set out in Table I (two-banded heterozygotes implying a monomeric structure, three-banded ones a dimeric structure, *etc.*), exceptional heterozygote patterns may result from one of a number of possibilities. It may mean that *in vitro* dissociation has occurred into enzymatically active forms of atypically low subunit number (which is perhaps likely to be the most common explana-

TABLE I. Subunit Number n and Subunit Molecular Weight SMW ($\times 10^{-3}$)[a]

Enzyme and abbreviation	EC number	Invertebrates n	Invertebrates SMW	Vertebrates n	Vertebrates SMW
Oxidoreductases					
Alcohol dehydrogenase (ADH)	1.1.1.1	2	27.6[1]	2	40.3[1]
Glycerol-3-phosphate dehydrogenase (G3PDH)	1.1.1.8	2	33.0[1]	2	36.0[1]
Iditol dehydrogenase (IDDH)[b]	1.1.1.14	4[35]	—	4	38.0[3]
l-Lactate dehydrogenase (*l*-LDH)	1.1.1.27	4	35.0[4]	4	34.0[3–6]
d-Lactate dehydrogenase (*d*-LDH)	1.1.1.28	2[52]	35.4[7]	—	—
3-Hydroxybutyrate dehydrogenase (HBDH)	1.1.1.30	2	31.5[50]	—	—
Malate dehydrogenase (MDH)	1.1.1.37	2	29.0[2,8]	2	35.0[3,4]
NADP-malate dehydrogenase (ME)[c]	1.1.1.40	4	63.1[1]	4	63.0[1]
Isocitrate dehydrogenase (IDH)	1.1.1.42	2	50.0[1]	2	44.8[1]
Phosphogluconate dehydrogenase (6PGDH)	1.1.1.44	2	56.0[9,10]	2	51.0[3–5,10]
Glucose dehydrogenase (GLUD)[d]	1.1.1.47	2[?]	90[42]	2	111.0[3]
Glucose-6-phosphate dehydrogenase (G6PDH)	1.1.1.49	2/4	58.9[1]	2/4	56.5[1]
Octanol dehydrogenase (ODH)	1.1.1.73	2	55.0[2]	2[11]	?
Xanthine dehydrogenase (XDH)	1.1.1.204[e]	2	140.0[2]	2	144.0[12,36]
Glyceraldehyde-3-phosphate dehydrogenase (GAPDH)	1.2.1.12	4	36.6[1]	4	36.0[1]
Aldehyde oxidase (AO)	1.2.3.1	2	132.0[2]	2	149[44]
Glutamate dehydrogenase (GDH)	1.4.1.2–4	6	52.5[13]	6	54.0[14]
Diaphorase (DIA)[f]	1.8.1.4	1	39.0[33]	1	30.0[3]
Catalase (CAT)	1.11.1.6	4	58.0[15]	4	58.8[3,4]
Superoxide dismutase (SOD)[g]	1.15.1.1	2	16.0[1]	2	16.3[1]
Transferases					
Purine nucleoside phosphorylase (NP)	2.4.2.1	3[16]	—	3	28.0[3,17]
Aspartate aminotransferase (AAT)	2.6.1.1	2	44.0[1]	2	47.4[1]
Alanine aminotransferase (ALAT)[h]	2.6.1.2	2	43.5[31]	2	50.0[3]
Hexokinase (HK)	2.7.1.1	1	50–100[40]	1	103.0[3,18]
Pyruvate kinase (PK)	2.7.1.40	4	57.0[19]	4	56.3[3–5]
Creatine kinase (CK)	2.7.3.2	—	—	2	40.9[1]
Arginine phosphokinase (APK)	2.7.3.3	1	40.0[20]	—	—
Adenylate kinase (AK)	2.7.4.3	1	27.0[41]	1	21.5[3,21]
Hydrolases					
Esterase (EST)	3.1.1.-	V[i]	V	1	V
Esterase-D (EST-D)	3.1.1.-	2[53]	—	2	28.0[3]
Alkaline phosphatase (ALP)	3.1.3.1	V	V	2	68.3[3,24]
Acid phosphatase (ACP)	3.1.3.2	2	55.0[25]	1	V
Amylase (AMY)	3.2.1.1	1	50.0[26]	1	55.0[3,37]
Leucine aminopeptidase (LAP)	3.4.11.1	1[54]	—	1[54]	—
Peptidase (PEP)	3.4.11.-	V	V	1	V
Adenosine deaminase (ADA)	3.5.4.4	1	35.0[39]	1	37.5[3,38]

TABLE I. Continued

Enzyme and abbreviation	EC number	Invertebrates		Vertebrates	
		n	SMW	*n*	SMW
Lyases					
Aldolase (ALD)	4.1.2.13	4	39.9[1]	4	40.0[1]
Fumarase (FUM)	4.2.1.2	4	43.5[22,48]	4	48.8[3,4,49]
Aconitase (ACO)	4.2.1.3	1[55]	—	1	76.8[3,43]
Isomerases					
Triosephosphate isomerase (TPI)	5.3.1.1	2	25.5[22,27]	2	27.0[3,28]
Mannosephosphate isomerase (MPI)	5.3.1.8	1[32]	—	1	43.0[3]
Glucosephosphate isomerase (GPI)	5.3.1.9	2	56.3[22,29]	2	61.5[3,5,30]
Phosphoglucomutase (PGM)	5.4.2.2[j]	1	53.5[34]	1	58.6[3,23]
Nonenzymic proteins					
Hemoglobin (HB)		V	V	4	16.0[5,51]
Albumin (ALB)		—	—	1	66.5[46]
Transferrin (TF)		—	—	1	77.5[47]
Unspecified proteins (PROT)		V	V	1	V

[a] EC numbers follow IUBNC (1984). Superscript numerals indicate references: 1, see Table II; 2, Koehn and Eanes, 1977; 3, Hopkinson *et al.*, 1976; 4, Dixon and Webb, 1979; 5, Darnall and Klotz, 1975; 6, Onoufriou and Alahiotis, 1982; Dell'Agata *et al.*, 1990; Rehse and Davidson, 1986; 7, Long and Kaplan, 1968, 1973, Selander and Yang, 1970; Gleason *et al.*, 1971; 8, Weeda, 1981; 9, Hori and Tanda, 1980; 10, Williamson *et al.*, 1980b; 11, Ward and Beardmore, 1977; 12, Markley *et al.*, 1973; 13, Walsh, 1981; Ling *et al.*, 1986; Catmull *et al.*, 1987; 14, E. L. Smith *et al.*, 1975; 15, Nahmias and Bewley, 1984; 16, Knight and Ward, 1986; 17, Unemura *et al.*, 1982; 18, Colowick, 1975; 19, Carvajal *et al.*, 1986 (native molecular weight only given, 230,000, assumed here to consist of four equal subunits); 20, Morrison, 1975; 21, Noda, 1975; 22, Leigh Brown and Langley, 1979; 23, Ray and Peck, 1975; 24, Fosset *et al.*, 1974; Unakami *et al.*, 1987; 25, MacIntyre, 1971; 26, Panara, 1985; Janska *et al.*, 1988; Mayzaud, 1985; Doane *et al.*, 1975; 27, Gavilanes *et al.*, 1981; 28, Norton *et al.*, 1970; 29, Widmer *et al.*, 1986; 30, Cini and Gracy, 1986; 31, Leigh Brown and Voelker, 1980 (native weight only given, 87,000, assumed here to consist of two equal subunits); 32, Shaklee, 1983; Siegismund, 1985; 33, Ward and Warwick, 1980; Weller, 1982; 34, Leigh Brown and Langley, 1979; Fucci *et al.*, 1979; 35, Buroker *et al.*, 1975; Harrison, 1979; 36, Rajagopalan and Handler, 1967; 37, Karn and Malacinski, 1978; 38, Ingolia *et al.*, 1985; 39, Umemori-Aikawa and Aikawa, 1974; Aikawa *et al.*, 1977; 40, Komuniecki and Roberts, 1977; Mochizuki and Hori, 1977; Moser *et al.*, 1980; Sanchez *et al.*, 1983; 41, Storey, 1976; 42, Mochizuki and Hori, 1976 (native weights only given, mean of 180,280, assumed here to consist of two equal subunits); 43, Scholze, 1983; Ryden *et al.*, 1984; 44, Ohkubo *et al.*, 1983; Yoshihara and Tatsumi, 1985; 46, Peters, 1975; 47, Putnam, 1975; 48, Payne *et al.*, 1979; 49, Beeckmans and Kanarek, 1977; 50, Borack and Sofer, 1971; 51, Perutz, 1990; 52, see text; 53, Ahmad *et al.*, 1977; 54, Ward, 1977; Murphy *et al.*, 1990; 55, R. D. Ward, unpublished observations.

[b] Sorbitol dehydrogenase.

[c] Malic enzyme.

[d] Hexose-6-phosphate dehydrogenase.

[e] Former EC number 1.2.1.37.

[f] Dihydrolipoamide dehydrogenase; former EC number 1.6.4.3.

[g] Tetrazolium oxidase; indophenol oxidase.

[h] Glutamate pyruvate transaminase.

[i] Variable.

[j] Former EC number 2.7.5.1.

TABLE II. Subunit Number and Size for a Range of Enzymes

Enzyme	Species	Group	Size	Reference[a]
ADH	Fruit fly (7 spp.)	Insect	2*27,200–28,000	11
ADH	Human	Mammal	2*40,000	3
ADH	Horse	Mammal	2*41,000	4
ADH	Shrew	Mammal	2*40,000	19
G3PDH	Fruit fly	Insect	2*32,067	1
G3PDH	Bee	Insect	2*34,000	2
G3PDH	Human	Mammal	2*36,000	3
G3PDH	Rat	Mammal	2*32,000	4
G3PDH	Hen	Bird	2*40,000	4
ME	Fruit fly	Insect	4*58,000	25
ME	Fruit fly	Insect	4*67,250	14
ME	*Ascaris*	Nematode	4*64,000	2
ME	Pigeon	Bird	4*65,000	2
ME	Mouse	Mammal	4*67,000	25
sME	Human	Mammal	4*60,000	3
mME	Human	Mammal	4*60,000	3
IDH	Fruit fly	Insect	2*56,000	5
IDH	Fruit fly	Insect	2*55,000	27
IDH	Silkmoth	Insect	2*44,000	32
sIDH	Mussel	Mollusk	2*45,000	28
sIDH	Pig	Mammal	2*37,500	31
sIDH	Human	Mammal	2*48,000	3
sIDH	Cow	Mammal	2*48,000	30
sIDH	Killifish	Fish	2*45,000	6
mIDH	Killifish	Fish	2*45,000	6
mIDH	Ox	Mammal	2*45,000	29
G6PDH	Fruit fly	Insect	4*58,000	25
G6PDH	Fruit fly	Insect	2 or 4*69,000	12
G6PDH	*Hydra*	Coelenterate	2 or 4*57,500	12
G6PDH	Earthworm	Annelid	2 or 4*55,000	12
G6PDH	Crayfish	Crustacea	2 or 4*55,000	12
G6PDH	Human	Mammal	2*53,000	3
G6PDH	Rat	Mammal	2*63,000	4
G6PDH	Mouse	Mammal	4*52,000	25
G6PDH	Rat	Mammal	2*55,000	12
G6PDH	Cow	Mammal	4*64,600	2
G6PDH	Chicken	Bird	2*55,000	12
G6PDH	Snake	Reptile	2*55,000	12
G6PDH	Turtle	Reptile	2*50,000	12
G6PDH	Frog	Amphibia	4*65,000	12
G6PDH	Salamander	Amphibia	4*52,500	12
G6PDH	Trout	Fish	4*56,250	13

TABLE II. Continued

Enzyme	Species	Group	Size	Reference[a]
GAPDH	Fruit fly	Insect	?*35,500–37,000	17
GAPDH	Lobster	Crustacean	4*37,000	2
GAPDH	Human	Mammal	4*36,000	3
GAPDH	Rabbit	Mammal	4*36,000	4
GAPDH	Pig	Mammal	4*36,000	4
GAPDH	Fish (3 spp.)	Fish	4*36,250	16
GAPDH	Duck	Bird	4*36,000	15
GAPDH	Caiman	Reptile	4*36,000	18
sSOD	Fruit fly	Insect	2*16,000	8
sSOD	Housefly	Insect	2*16,000	9
sSOD	Human	Mammal	2*16,000	3
sSOD	Horse	Mammal	2*16,500	10
sSOD	Ox	Mammal	2*16,250	10
mSOD	Human	Mammal	4*18,000	3
mSOD	Chicken	Bird	4*20,000	10
AAT	Mussel	Mollusk	2*44,000	20
AAT	Pig	Mammal	2*50,000	4
AAT	Chicken	Bird	2*50,000	2
sAAT	Human	Mammal	2*46,000	3
mAAT	Human	Mammal	2*46,000	3
mAAT	Chicken	Bird	2*45,000	7
CK	Human	Mammal	2*41,000	3
CK	Rabbit	Mammal	2*40,000	4
CK	Bat	Mammal	2*42,500	24
CK	Chicken	Bird	2*40,000	4
ALD	*Eimeria*	Protozoan	4*40,000	26
ALD	Fruit fly	Fly	4*39,750	1
ALD	Snail	Mollusk	4*40,000	21
ALD	Lobster	Crustacean	4*40,000	22
ALD	Human	Mammal	4*40,000	3
ALD	Rabbit	Mammal	4*40,000	4
ALD	Fish (3 spp.)	Fish	4*40,000	23

[a] 1, Koehn and Eanes, 1978; 2, Darnall and Klotz, 1975; 3, Hopkinson *et al.*, 1976; 4, Dixon and Webb, 1979; 5, Leigh Brown and Langley, 1979; 6, Gonzalez-Villasenor and Powers, 1985; 7, Jaussi *et al.*, 1985; 8, Lee *et al.*, 1981; 9, Bird *et al.*, 1986; 10, Tegelstrom, 1975; 11, Atrian and Gonzalez-Duarte, 1985; Winberg *et al.*, 1986; 12, Hori and Tanda, 1980; 13, Cederbaum and Yoshida, 1976; 14, Geer *et al.*, 1980; 15, Barbosa and Nakano, 1987; 16, Nakagawa and Nagayama, 1989*a*; 17, Sullivan *et al.*, 1985; 18, Vieira *et al.*, 1983; 19, Keung *et al.*, 1989; 20, McCormick *et al.*, 1986; 21, Buczylko *et al.*, 1980; 22, Guha *et al.*, 1971; 23, Nakagawa and Nagayama, 1989*b*; 24, Afolayan and Daini, 1986; 25, Lee *et al.*, 1978; 26, Mitchell *et al.*, 1982; 27, Williamson *et al.*, 1980a (molecular weight estimated at 110,000, with apparently different size subunits of 60,000 and 50,000); 28, Head, 1980; 29, MacFarlane *et al.*, 1977; 30, Carlier and Pantaloni, 1973; 31, Illingworth and Tipton, 1970; 32, Miake *et al.*, 1977.

tion, especially after considering that the great majority of such cases imply lower subunit numbers than those of Table I), that the enzyme really does have a different quaternary structure, that heteromultimeric enzymes are not being formed (see CK, below), that a different enzyme is in fact being scored, or that the heterozygote is not truly a heterozygote but rather that posttranslational modifications are being incorrectly scored as genetic variants. Whatever the explanation of these inconsistencies, we have assumed that the described variation is in fact genetic and that the true *in vivo* structure of the enzyme is as in Table I.

The structure and subunit molecular weights of most enzymes routinely screened in electrophoretic studies are well defined and are given in Tables I and II. However, the structures of some enzymes merit further comment and, in enzyme alphabetical order, are discussed below.

Acid Phosphatase. Drosophila ACP has been characterized as a dimer with a subunit molecular weight of 55,000 (MacIntyre, 1971), but in vertebrates there is a variety of acid phosphatases with differing molecular weights (e.g., Panara, 1985; Janska *et al.,* 1988). In humans, Hopkinson *et al.* (1976) describe three isozymes, a monomer with a molecular weight of 15,000 and two dimers with subunit molecular weights of 60,000 and 48,000. Four acid phosphatases were found in frog liver tissue, with apparent molecular weights of about 240,000, 110,000, 38,000, and 17,000, and associated respectively with the microsomal, mitochondrial-lysosomal, nuclear, and soluble fractions (Panara *et al.,* 1989). Janska *et al.* (1989) describe the 38,000-molecular-weight ACP as a monomer, and further identify a 140,000-molecular-weight fraction (presumably corresponding to Panara *et al.*'s 110,00-molecular-weight ACP) as a dimer. High- and low-molecular-weight enzymes also exist in fish (118,000 and 16,600) (Panara and Pascolini, 1989).

Adenosine Deaminase. In vertebrates, multiple-molecular-weight forms of this enzyme have been isolated (e.g., Ma and Fisher, 1968). The low-molecular-weight form is a monomer with a molecular weight of around 35,000, but it can bind to complexing proteins to yield high-molecular-weight aggregates (Daddona and Kelley, 1979; Weisman *et al.,* 1988). Bivalve mollusks also display two forms of ADA, a high-molecular-weight form in the midgut gland and a low-molecular-weight form (35,000) in the adductor muscle (Umemori-Aikawa and Aikawa, 1974; Aikawa *et al.,* 1977). We have assumed here that in invertebrates, as in vertebrates, the low-molecular-weight form represents the unaggregated enzyme.

Alcohol Dehydrogenase. Drosophila ADH is not closely related to mammalian (horse liver) ADH. While both are dimers, *Drosophila* ADH has a smaller subunit size (Table II), and mammalian ADH is in fact structurally more similar to mammalian IDDH than to *Drosophila* ADH (Jornvall *et al.,* 1981).

Alkaline Phosphatase. This enzyme has been most intensively studied in mammals, where it occurs as a dimer with a subunit molecular weight of around 68,000 (e.g., Hopkinson *et al.,* 1976; Fosset *et al.,* 1974; Unakami *et al.,* 1987). In *Drosophila* it is a dimer (Beckman and Johnson, 1964), with a native molecular weight "tentatively" estimated by Harper and Armstrong (1972) at 300,000, giving an estimated subunit size of around 150,000. However, in the silkworm, ALP was characterized as a monomer with a molecular weight of around 60,000 (Okada *et al.,* 1989), and in *Glossina,* another insect, heterozygotes were two-banded, again implying a monomeric structure (Gooding and Rolseth, 1978). This enzyme has not been included in our invertebrate analyses of structure because of its apparently variable subunit number and weight.

Arginine Phosphokinase. This enzyme is a monomer in arthropods and mollusks, but in annelids and echinoderms it appears to be a dimer. In each case the subunit molecular weight is around 40,000 (Morrison, 1975). For the purposes of the present analysis, since about 96% of the invertebrates covered in the present analysis are arthropods or mollusks, and since all the data contributing to this enzyme in our data base come from arthropods and mollusks, we are considering this enzyme as a monomer. Electrophoretic patterns of heterozygotes in crustaceans [*Daphnia* (personal observations), *Gammarus* (Bulnheim and Scholl, 1981)], mollusks [*Littorina* (personal observations)], and insects [*Glossina* (Gooding and Rolseth (1979)] are two-banded, consistent with a monomeric structure.

Creatine Kinase. Several creatine kinase loci exist in vertebrate species, and the enzyme is a dimer with a subunit molecular weight of 40,000 (Hopkinson *et al.,* 1976; Dixon and Webb, 1979). Interestingly, the skeletal muscle form of teleost fishes does not form the expected heterodimer in heterozygous individuals (Ferris and Whitt, 1978); thus, heterozygotes show just two bands following electrophoresis. This is one of very few examples where subunit number may be misinterpreted from the results of gel electrophoresis alone (another example being G6PDH in mammals, where the X-linked nature of the locus coupled with inactivation of one X chromosome in each cell of females means that heterozygous females cannot produce the heterodimer *in vivo*).

Esterase. Esterases acting on α-naphthyl acetates produce both two- and three-banded heterozygotes in invertebrates, implying a variable quaternary structure. In *Drosophila,* polymorphism for dimerizing ability has been reported (Cobbs, 1976). Electrophoretic surveys of naphthyl acetate esterases in vertebrates almost always describe heterozygotes as having a two-banded phenotype, consistent with a monomeric structure, although a few surveys have identified such esterases active both as monomeric and trimeric forms (e.g., Searle, 1986; see also Inkerman *et al.*, 1975). Here we have assumed a monomeric structure for vertebrate esterase but have not assigned a fixed structure for the invertebrate enzyme. Molecular weights are variable (e.g., Hopkinson *et al.*, 1976). The enzyme listed as esterase-D in Table I hydrolyzes methyl umbelliferyl esters: it appears to be a dimer in a wide range of vertebrate and invertebrate species and is identified as EST-D in humans (Hopkinson *et al.*, 1973).

Glucose 6-Phosphate Dehydrogenase. Mammalian, avian, and reptilian G6PDH is generally classified as a dimer (Hopkinson *et al.*, 1976; Hori *et al.*, 1975), whereas in the lower vertebrates (fish and amphibia) G6PDH has about twice the molecular weight and appears to be a tetramer (Hori *et al.*, 1975; Ohnishi and Hori, 1977). We have used these structures in our analyses. Interconversion is possible *in vitro* by the use of SH reagents. Native trout enzyme has the molecular weight of a tetramer, but at high pH seems to dissociate into an active dimeric or even monomeric form (Cederbaum and Yoshida, 1976). See the creatine kinase section for further comment on mammalian G6PDH. A variety of invertebrate species appear to have both dimeric and tetrameric forms (Ohnishi and Hori, 1977; Hori and Tanda, 1980), and in *Drosophila* three alleles have been described which yield respectively monomeric, dimeric, and tetrameric forms (Williamson and Bentley, 1983). We have not designated a quaternary structure for this enzyme in our invertebrate analyses. Subunit molecular weight is similar among invertebrates and vertebrates (Table II).

Glutamate Dehydrogenase. Bovine liver glutamate dehydrogenase has been extensively studied, and from amino acid sequence data is known to be a hexamer of identical subunits with a total molecular weight of 332,000 (E. L. Smith *et al.*, 1975). The mitochondrial form of a nematode GDH had a molecular weight of 335,000, implying a similar structure, whereas gel filtration of the cytosolic GDH from the same species showed four peaks of activity with molecular weights of approximately 610,000, 285,000, 180,000, and <100,000 (Turner *et al.*, 1986). GDH from both a coral (Catmull *et al.*, 1987) and a protozoan (Ling *et al.*, 1986) has been shown to be a hexamer, with subunit molecular weights of 56,000 and 49,000, respectively. Glutamate dehydrogenase is especially sensitive in terms of subunit aggregation to the concentration and composition of the solvent medium (Dixon and Webb, 1979), but it appears most commonly to aggregate as a hexamer and is so designated here.

Hemoglobin. Hemoglobin is perhaps the most extensively studied of all proteins, and in all bony vertebrates is a tetramer made up of a pair of closely linked $\alpha\beta$ dimers, each chain having a subunit molecular weight of around

16,000 daltons (Perutz, 1990). Invertebrate hemoglobins are, however, much more diverse both in quaternary structure and in subunit weight (Riggs, 1991). While most invertebrate HB chains have a relative molecular weight of around 16,000 (Perutz, 1990), the brine shrimp, *Artemia salina*, has HB chains with a subunit molecular weight of around 126,000 (Moens and Kondo, 1978). These long chains are made up of homologous, myoglobinlike domains linked in tandem (Manning *et al.*, 1990). The blood clam, *Scapharca inaequivalvis*, has both a heterotetrameric HB (A_2B_2) and a dimeric HB (C_2) (Royer *et al.*, 1985), the brine shrimp (Moens and Kondo, 1978) and the sea cucumber, *Molpaedia arenicola* (Kolatkar *et al.*, 1988) have dimeric HB, and the fat inn-keeper worm, *Urechis campo*, and the earthworm, *Lumbricus terrestris*, have tetrameric structures but with quite different compositions (the former is homomeric, the latter is an *abc* trimer linked to a *d* monomer) (Kolatkar *et al.*, 1988; Fushitani *et al.*, 1988). Hemoglobin is rarely included in invertebrate electrophoresis surveys, and in fact the data available in this group do not meet our minimal criteria for inclusion in Table IX and subsequent analyses. This is perhaps fortunate, considering the bewildering diversity of structures seen in the invertebrates. Hemoglobin is commonly screened in vertebrates, especially in mammals (Table VIII).

Hexokinase. This is a monomeric protein in vertebrates, with a molecular weight of around 100,000 (Colowick, 1975). However, the situation in invertebrates is not so clear-cut. HK from a tapeworm (Komuniecki and Roberts, 1977) and the most abundant HK from a mussel (Sanchez *et al.*, 1983) have a molecular weight around 100,000, but in *Drosophila* (Moser *et al.*, 1980) and a starfish (Mochizuki and Hori, 1977) HK has a molecular weight of about 50,000. It does appear to be a monomer in all taxa.

Lactate Dehydrogenase. Vertebrate LDH, which is *l*-lactate specific, is a tetramer, but two distinct LDH enzymes are found in invertebrates (Long and Kaplan, 1968; Scheid and Awapara, 1972; Ellington and Long, 1978). Mollusks, chelicerates, and polychaetes possess a *d*-specific form. This is a dimer both in the chelicerates *Limulus polyphemus* and tarantula (both species with a subunit molecular weight of 35,000) (Long and Kaplan, 1973; Selander and Yang, 1970; Gleason *et al.*, 1971) and in the polychaete *Nereis virens* (subunit weight 36,200) (Long and Kaplan, 1973), and is also typically dimeric in gastropods (Morris, 1979; Hill *et al.*, 1983). However, five-banded heterozygotes have been reported from three bivalve species, implying a tetrameric structure (Fujio *et al.*, 1983; Hebert *et al.*, 1989). Species contributing data for *d*-LDH to the present analysis were essentially gastropods (and *Limulus*), bivalve mollusks not being scored for this enzyme in any incorporated surveys, and thus we feel justified in scoring this enzyme as a dimer in ensuing analyses. Oligochaetes, mandibulates (a subphylum including the crustaceans and insects), and echinoderms generally have an *l*-specific form similar in subunit number and size to vertebrate *l*-LDH. One confusing exception seems to be in the barnacles, a crustacean group which has a *d*-specific LDH with a molecular weight of around 140,000, implying a tetrameric structure (Gleason *et al.*, 1971; Ellington and Long, 1978). However, no barnacles contributed LDH data to our data base.

Leucine Aminopeptidase. Electrophoretic surveys of both invertebrates and vertebrates invariably describe heterozygotes for this enzyme as two-banded, implying a monomeric structure. Yet some biochemical investigations have identified mammalian forms as being tetrameric or hexameric (Dixon and Webb, 1979). We have assumed here that the LAP of electrophoretic surveys is a monomer.

Malic Enzyme. The NADP-dependent malate dehydrogenase or malic enzyme is generally recognized as a tetramer in both invertebrates (e.g., Lee *et al.,* 1978; Darnall and Klotz, 1975; May and Holbrook, 1978) and vertebrates (e.g., Shows *et al.,* 1970; Hopkinson *et al.,* 1976; Nevaldine *et al.,* 1974; Stoneking *et al.,* 1979; Cross *et al.,* 1979), yet the electrophoresis literature contains several accounts of this enzyme behaving as a monomer in some invertebrate (Tracey *et al.,* 1975; Fujio *et al.,* 1983) and vertebrate taxa (Hedgecock and Ayala, 1974; Frydenberg and Simonsen, 1973). We assume that the true native state of ME is tetrameric, but that sometimes it degrades to a monomeric form which retains significant activity.

Peptidase. Enzymes categorized in this survey as PEP are peptidases which act on one or more of a wide range of di- and tripeptides (excluding those specific to leucine naphthylamide), and which exhibit a bewildering range of structures ranging from monomers and dimers to tetramers and hexamers. The subunit numbers of monomorphic enzymes cannot be known without extensive biochemical investigation, and no attempt has been made here to relate genetic variation in PEP enzymes to structural features.

Superoxide Dismutase. The cytoplasmic enzyme (Cu–Zn SOD) is a dimer, while the mitochondrial form (Mn SOD) is a tetramer (McCord, 1979). While the subunits of both enzymes are small, that of the cytoplasmic form seems to be slightly smaller than that of the mitochondrial enzyme, with mean molecular weights of around 16,000 and 19,000, respectively (see Table II). The surveys included in the present analysis scored an average of 1.21 (invertebrates) and 1.25 (vertebrates) loci per species. Thus the majority of surveys describe a single locus for this enzyme: this can be assumed to be the supernatant form (which is the most active isozyme), since whenever heterozygotes are described they show the three bands typical of a dimeric protein. The extra locus identified in a minority of surveys may represent a monomorphic mitochondrial isozyme or may represent a charge variant of the cytoplasmic form. Three Cu–Zn SOD isozymes were described in the housefly (Bird *et al.,* 1986), but whether these represent allelic forms, forms encoded by different loci, or posttranslational modifications of a single isozyme is uncertain. For simplicity, we have assumed that all the SOD loci covered in this survey code for the cytoplasmic type, with the exception of human Sod_1, which is known to be the mitochondrial locus and is thus excluded from our analysis. Since tetrameric enzymes are on average less variable than dimers (Ward, 1977; Harris *et al.,* 1977; and present work), the true mean heterozygosity of dimeric SOD may be slightly underestimated.

Conservation of Enzyme Structure among Taxa

Subunit numbers and subunit molecular weights are given in Tables I and II. Some enzymes still have not been extensively studied, but all those with a known and fixed quaternary structure in both invertebrates and vertebrates have the same structure in the two groups. Twenty eight enzymes have defined subunit molecular weights in both vertebrates and invertebrates, and these are plotted against one another in Fig. 1. The nonspecific and structurally variable hydrolases such as esterases, phosphatases, and peptidases are excluded here. Note that in fact there are 29 points, as HK in invertebrates appears to exist as two molecular-weight forms and was entered into this figure as two points, with coordinates (50, 100) and (100, 100). The

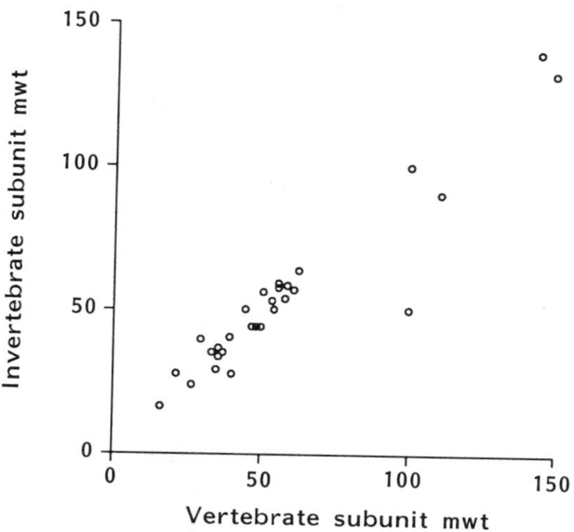

FIG. 1. Subunit molecular weights of invertebrate enzymes plotted against subunit molecular weights of the corresponding vertebrate enzymes.

correlation coefficient for the 29 points is 0.95 ($P < 0.001$), and omission of the data for ADH (where there is little homology between the insect and vertebrate enzyme; see earlier) and HK raises the correlation to 0.99 ($n = 26$, $P < 0.001$). The conservation of structure of homologous enzymes among animal taxa is very impressive, and means that for most enzymes with defined and specific functions a well-validated subunit molecular-weight estimate from one animal species will serve as a good guide to molecular weight in any other animal species. The small differences in subunit sizes between such enzymes from vertebrates and invertebrates may in large part reflect experimental errors in molecular weight determination rather than real variation. Interestingly, where variation of structure is restricted to a single subkingdom, the variation is mainly found within the invertebrates (e.g., HK, HB, *d*-LDH, APK). This may reflect the greater evolutionary time available for divergence in the invertebrates compared with the bony vertebrates (around 650 million years compared with 400 million years) or the greater range of habitats occupied by invertebrates.

VARIATION IN MEAN HETEROZYGOSITY AMONG SPECIES AMONG TAXA

For each species in the ENTIRE data base, an average heterozygosity across all loci was estimated as H (equal to $\sum H_j/k$, where k, the number of loci, varied from 15 to 69). Table III lists the mean values of H for the major taxonomic groupings considered, and also gives the number of species contributing data (908 species in total) and the average number of loci screened per species. Date obtained by Nevo *et al.* (1984) are given for comparison. There is considerable overlap in the distributions of H in vertebrates and invertebrates, but most species with high H values are invertebrates (Fig. 2). The modal value of H is around 0.04 for vertebrates and 0.10 for invertebrates.

These data were initially analyzed by a nested analysis of variance where the two major groups are vertebrates and invertebrates (Table IV). The subgroups are, for the vertebrates, the phyla of mammals, birds, reptiles, amphibians, and fish, and for the invertebrates, two classes of mandibulate members of the phylum Arthropoda (crustacea and insects), the phylum Mollusca, and a heterogeneous collection of other species. Clearly there are highly significant differences in heterozygosity both among groups (with invertebrates being significantly more heterozygous than vertebrates) and among subgroups within each of the two groups. The source of this variation among subgroups within each of the two groups was examined using the Tukey

TABLE III. Estimates of Average Heterozygosity in Major Taxonomic Groupings[a]

	Present analysis			Nevo *et al.* (1984)	
	$\bar{H}$	$\bar{L}$	n	$\bar{H}$	n
Vertebrates					
Total	0.071 ± 0.002	24.76 ± 0.30	648	0.054 ± 0.003	551
Mammals	0.067 ± 0.004	24.48 ± 0.52	172	0.041 ± 0.003	184
Birds	0.068 ± 0.005	28.70 ± 0.90	80	0.051 ± 0.004	46
Reptiles	0.078 ± 0.007	22.72 ± 0.54	85	0.055 ± 0.006	70
Amphibians	0.109 ± 0.006	21.93 ± 0.44	116	0.067 ± 0.007	61
Fish	0.051 ± 0.003	25.97 ± 0.65	195	0.051 ± 0.003	183
Invertebrates					
Total	0.122 ± 0.004	21.71 ± 0.39	370	0.100 ± 0.005	361
Insects	0.137 ± 0.005	21.18 ± 0.70	170	0.096[b]	156
Crustaceans	0.052 ± 0.005	23.01 ± 0.82	80	0.082 ± 0.007	122
Mollusks	0.145 ± 0.010	21.78 ± 0.36	105	0.148 ± 0.025	46
Others	0.160 ± 0.016	20.33 ± 1.77	15	0.112[b]	35

[a] $\bar{H}$ is the average heterozygosity per species (±SE), $\bar{L}$ is the average number of loci screened per species (±SE), and n is the number of species.

[b] Nevo *et al.* (1984) split the insects into two categories, *Drosophila* and insects excluding *Drosophila,* and listed separately invertebrate taxa with small sample sizes such as brachiopods and coelenterates.

HSD method (T-method) and the Tukey–Kramer adjustment for unequal sample sizes (Table IV). This shows that for the vertebrates, mammals, birds, and reptiles have average heterozygosities that do not differ from each other, while amphibians and fish are, respectively, more and less heterozygous than these three taxa. For the invertebrates, mollusks, insects, and "others" have similar mean heterozygosities, while crustaceans are less heterozygous.

It will be apparent from Table III that, while Nevo *et al.* (1984) also found the amphibia to be the most heterozygous vertebrate taxon and the crustaceans the least variable invertebrate taxon, our estimates of $\bar{H}$ are higher than those of Nevo *et al.* by a factor of about 1.3 for vertebrates and about 1.2 for invertebrates. To what might this difference in mean heterozygosities be attributed? The requirements for inclusion in Nevo *et al.*'s data base were slightly less restrictive than ours, being a minimum of 14 loci and 10 individuals per species (versus 15 and 15, respectively, for our analyses), but mean numbers of loci studied per species were similar in the two surveys (Nevo *et al.* give a global mean of 23, compared with means of 24.8 and 21.7 for vertebrates and invertebrates, respectively, in our survey).

There is some suggestion of an inverse relationship between the number of loci scored in a species and its mean heterozygosity (Table V), as judged by

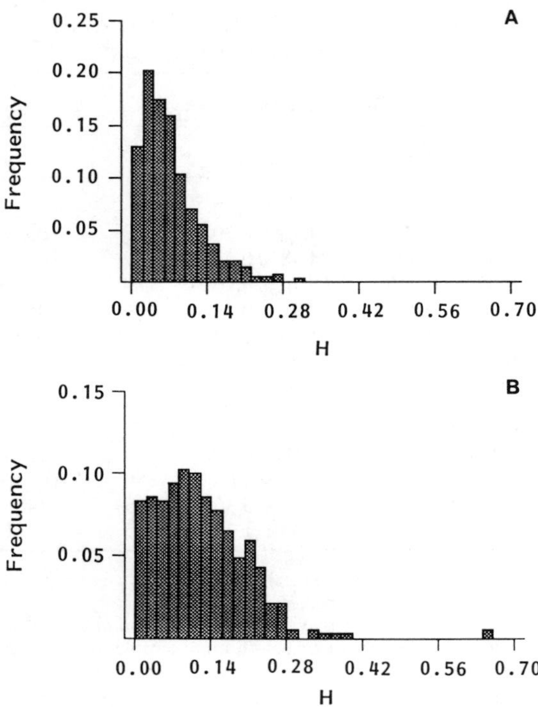

FIG. 2. Distributions of average heterozygosities per species among vertebrates and invertebrates. (A) Vertebrates (648 species). (B) Invertebrates (370 species).

the significant negative regression slopes for the pooled vertebrate and pooled invertebrate data. Such a relationship would imply that those investigators using smaller numbers of loci tend to choose preferentially those loci with a higher probability of being heterozygous. However, this is a simplistic view. Among the vertebrates, the bird surveys show a very striking and significant negative regression of H on number of loci. But this result is completely attributable to just two studies. The study surveying the smallest number of loci (15) analyzed 7 species of blackbirds and gave an average H for these 7 species of 0.177 ± 0.016 (J. K. Smith and Zimmerman, 1976), whereas a study analyzing 39 loci in 6 species of sparrows gave $\bar{H}$ for these species of 0.032 ± 0.003 (Zink, 1982). Deleting these two studies from the analysis leaves 67 avian species and completely removes any significant linear regression component ($P = 0.820$). The significant overall negative regression in vertebrates between H and number of loci seems primarily to result from inclusion of the amphibia, a taxon with high average H and (relatively) small

TABLE IV. Analysis of Variation in Heterozygosity H among Taxa

Nested analysis of variance

Source of variation	df	F	P
Among groups (vertebrates and invertebrates)	1	74.74	<0.001
Among subgroups within groups	7	31.44	<0.001
Among subgroups within vertebrates	4	19.10	<0.001
Among subgroups within invertebrates	3	47.89	<0.001
Within subgroups (error; between species within subgroups)	1009		

Tukey HSD test[a]

Vertebrates	Mammals	Birds	Reptiles	Amphibians	Fish
Mammals	—	+<0.001	+0.011	+0.043	−0.016
Birds	0.994	—	+0.012	+0.043	−0.016
Reptiles	0.670	0.939	—	+0.031	−0.027
Amphibians	<0.001	<0.001	<0.001	—	−0.059
Fish	0.030	0.051	0.002	<0.001	—

Invertebrates	Insects	Crustaceans	Mollusks	Others
Insects	—	−0.086	+0.008	+0.023
Crustaceans	<0.001	—	+0.093	+0.108
Mollusks	0.999	<0.001	—	+0.015
Others	0.685	<0.001	0.717	—

[a] Above the diagonal are pairwise $\bar{H}$ differences, where a plus sign indicates that the column $\bar{H}$ exceeds row $\bar{H}$ and a minus sign indicates the converse. Below the diagonal are the P values derived from tests on arcsine-transformed H estimates.

average number of screened loci. Omitting the amphibia (which themselves show no significant regression) from the analysis removes the association ($P = 0.322$). The significant overall negative regression in invertebrates is, in a rather similar manner, dependent on the inclusion of the crustaceans, a group with a low average H but a high average number of screened loci. Omitting this group removes the significance of the invertebrate regression ($P = 0.305$). The crustaceans themselves show a significant negative regression, but this significance is generated by a single study which analyzed large numbers of loci (37) in 12 species of penaeid prawns, with a low average heterozygosity of 0.019 ± 0.003 (Mulley and Latter, 1980). Removing this study removes the significance of the regression ($P = 0.774$).

The difference in H estimates between our study and Nevo *et al.* (1984) is probably related to differences in the definition of H. As stated earlier, we calculate H_j and H as total expected heterozygosity for that locus and species, respectively, H_T in Nei's (1973) terminology. Nevo *et al.*, in contrast, used,

TABLE V. Relationships between H (Arcsine Transformed) and Number of Loci Screened per Species[a]

		Regression analysis			
Taxon	n	Intercept	Slope	P	r^2
Vertebrates	648	16.139	−0.074	0.022	0.008
Mammals	172	13.463	0.017	0.800	<0.001
Birds	80	21.013	−0.237	<0.001	0.154
Reptiles	85	13.043	0.082	0.597	0.003
Amphibians	116	18.952	−0.029	0.823	<0.001
Fish	195	11.706	0.016	0.689	0.001
Invertebrates	370	21.960	−0.127	0.017	0.015
Insects	170	21.386	−0.014	0.776	<0.001
Crustaceans	80	16.110	−0.180	0.038	0.054
Mollusks	105	26.433	−0.243	0.282	0.011
Others	15	33.295	−0.502	0.010	0.414

[a] H and number of loci are the Y and X variables, respectively; n is the number of species.

where possible, observed heterozygosity per individual estimates. In a sexually reproducing species, these will generally be very similar to Hardy–Weinberg expected heterozygosities, but using observed heterozygosities means that the H estimates of Nevo *et al.* will be analogous to the H_S parameter of Nei (1973), that is, an average H per subpopulation. H_T can never be less than H_S, and the amount by which H_T exceeds H_S depends on the amount of interpopulation differentiation D_{ST}. Thus,

$$H_T = H_S + D_{ST}$$

It is therefore to be expected that our estimates of H (H_T) would exceed those of Nevo *et al.* (H_S). Note also that Nevo *et al.* incorporated data from some parthenogenetically reproducing species in their analysis.

We have estimated the sizes of $\bar{\bar{H}}_S$ and $\bar{\bar{H}}_T$ (averages of $\bar{H}_S$ and $\bar{H}_T$ over species) (Table VI) for the WITHIN species data base (which, unlike the ENTIRE data base, is so structured as to allow this analysis), and these $\bar{\bar{H}}_S$ estimates are more similar to the $\bar{H}$ estimates of Nevo *et al.* than are our earlier $\bar{H}$ (H_T) figures given in Table III, with the exception of the small reptile dataset (22 species). However, it is interesting to note that the $\bar{\bar{H}}_T$ estimates of Table VI from the WITHIN data base are higher than the $\bar{H}$ (H_T) estimates of Table III derived from the ENTIRE data base (t-tests on arcsine-transformed H values: vertebrates $P = 0.001$, invertebrates $P = 0.039$). This probably arises from the inclusion in the ENTIRE data base of

TABLE VI. Comparison of $\bar{\bar{H}}_T$, $\bar{\bar{H}}_S$, $\bar{G}_{ST}$, and $\bar{D}_m$ Estimates (Nei, 1973) with SE, for Different Taxonomic Groupings in the WITHIN Dataset[a]

	$\bar{\bar{H}}_T$	$\bar{\bar{H}}_S$	$\bar{G}_{ST}$	$\bar{D}_m$	n
Vertebrates					
Total	0.086 ± 0.004	0.064 ± 0.003	0.202 ± 0.015	0.031 ± 0.004	207
Mammals	0.078 ± 0.007	0.054 ± 0.004	0.242 ± 0.030	0.031 ± 0.005	57
Birds	0.059 ± 0.007	0.054 ± 0.006	0.076 ± 0.020	0.007 ± 0.002	16
Reptiles	0.124 ± 0.016	0.090 ± 0.013	0.258 ± 0.050	0.048 ± 0.012	22
Amphibians	0.136 ± 0.013	0.094 ± 0.011	0.315 ± 0.040	0.061 ± 0.010	33
Fish	0.067 ± 0.005	0.054 ± 0.003	0.135 ± 0.021	0.018 ± 0.006	79
Invertebrates					
Total	0.138 ± 0.009	0.113 ± 0.009	0.171 ± 0.020	0.038 ± 0.005	114
Insects	0.138 ± 0.009	0.122 ± 0.007	0.097 ± 0.015	0.023 ± 0.005	46
Crustaceans	0.088 ± 0.016	0.063 ± 0.011	0.169 ± 0.061	0.036 ± 0.019	19
Mollusks	0.157 ± 0.020	0.121 ± 0.019	0.263 ± 0.036	0.057 ± 0.009	44
Others	0.150 ± 0.031	0.141 ± 0.030	0.060 ± 0.021	0.014 ± 0.008	5

[a] n = is the number of species.

some species which only have data contributed from a single subpopulation, and thus will not show the elevated H_T values which occur when allele frequencies are averaged over subpopulations.

THE EXTENT OF POPULATION SUBDIVISION AMONG TAXA

Table VI gives $\bar{\bar{H}}_S$ and $\bar{\bar{H}}_T$ and the derived parameters $\bar{G}_{ST}$ and $\bar{D}_m$ estimated as means over the n species contributing data. G_{ST} represents that proportion of the genetic variation that can be attributable to subpopulation differentiation in gene frequencies, and is the parameter most commonly used for measuring degrees of genetic population subdivision. It can be seen that the mean value of G_{ST} is highest for the amphibians and mollusks, and lowest for birds and insects (excluding the heterogeneous and small group of invertebrate "other" species). The striking contrasts between the high G_{ST} estimates of amphibia and mollusks and the low values of birds and insects, and the extensive variation in G_{ST} among different amphibia and mollusk species, are shown clearly in Fig. 3. About 30% of genetic variation in amphibia and mollusks ($\bar{G}_{ST}$ = 0.315 and 0.263, respectively) can be attributed to variation among subpopulations, compared with around 10% for birds and insects ($\bar{G}_{ST}$ = 0.076 and 0.097, respectively). Particularly noteworthy is the comparative lack of amphibian species showing G_{ST} values less than 0.05.

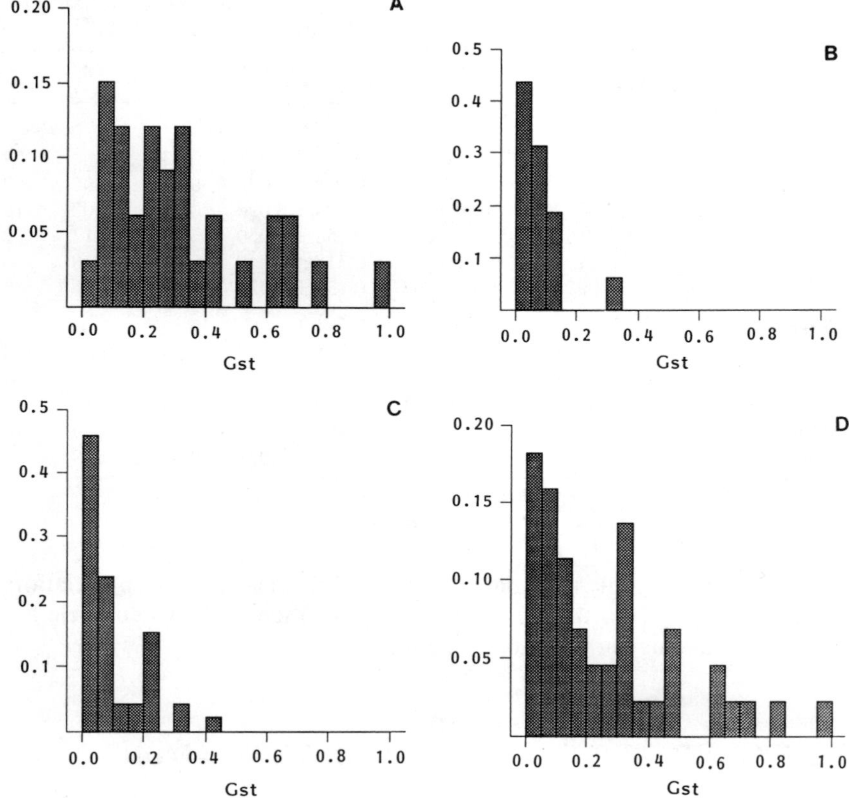

FIG. 3. G_{ST} distributions of amphibia, birds, insects, and mollusks. (A) Amphibia (33 species). (B) Birds (16 species). (C) Insects (46 species). (D) Mollusks (44 species).

An analysis of variation in G_{ST} is given in Table VII, which confirms the significant heterogeneity of $\bar{G}_{ST}$ among subgroups. Tukey's HSD tests show that the significant differences in the vertebrates arise from the amphibia, mammals and reptiles having higher $\bar{G}_{ST}$ values than birds and fish. With respect to invertebrates, the one significant test shows that mollusks have a higher $\bar{G}_{ST}$ than insects.

Values of G_{ST} are dependent on values of $\bar{H}_T$, such that when $\bar{H}_T$ is small, G_{ST} may be large even if the absolute differentiation is small. In order to adjust for any effects of variation in H_T, Nei (1973) suggested that the parameter D_m be used to compare the degree of genetic differentiation of different species. D_m values were analyzed in a similar fashion to the G_{ST} values. Results and conclusions are very similar, except that while there is (nearly) significant difference in $\bar{G}_{ST}$ between vertebrates and invertebrates

TABLE VII. Analysis of Variation in G_{ST} among Taxa

Nested analysis of variance

Source of variation	df	F	P
Among groups (vertebrates and invertebrates)	1	3.84	0.051
Among subgroups within groups	7	7.74	<0.001
Among subgroups within vertebrates	4	9.02	<0.001
Among subgroups within invertebrates	3	6.03	<0.001
Within subgroups (error; between species within subgroups)	312		

Tukey HSD test[a]

Vertebrates	Mammals	Birds	Reptiles	Amphibians	Fish
Mammals	—	−0.167	+0.016	+0.073	−0.107
Birds	0.028	—	+0.182	+0.240	+0.060
Reptiles	0.996	0.041	—	+0.058	−0.123
Amphibians	0.292	<0.001	0.741	—	−0.180
Fish	0.006	0.909	0.037	<0.001	—

Invertebrates	Insects	Crustaceans	Mollusks	Others
Insects	—	+0.072	+0.166	−0.037
Crustaceans	0.728	—	+0.094	−0.109
Mollusks	0.001	0.160	—	−0.202
Others	0.976	0.776	0.120	—

[a] Above the diagonal are pairwise G_{ST} differences, where a plus sign indicates that the column G_{ST} exceeds row G_{ST} and a minus sign indicates the converse. Below the diagonal are the P values derived from tests on arcsine-transformed G_{ST} estimates.

(P = 0.051), $\bar{D}_m$ shows no significant difference (P = 0.85). Thus, after correcting for the effects of differences in $\bar{H}_T$, vertebrates and invertebrates show similar overall levels of genetic population subdivision. However, the variation in $\bar{G}_{ST}$ and $\bar{D}_m$ among taxa means that this conclusion would not be robust to changes in the relative proportions of taxa within vertebrates and invertebrates.

ENZYME HETEROZYGOSITY AND TAXONOMIC CLASSIFICATION

Vertebrates

Mean heterozygosities H_E of enzymes and some nonenzymatic proteins are given, for each vertebrate class, in Table VIII. Fourteen proteins

TABLE VIII. Mean Heterozygosities ($H_E \pm$ SE) of Enzymes and Nonenzymatic Proteins in Vertebrate Taxa[a]

Enzyme	Birds	Mammals	Reptiles	Amphibia	Fish	Total
Oxidoreductases						
ADH	0.018 ± 0.009	0.075 ± 0.017	0.058 ± 0.019	0.157 ± 0.047	0.058 ± 0.014	0.065 ± 0.009
	38 38	71 71	35 35	22 13	101 92	267 249
G3PDH	0.043 ± 0.011	0.072 ± 0.012	0.032 ± 0.010	0.097 ± 0.019	0.061 ± 0.101	0.062 ± 0.006
	90 67	121 118	82 74	83 78	214 137	590 474
IDDH	0.055 ± 0.018	0.063 ± 0.016	0.074 ± 0.062	0.129 ± 0.030	0.087 ± 0.016	0.080 ± 0.010
	53 51	82 82	9 7	43 43	100 81	287 264
l-LDH	0.010 ± 0.003	0.024 ± 0.004	0.033 ± 0.007	0.078 ± 0.011	0.024 ± 0.004	0.032 ± 0.003
	139 76	366 172	168 85	218 116	507 193	1398 642
MDH	0.002 ± 0.001	0.014 ± 0.003	0.045 ± 0.011	0.061 ± 0.009	0.029 ± 0.004	0.029 ± 0.003
	143 78	315 171	151 83	211 115	478 186	1298 633
ME	0.087 ± 0.028	0.079 ± 0.017	0.056 ± 0.037	0.194 ± 0.027	0.035 ± 0.008	0.083 ± 0.009
	38 32	82 63	22 22	71 54	149 99	362 270
IDH	0.073 ± 0.013	0.048 ± 0.008	0.065 ± 0.015	0.087 ± 0.011	0.050 ± 0.008	0.062 ± 0.005
	117 67	280 158	75 45	202 114	248 144	922 528
6PGDH	0.109 ± 0.019	0.100 ± 0.012	0.182 ± 0.023	0.166 ± 0.023	0.045 ± 0.010	0.110 ± 0.007
	77 76	157 155	77 70	84 83	136 123	531 507
GLUDH	0.000 ± 0.000	0.213 ± 0.050	—	0.000 ± 0.000	0.022 ± 0.020	0.129 ± 0.035
	2 2	23 19		2 2	17 14	44 37
G6PDH	0.046 ± 0.030	0.066 ± 0.015	0.234 ± 0.062	0.058 ± 0.026	0.001 ± 0.001	0.060 ± 0.010
	19 19	92 90	13 13	21 20	43 34	188 176
ODH	0.000 ± 0.000	0.000	0.054 ± 0.035	0.091 ± 0.037	0.088 ± 0.041	0.071 ± 0.020
	3 3	1 1	15 9	14 14	15 14	48 41
XDH	0.078 ± 0.033	0.074 ± 0.028	0.056 ± 0.020	0.123 ± 0.047	0.028 ± 0.014	0.061 ± 0.012
	7 7	27 27	48 46	22 18	50 50	154 148

GAPDH	0.000 ± 0.000	0.000 ± 0.000	0.000 ± 0.000	0.036 ± 0.024	0.031 ± 0.012	0.025 ± 0.009
	16 16	17 16	2 2	29 29	88 48	152 111
GDH	0.000 ± 0.000	0.074 ± 0.028	0.085 ± 0.054	0.025 ± 0.011	0.033 ± 0.021	0.031 ± 0.008
	47 39	26 25	15 13	84 82	35 29	207 188
DIA	0.125 ± 0.125	0.042 ± 0.021	—	0.000	0.008 ± 0.008	0.033 ± 0.014
	2 1	31 23		1 1	16 16	50 41
CAT	0.000	0.020 ± 0.020	0.000 ± 0.000	—	0.097 ± 0.097	0.025 ± 0.018
	1 1	23 23	3 3	—	3 3	30 30
SOD	0.003 ± 0.002	0.019 ± 0.005	0.069 ± 0.017	0.064 ± 0.016	0.028 ± 0.007	0.032 ± 0.004
	97 56	170 142	76 54	82 77	196 168	620 497
Transferases						
NP	0.176 ± 0.033	0.069 ± 0.026	0.032 ± 0.032	0.019 ± 0.019	0.155 ± 0.084	0.120 ± 0.020
	33 32	26 24	4 4	5 5	8 8	76 73
AAT	0.022 ± 0.006	0.030 ± 0.006	0.087 ± 0.014	0.130 ± 0.014	0.054 ± 0.008	0.061 ± 0.004
	141 71	305 156	130 93	193 108	284 140	1053 561
ALAT	0.000 ± 0.000	0.037 ± 0.037	0.596 ± 0.026	0.125 ± 0.125	0.080 ± 0.072	0.056 ± 0.024
	23 23	12 12	2 2	4 3	7 5	48 45
HK	0.199 ± 0.083	0.024 ± 0.024	—	—	0.000 ± 0.000	0.043 ± 0.021
	5 5	11 6			13 13	29 24
CK	0.002 ± 0.001	0.018 ± 0.016	0.000 ± 0.000	0.041 ± 0.019	0.029 ± 0.007	0.022 ± 0.005
	80 46	31 22	13 10	36 28	190 106	350 212
AK	0.002 ± 0.001	0.029 ± 0.014	—	0.124 ± 0.084	0.012 ± 0.006	0.019 ± 0.006
	26 22	56 44		5 5	95 80	182 151
Hydrolases						
EST	0.200 ± 0.020	0.152 ± 0.011	0.180 ± 0.017	0.216 ± 0.023	0.115 ± 0.011	0.160 ± 0.007
	157 61	401 131	169 73	127 59	302 116	1156 440

(continued)

TABLE VIII. Continued

Enzyme	Birds	Mammals	Reptiles	Amphibia	Fish	Total
EST-D	0.032 ± 0.011 15 12	0.000 ± 0.000 3 3	0.129 ± 0.103 2 2	—	0.018 ± 0.009 16 15	0.029 ± 0.008 36 32
ALP	0.207 ± 0.096 5 3	0.000 ± 0.000 21 19	0.000 ± 0.000 18 6	—	0.041 ± 0.030 17 9	0.029 ± 0.013 61 37
ACP	0.017 ± 0.007 57 40	0.059 ± 0.018 53 44	0.243 ± 0.086 8 5	0.122 ± 0.033 23 18	0.044 ± 0.020 54 42	0.058 ± 0.010 195 149
AMY	0.098 ± 0.072 3 2	0.118 ± 0.064 14 14	0.361 ± 0.067 8 8	—	0.380 1 1	0.200 ± 0.046 26 25
LAP	0.081 ± 0.019 63 42	0.018 ± 0.007 53 49	0.036 ± 0.010 69 39	0.059 ± 0.017 69 69	0.061 ± 0.027 29 26	0.051 ± 0.007 283 225
PEP	0.119 ± 0.013 175 63	0.069 ± 0.015 91 50	0.099 ± 0.017 101 59	0.164 ± 0.017 160 89	0.044 ± 0.007 288 102	0.093 ± 0.006 815 363
ADA	0.119 ± 0.022 44 44	0.160 ± 0.034 36 36	0.000 ± 0.000 9 3	0.085 ± 0.033 22 17	0.216 ± 0.036 44 37	0.144 ± 0.016 155 137
Lyases						
ALD	0.038 ± 0.031 16 16	0.012 ± 0.009 28 23	—	0.209 ± 0.123 4 4	0.033 ± 0.017 35 25	0.035 ± 0.012 83 68
FUM	0.000 ± 0.000 5 3	0.029 ± 0.026 19 18	0.079 ± 0.023 37 37	0.066 ± 0.021 37 37	0.025 ± 0.011 59 49	0.047 ± 0.009 157 144
ACO	0.030 ± 0.016 32 28	0.091 ± 0.051 14 8	0.070 ± 0.035 21 11	0.152 ± 0.027 44 24	0.095 ± 0.045 18 15	0.094 ± 0.015 129 86

Isomerases						
TPI	0.000 1 1	0.039 ± 0.031 16 14	—	0.000 ± 0.000 4 4	0.003 ± 0.003 14 12	0.019 ± 0.014 35 31
MPI	0.143 ± 0.021 70 61	0.193 ± 0.032 50 47	0.033 ± 0.019 16 13	0.168 ± 0.026 60 60	0.061 ± 0.016 68 65	0.130 ± 0.011 264 246
GPI	0.047 ± 0.009 70 66	0.030 ± 0.007 128 116	0.080 ± 0.016 71 70	0.183 ± 0.022 108 97	0.100 ± 0.009 364 181	0.093 ± 0.006 741 530
PGM	0.096 ± 0.015 108 70	0.114 ± 0.012 231 152	0.125 ± 0.018 126 84	0.109 ± 0.014 162 111	0.111 ± 0.011 263 192	0.112 ± 0.006 890 609
Nonenzymatic proteins						
HB	0.000 ± 0.000 32 25	0.075 ± 0.012 177 116	0.030 ± 0.027 18 10	0.002 ± 0.002 16 15	0.026 ± 0.017 40 16	0.053 ± 0.009 283 182
TF	0.149 ± 0.050 14 13	0.150 ± 0.020 104 104	0.260 ± 0.061 18 18	0.241 ± 0.044 31 31	0.077 ± 0.065 5 5	0.176 ± 0.017 172 171
ALB	0.042 ± 0.016 35 35	0.096 ± 0.016 132 129	0.097 ± 0.049 19 19	0.137 ± 0.031 39 39	0.000 ± 0.000 2 2	0.094 ± 0.012 227 224
PROT[b]	0.010 ± 0.017 99 48	0.048 ± 0.009 179 65	0.014 ± 0.004 262 73	0.058 ± 0.011 184 73	0.011 ± 0.003 371 102	0.025 ± 0.003 1095 361

[a] The first number given under the H_E estimate is the number of loci scored, the second is the number of species. Proteins scored from fewer than 25 loci are not included.

[b] Unspecified nonenzymatic proteins.

(G3PDH, *l*-LDH, MDH, ME, IDH, 6PGDH, SOD, AAT, PGM, EST, LAP, PEP, GPI, and PROT) have been screened in 20 or more species in each class, and the arcsine-transformed mean heterozygosities in this subtable of 14 rows and five columns were analyzed by a randomized complete-blocks ANOVA. This revealed highly significant differences both among class ($P < 0.001$, attributable largely to the higher heterozygosities of amphibia) and among proteins ($P < 0.001$). A total of 51.9% of the variance in (untransformed) H_E was attributable to protein-specific effects, and 21.2% to among-class effects. The 14 measures of H_E can be ranked for each vertebrate class and compared, a procedure which yields a highly significant value (Friedman test statistic = 44.13, $P < 0.001$) of 0.68 for Kendall's coefficient of concordance (which takes a range of 0–1). Thus, the rankings of the different proteins broadly agree among classes, indicating that patterns of variation in H_E are taxon independent. Esterases were the most variable of these 14 proteins, followed by PGM and 6PGDH, with MDH and PROT the least variable.

Invertebrates

Mean heterozygosities of each enzyme and a general nonenzymatic protein are given, for the invertebrate taxa, in Table IX. Nineteen proteins (MDH, ME, IDH, 6PGDH, XDH, GAPDH, SOD, AAT, HK, PGM, EST, ALP, ACP, LAP, PEP, ALD, MPI, GPI, and PROT) have been screened in 20 or more species of each of the three major taxa. A randomized complete-blocks ANOVA revealed highly significant differences in arcsine-transformed H_E both among taxa ($P < 0.001$, attributable largely to the lower heterozygosities of crustaceans) and among proteins ($P < 0.001$). A total of 41.06% of the variance in (untransformed) H_E was attributable to protein-specific effects, and 33.77% to taxa effects. The 19 measures of H_E were ranked for each invertebrate taxon and compared, giving a highly significant value (Friedman test statistic = 41.28, $P = 0.0014$) of 0.76 for Kendall's coefficient of concordance. Thus, the rankings of the different proteins broadly agree among classes, indicating that, as with the vertebrates, the patterns of variation in H_E are not taxon dependent. Phosphoglucomutase was the most variable protein, followed by MPI, GPI, and EST, with SOD and PROT the least variable. One notable exception to this general agreement across invertebrate taxa is provided by XDH. This enzyme is one of the most variable in insects, but one of the least variable in both mollusks and crustacea.

TABLE IX. Mean Heterozygosities per Locus ($H_E \pm$ SE) of Enzymes and Nonenzymatic Proteins in Invertebrate Taxa[a]

Enzyme	Crustacea	Mollusks	Insects	"Others"	Total
Oxidoreductases					
ADH	0.000 ± 0.000	0.064 ± 0.062	0.131 ± 0.022	0.539 ± 0.091	0.130 ± 0.021
	3 3	8 7	76 58	2 2	89 70
G3PDH	0.007 ± 0.004	0.088 ± 0.040	0.034 ± 0.008	0.005 ± 0.005	0.034 ± 0.007
	22 22	15 14	148 139	2 2	187 177
IDDH	0.102 ± 0.037	0.149 ± 0.028	0.141 ± 0.062	0.330 ± 0.074	0.149 ± 0.021
	16 16	49 47	15 12	5 5	85 80
l-LDH	0.032 ± 0.006	—	0.036 ± 0.019	0.000	0.033 ± 0.010
	44 39		30 29	1 1	75 68
d-LDH	—	0.137 ± 0.031	—	0.019 ± 0.019	0.131 ± 0.028
		37 32		3 2	40 34
HBDH	0.000 ± 0.000	—	0.102 ± 0.022	—	0.098 ± 0.021
	2 2		47 42		49 44
MDH	0.049 ± 0.010	0.074 ± 0.010	0.063 ± 0.008	0.068 ± 0.027	0.063 ± 0.005
	133 73	224 98	220 156	20 15	597 342
ME	0.029 ± 0.009	0.096 ± 0.028	0.095 ± 0.013	0.118 ± 0.038	0.076 ± 0.009
	74 53	49 47	146 137	10 8	279 245
IDH	0.016 ± 0.008	0.075 ± 0.011	0.124 ± 0.012	0.155 ± 0.043	0.094 ± 0.008
	44 38	175 88	197 141	13 11	429 278
6PGDH	0.055 ± 0.018	0.131 ± 0.021	0.094 ± 0.017	0.334 ± 0.124	0.108 ± 0.012
	33 33	90 90	85 75	5 5	213 203
G6PDH	0.001 ± 0.001	0.066 ± 0.023	0.198 ± 0.027	0.050 ± 0.036	0.117 ± 0.017
	11 9	55 46	54 53	5 5	125 113
ODH	0.049 ± 0.012	0.237 ± 0.086	0.096 ± 0.019	0.152 ± 0.143	0.101 ± 0.016
	26 26	10 10	55 48	3 2	94 86
XDH	0.012 ± 0.006	0.017 ± 0.010	0.180 ± 0.034	0.046 ± 0.032	0.077 ± 0.015
	28 28	51 51	49 48	6 6	134 133
GAPDH	0.000 ± 0.000	0.096 ± 0.034	0.058 ± 0.016	0.038 ± 0.019	0.053 ± 0.012
	24 24	31 31	52 52	10 9	117 116
GDH	0.000 ± 0.000	0.019 ± 0.019	0.126 ± 0.096	0.000	0.020 ± 0.014
	20 20	3 3	4 4	1 1	28 28

(continued)

TABLE IX. Continued

Enzyme	Crustacea	Mollusks	Insects	"Others"	Total
DIA	0.040 ± 0.037	0.402 ± 0.203	0.044 ± 0.019	0.014	0.068 ± 0.026
	10 8	3 3	20 8	1 1	34 20
SOD	0.021 ± 0.010	0.008 ± 0.005	0.018 ± 0.005	0.126 ± 0.062	0.021 ± 0.005
	54 43	99 83	143 117	17 15	313 258
AO	0.053 ± 0.016	0.272 ± 0.066	0.306 ± 0.029	—	0.196 ± 0.020
	66 43	8 8	80 61		154 112
Transferases					
NP	—	0.131 ± 0.036	—	0.080 ± 0.061	0.126 ± 0.033
		28 28		3 3	31 31
AAT	0.059 ± 0.014	0.143 ± 0.016	0.134 ± 0.013	0.153 ± 0.056	0.121 ± 0.009
	95 67	155 100	187 121	10 8	447 296
ALAT	0.121 ± 0.046	—	0.164 ± 0.043	0.030	0.141 ± 0.030
	17 17		20 20	1 1	38 38
HK	0.031 ± 0.014	0.127 ± 0.048	0.097 ± 0.011	0.323 ± 0.051	0.104 ± 0.010
	45 34	23 20	199 115	20 12	287 181
PK	0.000 ± 0.000	—	0.031 ± 0.019	—	0.029 ± 0.017
	2 1		28 28		30 29
APK	0.054 ± 0.042	0.164 ± 0.164	0.000 ± 0.000	—	0.028 ± 0.017
	11 11	2 2	20 20		33 33
AK	0.036 ± 0.031	0.137 ± 0.033	0.096 ± 0.018	0.079 ± 0.040	0.101 ± 0.014
	14 13	50 38	101 61	14 8	179 120
Hydrolases					
EST	0.074 ± 0.012	0.188 ± 0.016	0.274 ± 0.014	0.115 ± 0.032	0.195 ± 0.009
	178 74	231 98	329 112	35 11	773 295
ALP	0.019 ± 0.009	0.213 ± 0.041	0.138 ± 0.028	0.037 ± 0.037	0.106 ± 0.016
	59 44	38 38	37 24	3 3	137 109
ACP	0.050 ± 0.013	0.061 ± 0.019	0.199 ± 0.022	0.064 ± 0.038	0.115 ± 0.012
	79 58	62 47	112 98	16 7	269 210

AMY	0.068 ± 0.037 26 21	—	0.190 ± 0.032 45 40	—	0.145 ± 0.025 71 61
LAP	0.065 ± 0.021 57 54	0.216 ± 0.022 129 81	0.221 ± 0.020 136 94	0.165 ± 0.043 22 10	0.188 ± 0.013 344 239
PEP	0.013 ± 0.012 32 20	0.232 ± 0.020 145 80	0.111 ± 0.016 104 48	0.050 ± 0.050 3 1	0.159 ± 0.013 284 149
ADA	0.010 ± 0.007 30 18	—	—	—	0.010 ± 0.007 30 18
Lyases					
ALD	0.007 ± 0.005 28 24	0.030 ± 0.016 22 22	0.086 ± 0.013 100 84	—	0.063 ± 0.010 150 130
FUM	0.040 ± 0.019 35 31	0.000 1 1	0.054 ± 0.012 92 92	0.007 ± 0.007 4 4	0.048 ± 0.010 132 128
Isomerases					
TPI	0.041 ± 0.021 24 23	—	0.051 ± 0.027 20 20	0.271 ± 0.078 8 7	0.079 ± 0.022 52 50
MPI	0.142 ± 0.026 56 56	0.225 ± 0.025 79 72	0.249 ± 0.050 26 24	0.338 ± 0.104 5 5	0.201 ± 0.017 166 157
GPI	0.132 ± 0.021 86 77	0.204 ± 0.020 119 102	0.229 ± 0.017 144 140	0.244 ± 0.049 14 14	0.196 ± 0.011 363 333
PGM	0.121 ± 0.019 81 65	0.284 ± 0.021 153 105	0.274 ± 0.018 165 141	0.241 ± 0.055 20 14	0.246 ± 0.012 419 325
Nonenzymatic proteins					
PROT[b]	0.004 ± 0.002 186 55	0.054 ± 0.016 80 33	0.045 ± 0.008 202 40	0.026 ± 0.020 8 2	0.029 ± 0.004 476 130

[a] The first number given under the H_E estimate is the number of loci scored, the second is the number of species. Proteins scored from fewer than 25 loci are not included.

[b] Unspecified nonenzymatic proteins.

Considering Vertebrates and Invertebrates Together

The distributions of H_E values for the different vertebrate and invertebrate proteins are summarized in Fig. 4. The vertebrate distribution is clearly unimodal. The invertebrate distribution shows evidence of three peaks, but sample sizes are small and the general shape of the distribution is unimodal.

The 15 most heterozygous vertebrate proteins, of the 42 listed in Table VIII, in rank order, are AMY (most variable), TF, EST, ADA, MPI, GLUDH, NP, PGM, 6PGDH, ALB, ACO, PEP, GPI, ME, and SDH. The 15 most heterozygous invertebrate proteins, of 39 in total, again in rank order, are PGM, MPI, GPI, AO, EST, LAP, PEP, SDH, AMY, GPT, *d*-LDH, ADH, NP, AAT, and G6PDH. These proteins would be the ones to examine in any initial screening to find polymorphic enzymes in a previously uncharacterized species. However, the distributions of H_j values for

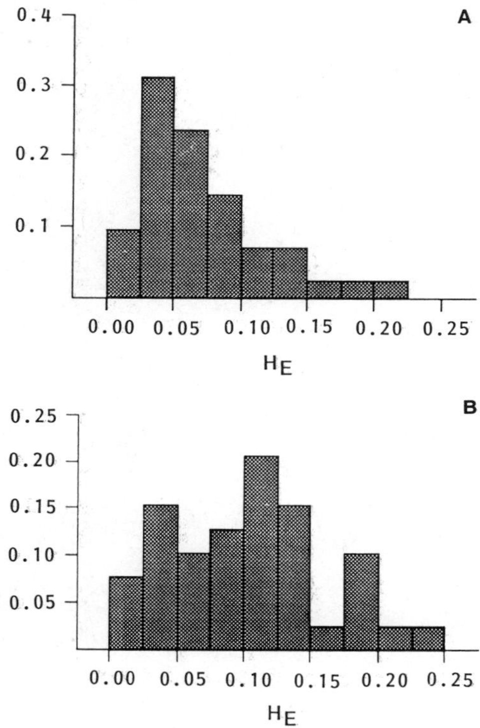

FIG. 4. H_E distributions of vertebrate and invertebrate proteins. (A) Vertebrate proteins (39 enzymes and 3 nonenzymatic proteins). (B) Invertebrate proteins (38 enzymes and 1 nonenzymatic protein).

MPI and EST (Fig. 5), two of the best sampled and most heterozygous enzymes in both vertebrates and invertebrates, indicate that even in these enzymes between 40 and 60% of loci are effectively monomorphic ($H_j < 0.05$). Note that some of the highly heterozygous proteins have only been analyzed in 20 or more species from one or the other of these two groups of animals (AO and dimeric LDH only in invertebrates, TF, ALB, ADA, and ACO only in vertebrates), and thus a direct comparison of the rankings of these two lists is inappropriate.

However, 32 enzymes and one general category of nonenzymatic protein have been screened in more than 20 species of both vertebrates and invertebrates. H_E values of these two taxa for each protein are plotted against one another in Fig. 6, and give a Spearman correlation coefficient of 0.69 ($P < 0.001$) and a Pearson correlation, using arcsine-transformed values, of 0.66 ($P < 0.001$). Twenty proteins have been screened in more than 100 species of vertebrates and more than 100 species of invertebrates. These generate Spearman and Pearson correlation coefficients of 0.66 and 0.71, respectively (the latter using transformed H_E; both correlations give $P < 0.001$). Clearly, this result demonstrates broad agreement between H_E values of invertebrates and vertebrates, and it can be concluded that while values of $\bar{H}$ differ among taxa (Tables III and IV), relative values of H_E are similar.

Standardizing Heterozygosity Estimates in Species Comparisons

One problem in comparing average species heterozygosities H of two species assayed in independent surveys is that the two species will probably have been assayed for different sets of enzymes. It is likely that while some enzymes will be common to both surveys, others will have been screened in only one of the species. This may confound any interpretation of differences in H. For example, if one species has been assayed mainly for enzymes of high average variability (such as EST, PGM, or GPI) while the other species has been assayed mainly for low-variability enzymes (such as GDH, GAPDH, or SOD), then differences in H are to be expected and need not reflect true differences in genomic heterozygosity.

The H_E estimates of Tables VIII and IX allow weighting factors to be calculated which permit the derivation of adjusted or standardized H_j estimates ($H_{j,\text{adj}}$) at each locus. For locus j in species or population x

$$H_{j,\text{adj}} = H_j(\bar{H}_E/H_E) \qquad \text{(method 1)}$$

For example, $\bar{H}_E$ of all 43 vertebrate proteins listed in Table VIII is 0.070 (close to $\bar{H}$ of Table III, 0.071), while H_{Est} is 0.160. Esterase loci are

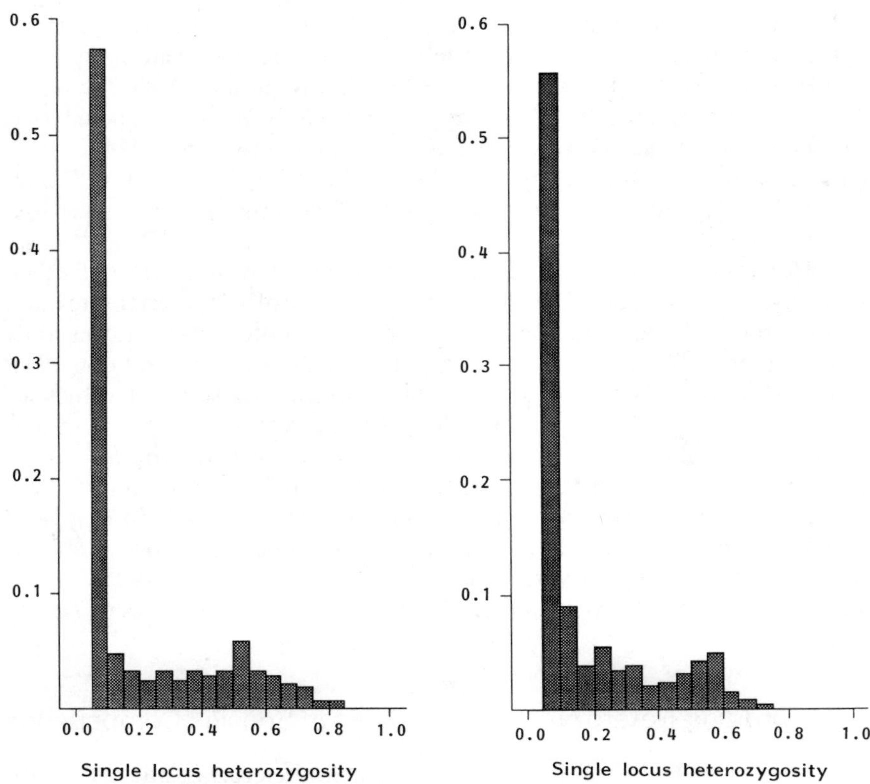

FIG. 5. Distribution of single-locus heterozygosity values of two of the most heterozygous enzymes in vertebrates and invertebrates. Left: EST, vertebrates (1156 loci); right: MPI, vertebrates (264 loci).

therefore expected to show, on average, high levels of variation, and an appropriate weighting factor for vertebrate esterase would be 0.438 (estimated from 0.070/0.160). If the observed (or Hardy–Weinberg expected) heterozygosity at an esterase locus is 0.100 (i.e., $H_j = 0.100$), then $H_{j,\text{adj}} = 0.0438$.

The mean of the $H_{j,\text{adj}}$ values of the different loci screened in a particular species gives H^*, the adjusted or standardized species heterozygosity.

The mean H_E of all 39 invertebrate proteins (Table IV) is 0.103 (a little lower than $\bar{H}$ of Table III, 0.122), and this value is incorporated into invertebrate weighting factors. Most H_E values have been derived from studies of more than 50 species and the resulting weighting factors are likely to be quite accurate. Where the number of species studied is currently low, the H_E val-

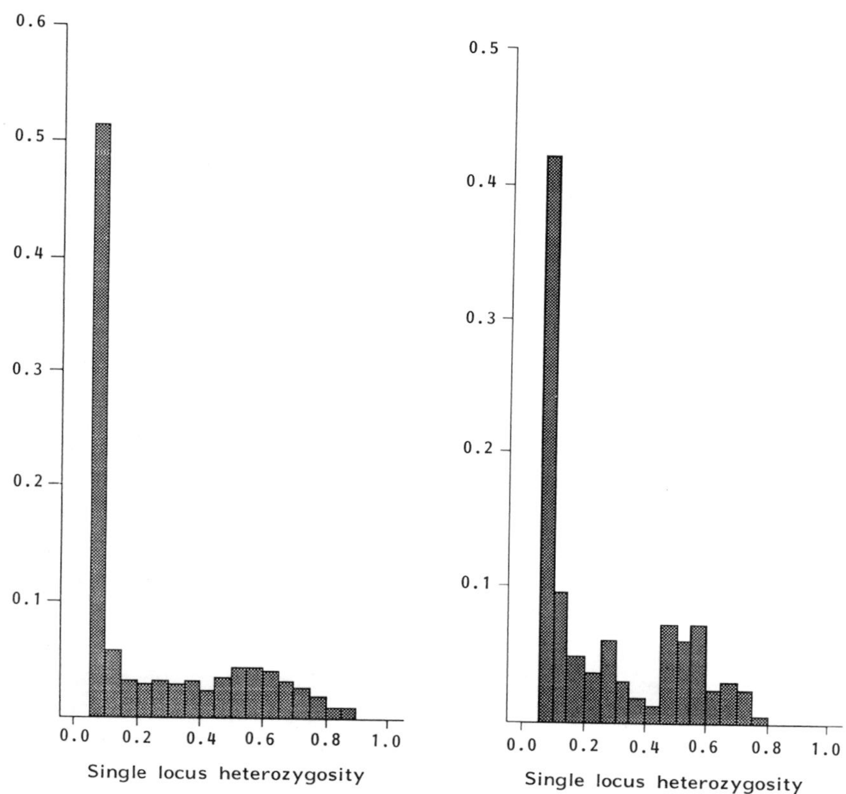

FIG. 5 (*Continued*). Left: EST, vertebrates (773 loci); right: MPI, invertebrates (166 loci).

ues and the weighting factors will be less reliable. In particular, ADA in invertebrates has a very high weighting factor but has only been studied in 18 species.

A simple hypothetical example of these weighting factors in use is given in Table X. Vertebrate species A appears to be substantially less variable than vertebrate species B ($H = 0.029$ and 0.086, respectively), but the suites of enzymes screened in the two species have only one enzyme in common. In fact, species A was screened for enzymes that tend toward low heterozygosity, whereas species B was mostly screened for enzymes that tend toward high heterozygosity. Standardizing for enzymes removes this confounding effect, and shows that now the two species show similar levels of variability ($H^* = 0.045$ and 0.057, respectively).

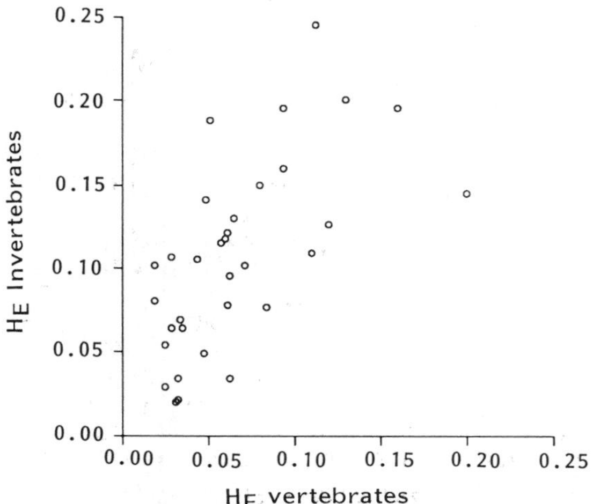

FIG. 6. H_E values of 33 proteins in invertebrates plotted against H_E values of the same enzymes in vertebrates.

One problem with this method of standardizing heterozygosities is that the adjusted values can exceed 1.0. This can occur when a species has been studied primarily for enzymes which normally show low-level variation but which in that species show extensive variation. An example is given as species C in Table X. In practice, such a situation is unlikely to occur, but an improved method of adjusting H_j values is clearly desirable.

Under neutral theory, heterozygosity is related to the product of the effective population size N_e and mutation rate μ. The weighting factor should then take account of the fact that differences in H_E among proteins are due only to differences in μ_E (N_e being similar, on average, for all proteins). Thus, when, for example, μ_{Est} is multiplied by a weighting factor, the result should be $\bar{\mu}$ (the average of μ_E over all proteins). The weight should be $\bar{\mu}/\mu_{\text{Est}}$ or, more generally, $\bar{\mu}/\mu_E$.

The correction equation can be represented as

$$\mu_{j,\text{adj}} = \mu_j(\bar{\mu}/\mu_E) \tag{1}$$

where $\mu_{j,\text{adj}}$ is the adjusted mutation rate of locus j in population x.

By the stepwise model of neutral variation (Ohta and Kimura, 1973),

$$H = 1 - \frac{1}{(8N_e\mu + 1)^{1/2}}$$

TABLE X. Weighting H_j Values to Facilitate Comparisons of Genetic Variability of Three (Hypothetical) Vertebrate Species[a]

Locus	H_E from Table VIII	Species A			Species B			Species C		
		H_j	$H_{j,adj}$ (1)	$H_{j,adj}$ (2)	H_j	$H_{j,adj}$ (1)	$H_{j,adj}$ (2)	H_j	$H_{j,adj}$ (1)	$H_{j,adj}$ (2)
Adh	0.065	0.052	0.056	0.056	—	—	—	0.650	0.700	0.663
l-Ldh	0.032	0.020	0.044	0.045	—	—	—	0.600	1.313	0.725
Mdh	0.029	0.015	0.036	0.037	0.036	0.087	0.086	0.650	1.569	0.773
Est-1	0.160	—	—	—	0.100	0.044	0.041	—	—	—
Est-2	0.160	—	—	—	0.150	0.066	0.065	—	—	—
Pgm	0.112	—	—	—	0.095	0.059	0.059	—	—	—
Mpi	0.130	—	—	—	0.050	0.027	0.025	—	—	—
Mean heterozygosity across loci H		0.029	—	—	0.086	—	—	0.633	—	—
Mean adjusted or standardized heterozygosity across loci H^*		—	0.045	0.046	—	0.057	0.055	—	1.094	0.720

[a] $H_{j,adj}(1)$ and $H_{j,adj}(2)$ refer to the two methods of adjustment detailed in the text. $H_{j,adj}(2)$ is based on the stepwise neutral model.

Rearranging gives

$$\mu = \frac{H(2 - H)}{8N_e(1 - H)^2}$$

Substituting for μ in the correction equation (1) gives, after canceling out N_e terms,

$$\frac{H_{j,\text{adj}}(2 - H_{j,\text{adj}})}{(1 - H_{j,\text{adj}})^2} = \frac{H_j(2 - H_j)}{(1 - H_j)^2} \frac{\bar{H}(2 - \bar{H})}{(1 - \bar{H})^2} \bigg/ \frac{H_E(2 - H_E)}{(1 - H_E)^2}$$

If the term on the right-hand side is set to z, then

$$H_{j,\text{adj}} = 1 - \frac{1}{(z + 1)^{1/2}} \qquad \text{(method 2)}$$

Thus, for example, if the observed heterozygosity H_j of enzyme EST in population or species of a vertebrate is 0.100, $H_{\text{Est}} = 0.160$ (this is H_E above), $\bar{H} = 0.070$, $z = 0.087817$, and $H_{j,\text{adj}} = 0.0412$.

As before, the mean of the $H_{j,\text{adj}}$ values of the different loci gives H^*.

Generally, the two methods of estimating $H_{j,\text{adj}}$ give similar values (see examples of species A and B given in Table X), but the second method is suggested by neutral theory and may be preferred. Furthermore, $H_{j,\text{adj}}$ estimated by the latter method does not exceed 1, even when a high degree of variability is recorded for loci typically showing little variation (see species C in Table X).

If required, $\bar{H}$ and H_E values can be taken from any of the subgroups given in Table VIII or Table IX, although it should be noted that some of the H_E estimates in these individual columns are based on small sample sizes and have high standard errors.

ENZYME STRUCTURE, FUNCTION, AND HETEROZYGOSITY

Multimeric Enzymes Forming Interlocus Hybrid Enzymes

Harris *et al.* (1977) reported that in humans, those multimeric enzymes that formed interlocus hybrid molecules were significantly less variable than those that did not. They emphasized that this result needed confirmation with more extensive data, and while we have not set out to test directly this finding, the data we have collected allow us to look at this subject from a different angle. This discussion will be restricted to vertebrates, because multimeric enzymes forming interlocus hybrid molecules appear to be a rarity in invertebrates.

One enzyme that is particularly instructive for examining this hypothesis is GPI. This dimeric enzyme is encoded by two loci in the vast majority of teleosts, and the products of the two loci hybridize to yield three well-separated bands in an electropherogram of a double homozygote (Avise and Kitto, 1973; Dando, 1980; Fisher *et al.,* 1980; Basaglia, 1989), but in other vertebrates GPI is almost always encoded by a single locus. This general finding receives substantial support from our data base analysis. One hundred and eighty one fish contribute data at 364 loci, an average of 2.01 loci per species (fish of tetraploid origin may show three or more loci), while 349 other species of vertebrate provide information on 377 loci, an average of 1.08 loci per species (mean values of 1.06, 1.10, 1.01, and 1.11 for birds, mammals, reptiles, and amphibia respectively).

If the finding of Harris *et al.* (1977) were to have general validity, GPI in fish should show reduced variation compared with GPI in other vertebrates. Such a result is not observed (Table XI). H_{Gpi} for fish is in fact a little higher (although not significantly so, Kruskal-Wallis test on individual locus values giving $P = 0.190$) than the mean value for all nonfish vertebrates. After correcting for class-related differences in average species heterozygosity across all loci (H), $H_{\mathrm{Gpi,adj}}$ is higher for fish than for any other vertebrate phylum.

This topic can also be addressed with reference to the dimeric enzyme MDH. Most fish have duplicate cytoplasmic loci and a single mitochondrial locus, and the products of the cytoplasmic loci form interlocus hybrids with one another but not with the product of the mitochondrial locus (Fisher *et al.,* 1980; Basaglia, 1989). Most other vertebrates have a single cytoplasmic locus and a single mitochondrial locus, again whose products do not hybridize with one another. This difference in numbers of MDH loci between fish and other vertebrates is borne out by our data base (mean numbers of MDH loci per species are fish, 2.57; birds, 1.83; mammals, 1.84; reptiles, 1.82; amphibia, 1.83). So fish have more MDH loci scored, on average, than other vertebrates, an observation consequent upon their possession of duplicate cytoplasmic MDH loci whose products hybridize. While it would be preferable to separate the cytoplasmic MDH data from mitochondrial MDH data, this cannot be done for most species since the majority of surveys do not give this information. Notwithstanding, it might be expected that H_{Mdh} for fish would be less than for other vertebrates, but in fact H_{Mdh} is equal to, and $H_{\mathrm{Mdh,adj}}$ higher than, values for other vertebrate phyla (Table XI).

Analysis of a subset of data which either give information on the subcel-

TABLE XI. Analysis of GPI and MDH Variation in Vertebrates[a]

Parameter	Fish	Vertebrates (excluding fish)	Birds	Mammals	Reptiles	Amphibia
$\bar{H}$	0.051 ± 0.003	0.081 ± 0.002	0.068 ± 0.005	0.067 ± 0.004	0.078 ± 0.007	0.109 ± 0.006
n_{sp}	195	453	80	172	85	116
GPI						
H_{Gpi}	0.100 ± 0.009	0.086 ± 0.004	0.047 ± 0.009	0.030 ± 0.007	0.080 ± 0.016	0.183 ± 0.022
$H_{Gpi,adj}$	0.139	0.075	0.049	0.032	0.073	0.119
n_{sp}	181	349	66	116	70	97
n_{loci}	364	377	70	128	71	108
MDH						
H_{Mdh}	0.029 ± 0.004	0.029 ± 0.002	0.002 ± 0.001	0.014 ± 0.003	0.045 ± 0.011	0.061 ± 0.009
$H_{Mdh,adj}$	0.040	0.025	0.002	0.015	0.041	0.040
n_{sp}	186	447	78	171	83	115
n_{loci}	478	820	143	315	151	211
Cytoplasmic MDH (sMDH)—subset of data						
H_{sMdh}	0.053 ± 0.015	0.007 ± 0.005	0.000 ± 0.000	0.002 ± 0.001	0.024 ± 0.024	0.020 ± 0.020
$H_{sMdh,adj}$	0.074	0.006	0.000	0.002	0.022	0.013
n_{sp}	35	45	9	24	9	3
n_{loci}	78	45	9	24	9	3

[a] H_{adj} takes into account variation in $\bar{H}$ among phyla. For example, if a phylum has a low $\bar{H}$ it is expected to have a reduced H_{Gpi}. In order to make values of H_{Gpi} more comparable among phyla, we take $H_{Gpi,adj} = (\bar{H}_{Total}/\bar{H}_{phylum})H_{Gpi}$. Thus, for fish, $H_{Gpi,adj} = (0.071/0.051)\ 0.100 = 0.139$, where the values of $\bar{H}_{Total}$ and $\bar{H}_{phylum}$ come from Table III.

lular location of the MDH isozymes or enable this to be determined allows a second analysis, specifically of the cytoplasmic or supernatant MDH (sMDH) isozymes, to be achieved. The nonfish vertebrate species in this dataset each show a single sMDH locus. Of the 35 fish species, all teleosts, 28 had two sMDH loci, 6 had three sMDH loci (one of the duplicated loci further duplicated), and 1 had four sMDH loci (both duplicate loci further duplicated). Both H_{sMdh} and $H_{\text{sMdh,adj}}$ are higher in fish than other vertebrates (Table XI), although the nonparametric Kruskal-Wallis test applied to individual locus values of H_{sMdh} indicates that the null hypothesis that the distributions of these values is the same among phyla cannot be rejected ($\chi^2 = 3.66$, df = 4, $P = 0.45$).

It is clear that the presence of interlocus hybrid molecules for both GPI and sMDH in fish and their absence from other vertebrates has not reduced either GPI or sMDH variability in fish.

Quaternary Structure and Heterozygosity

Figure 7 shows the relationship between H_E (from the Total columns of Tables VIII and IX) and quaternary structure for all proteins having a defined and fixed structure. Thus, the vertebrate figure, Fig. 7A, excludes ACP, PEP, and PROT (but includes dimeric and tetrameric LDH, transferrin, albumin, and hemoglobin) and the invertebrate figure, Fig. 7B, excludes G6PDH, EST, ALP, PEP, and PROT (but includes dimeric and tetrameric LDH). Both regressions are negative and significant, and in both instances

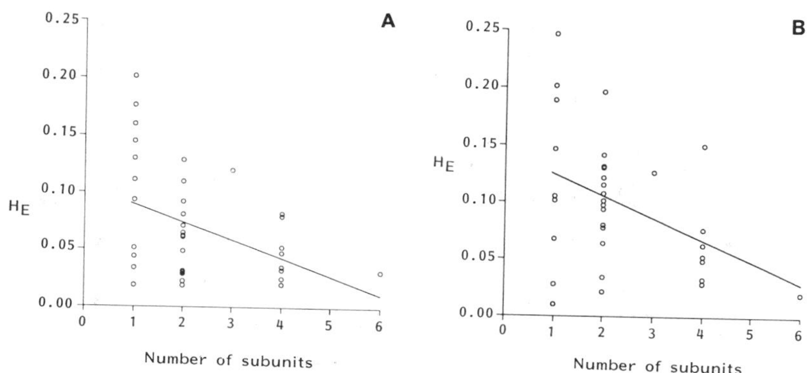

FIG. 7. Mean heterozygosity H_E per protein plotted against number of subunits. (A) Vertebrates: $H_E = 0.1063 - 0.0159 \times$ subunit number ($n = 40$, $P = 0.006$). (B) Invertebrates: $H_E = 0.1439 - 0.0192 \times$ subunit number ($n = 34$, $P = 0.019$).

TABLE XII. Mean H_E Values and ANOVA for Protein Quaternary Structures

Mean heterozygosities of different quatenary structures of proteins (±SE)[a]

	$\bar{H}_E$				
	Monomers	Dimers	Trimers	Tetramers	Hexamers
Birds	0.096 ± 0.025 [7] (0.089 ± 0.023 [8])	0.032 ± 0.012 [10]	0.176 [1]	0.051 ± 0.022 [3] (0.038 ± 0.020 [4])	0.000 [1]
Mammals	0.101 ± 0.027 [7] (0.106 ± 0.021 [9])	0.050 ± 0.009 [11]	0.069 [1]	0.040 ± 0.013 [5] (0.046 ± 0.012 [6])	0.074 [1]
Reptiles	0.114 ± 0.041 [3]	0.075 ± 0.015 [9]	—	0.056 ± 0.013 [3]	—
Amphibians	0.141 ± 0.027 [5] (0.155 ± 0.023 [7])	0.103 ± 0.018 [8]	—	0.093 ± 0.024 [6]	0.025 [1]
Fish	0.096 ± 0.029 [6]	0.048 ± 0.007 [10]	—	0.034 ± 0.010 [7]	0.033 [1]
All vertebrates	0.099 ± 0.019 [10] (0.105 ± 0.017 [12])	0.059 ± 0.008 [17]	0.120 [1]	0.043 ± 0.009 [8] (0.044 ± 0.008 [9])	0.031 [1]

Crustacea	0.085 ± 0.020 [5]	0.045 ± 0.010 [12]	—	0.020 ± 0.007 [5]	0.000 [1]
Mollusks	0.198 ± 0.029 [5]	0.094 ± 0.021 [9]	0.131 [1]	0.093 ± 0.024 [4]	—
Insects	0.161 ± 0.038 [7]	0.128 ± 0.020 [15]	—	0.060 ± 0.011 [6]	—
All invertebrates	0.121 ± 0.027 [9]	0.107 ± 0.012 [16]	0.126 [1]	0.065 ± 0.015 [7]	0.020 [1]

Analysis of variance in (arcsine-transformed) mean heterozygosity among quaternary structures[b]

Source of variation	Enzymatic proteins only			All proteins		
	df	F	P	df	F	P
Among groups (vertebrates and invertebrates)	1	3.63	0.059	1	4.03	0.047
Among subgroups (e.g., phyla) within groups	6	5.99	<0.001	6	7.19	<0.001
Among subsubgroups (structures) within subgroups	16	2.58	0.002	16	2.81	0.001
Within subsubgroups (error term)	144			151		

[a] Numerals in brackets are numbers of proteins contributing data. Minimum sample size: 20 species per protein per taxon. Figures in parentheses include the nonenzymatic proteins transferrin, hemoglobin, and albumin.

[b] The single trimeric and hexameric proteins are excluded from the analysis.

TABLE XIII. Analysis of Variation in (Arcsine-Transformed) Mean Protein Heterozygosity H_E in Vertebrates and Invertebrates Related to Quaternary Structure (Monomers, Dimers, and Tetramers), Using Data from the Totals Columns of Tables VIII and IX and Including Nonenzymatic Proteins in the Vertebrate Analyses

One-way ANOVAs

	All data			Minimum sample size 50 loci/protein		
Source of variation	df	*F*	*P*	df	*F*	*P*
Vertebrates						
Between structural classes	2	5.70	0.007	2	4.50	0.021
Within structural classes	35			27		
Invertebrates						
Between structural classes	2	1.73	NS	2	5.04	0.016
Within structural classes	29			22		

Tukey HSD test[a]

	Vertebrates			Invertebrates		
	Monomers	Dimers	Tetramers	Monomers	Dimers	Tetramers
Monomers	—	0.032	0.009	—	0.969	0.226
Dimers	0.074	—	0.618	0.075	—	0.237
Tetramers	0.024	0.723	—	0.014	0.396	—

[a] *P* values above the diagonal are using all data, below the diagonal using a minimum sample size of 50 loci per protein.

around 16% of the variation in H_E can be explained by subunit number (vertebrates, $r^2 = 0.183$; invertebrates, $r^2 = 0.161$). If attention is restricted to those enzymes screened for 50 or more loci in either vertebrates or invertebrates, then r^2 remains more or less unchanged for vertebrates ($r^2 = 0.196$, $n = 32$), but increases for invertebrates ($r^2 = 0.261$, $n = 25$). It seems that around 20% of variation in H_E can be attributed to variation in subunit number in both groups of animals.

Only two enzymes considered here of fixed quaternary structure are not monomers, dimers, or tetramers. NP is a trimer and GDH is a hexamer. Thus, GDH has the most complex quaternary structure of any enzyme routinely screened by electrophoresis,* and in terms of H_E ranks 31st of 37

* Note that peptidases have a very variable quaternary structure, and at least one hexameric peptidase has been recorded (Lavery and Shaklee, 1989).

enzymes in vertebrates (or 32nd of 38 if dimeric and tetrameric G6PDH are considered separately), and 37th of 38 enzymes in invertebrates.

Mean H_E values of the different structural groups of enzymes are given for the various taxa in Table XII. Several ANOVAs were carried out on these data. A three-level nested ANOVA (Table XII) showed significant variation in arcsine-transformed H_E among the three structural classes of monomers, dimers, and tetramers within subgroups (five vertebrate and three invertebrate subgroups) for enzymes alone ($P = 0.003$) and for enzymes and nonenzymatic proteins ($P = 0.001$). This analysis also revealed the expected variation in H_E among groups and among subgroups.

One-way ANOVAs on the total vertebrate and total invertebrate data were also carried out (Table XIII). Significant variation related to quaternary structure was shown for vertebrates, but for the invertebrates, using all the H_E values listed in the Total column of Table IX, this variation was not significant. However, if analysis is restricted to those proteins with sample sizes of 50 or more loci, then the analysis becomes significant. Two monomeric enzymes show atypically low levels of H_E in invertebrates, APK ($H_{\mathrm{Apk}} = 0.028$) and ADA ($H_{\mathrm{Ada}} = 0.010$)), and both these have relatively low sample sizes (numbers of loci scored being 33 and 30, respectively). It should also be pointed out that ADA has only been screened (in contributing datasets) from crustacean species of lower than average crustacean heterozygosity. Adjustment for this low average heterozygosity raises H_{Ada} to 0.030, and further adjustment to compensate for the fact that crustaceans as a group are less variable than other invertebrates raises H_{Ada} to 0.046. Note that while APK activity is restricted to invertebrate muscle, ADA has been screened in large numbers of vertebrates, and is here one of the most polymorphic enzymes (third only to esterase and amylase). It might be anticipated that as more data are collected for ADA and APK in invertebrates, so H_E for these enzymes will rise. Omitting these two proteins from the invertebrate ANOVA (all data) renders the analysis significant ($P = 0.013$).

Tukey's HSD test was performed on the data to see where the difference in arcsine-transformed $\bar{H}_E$ lay (Table XIII). In general terms, monomers were found to be more variable than dimers and tetramers, but no significant differences were observed between dimers and tetramers. Thus, the conclusion from this analysis appears to be that monomers are more variable than multimers, but that there are no differences among the multimeric classes. However, this conclusion may be premature. There are eight subgroups listed in Table XII, and for seven of these, $\bar{H}_E$ for monomers was greater than $\bar{H}_E$ for dimers, which was in turn greater than $\bar{H}_E$ for tetramers. The sole exception was in the subgroup birds, where tetramers had a higher $\bar{H}_E$ than dimers. Analysis of this ranked data by Friedman's nonparametric two-way ANOVA shows a very high concordance [Kendall coefficient of 0.89 (P

TABLE XIV. Correlation Coefficients between (Arcsine-Transformed) Mean Protein Heterozygosity H_E and Subunit Molecular Weights[a]

	All proteins		Monomers		Dimers		Tetramers	
	r	r^2	r	r^2	r	r^2	r	r^2
Birds	0.166 [20]	0.028	0.191 [5]	0.036	0.431 [10]	0.186	0.820 [3]	0.672
	(0.262 [22]	0.069)	(0.118 [6]	0.014)			(0.925 [4]	0.855)
Mammals	0.198 [23]	0.039	0.646 [5]	0.417	0.391 [11]	0.153	0.351 [5]	0.124
	(0.255 [26]	0.065)	(0.526 [7]	0.277)			(−0.131 [6]	0.017)
Reptiles	0.065 [13]	0.004	— [1]		0.005 [9]	0.000	0.556 [3]	0.309
Amphibia	0.450 [18]	0.202	0.197 [3]	0.039	0.729 [8]	0.532*	0.479 [6]	0.230
	(0.575 [20]	0.330**)	(0.357 [5]	0.127)				
Fish	−0.035 [22]	0.001	0.507 [4]	0.257	0.088 [10]	0.008	−0.418 [7]	0.174
All vertebrates	0.242 [34]	0.059	0.095 [8]	0.009	0.499 [16]	0.239*	0.081 [8]	0.006
	(0.292 [37]	0.085)	(0.198 [10]	0.039)			(0.062 [9]	0.004)
Crustacea	0.023 [21]	0.001	−0.419 [3]	0.175	0.040 [12]	0.002	0.255 [5]	0.065
Mollusks	−0.195 [17]	0.038	1.000 [3]	0.999*	0.223 [9]	0.050	0.034 [4]	0.001
Insects	0.510 [26]	0.260**	0.480 [5]	0.231	0.703 [15]	0.494**	0.232 [6]	0.054
All invertebrates	0.251 [32]	0.063	0.605 [7]	0.365	0.382 [16]	0.146	0.122 [7]	0.015

[a] Numerals in brackets are numbers of proteins contributing data. Minimum sample size: 20 species per taxon per protein. Figures in parentheses include, where sample sizes are sufficient, transferrin, hemoglobin, and albumin. *: $0.05 > P > 0.01$; **: $0.01 > P > 0.001$.

= 0.0008), appreciably higher than the value of 0.75 (P = 0.0025) expected for this data set if monomers were always more variable than multimers but dimers and tetramers did not differ systematically]. Comparing $\bar{H}_E$ of dimers and tetramers alone for the eight subgroups using the Wilcoxon signed ranks test gave a probability of no significant difference of 0.069 when considering enzymes alone, and 0.036 when hemoglobin is included in the tetramer estimate of $\bar{H}_E$ for birds and mammals (see Table XII). Thus, there is in fact evidence from this analysis and from the regression analysis (Fig. 7) that dimers are, on average, more variable than tetramers.

Subunit Molecular Weight and Heterozygosity

Correlation coefficients between subunit molecular weight and H_E, and coefficients of determination, were estimated for the various taxa for all proteins and, because of the known effects of subunit number, separately for monomers, dimers, and tetramers (Table XIV). A total of 46 correlation coefficients were estimated, and (although many of these are nonindependent) only 5 are negative. Six correlations were statistically significant at the 5% level, and all these were positive. Sample sizes are small after dividing the proteins into the different subunit categories, and few degrees of freedom are available for testing the significances of the r values. The dimeric category contains more enzymes than either of the monomeric or tetrameric classes, and two of its eight subgroups (amphibia and insects, Fig. 8) had significant r values. All values of r for dimers are positive. For the amphibia and insects, about 50% of variation in H_E for dimeric enzymes can be attributable to

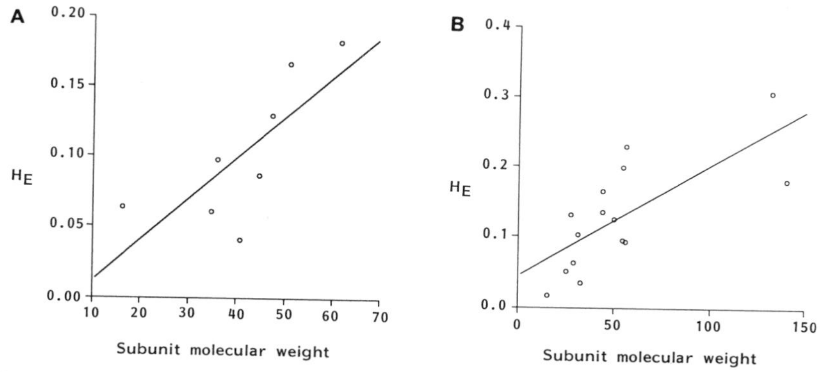

FIG. 8. Mean heterozygosity H_E per dimeric enzyme plotted against subunit molecular weight (SMW) $\times$ 10^{-3}. (A) Amphibia: $H_E = -0.0179 + 0.0000029\text{SMW}$ (n = 8, P = 0.030). (B) Insects: $H_E = 0.0449 + 0.0000016\text{SMW}$ (n = 15, P = 0.002).

subunit molecular weight. For pooled vertebrates and pooled invertebrates, r^2 for dimers falls to about 25% and 15%, respectively.

The evidence for such a relationship for monomers and tetramers is less strong. For these enzymes in vertebrates, less than 5% of the variation in H_E can be attributed to subunit molecular weight, and this also appears to be true of the invertebrate tetrameric class. Analysis of the invertebrate monomeric class is compromised by the very small sample sizes available, although two of the three subgroups show evidence of a positive relationship.

There are no significant differences in average subunit molecular weight among the three major quaternary structure classes (Table XV), and thus differences in heterozygosity among quaternary structure classes will not be significantly confounded by possible differences in subunit size.

Joint Consideration of Subunit Number and Subunit Size

Protein heterozygosity is dependent on at least two aspects of structure, subunit number and subunit size, and it is interesting to consider the relative influences of these two factors. This was achieved through a multiple regression analysis (Table XVI). Two groups were analyzed, insects, and all vertebrates pooled, and, as before, all proteins contributing data had to be scored from 20 or more species. The vertebrate analysis used 10 monomeric proteins (PGM, AK, ADA, MPI, DIA, ACO, AMY, HK, TF, ALB), 16 dimeric enzymes (ADH, MDH, G3PDH, IDH, 6PGDH, AAT, GPI, XDH, CK, TPI, ALP, GLUDH, ALAT, EST-D, SOD, dimeric G6PDH), 1 trimer (NP), 9 tetramers (SDH, LDH, ME, FUM, ALD, GAPDH, CAT, tetrameric G6PDH, HB), and 1 hexamer (GDH). The insect analysis used data from 5 monomeric enzymes (AK, AMY, APK, PGM, MPI), 15 dimers (ADH, MDH, G3PDH, IDH, 6PGDH, SOD, AAT, XDH, ACP, TPI, ODH, HBDH, AO, GPI, ALAT) and 6 tetramers (LDH, ME, FUM, ALD, GAPDH, PK).

No significant association between subunit number and size in either vertebrates or insects was observed, but both subunit number and subunit size are correlated with average heterozygosity in both groups (P values ranging between 0.088 and 0.008). About 22% (vertebrates) and 37% (insects) of the variation in H_E can be explained by the combined effects of subunit number and size. The conventional partial regression coefficient associated with subunit number in the vertebrate analysis was significant ($P = 0.023$), but that associated with size was not ($P = 0.148$). Both coefficients in the insect analysis were associated with low probabilities ($P = 0.054$ and $P = 0.006$, respectively). Since these two variables are measured in different

TABLE XV. Mean Subunit Molecular Weights of Vertebrate and Invertebrate Proteins of Differing Quaternary Structure[a]

All proteins	Monomers	Dimers	Tetramers
Vertebrates			
51,392 ± 4,289 [37]	56,940 ± 7,882 [10]	53,594 ± 8,071 [16] $F = 0.637$, df = 2, 32, $P = 0.536$	43,622 ± 4,991 [9]
Invertebrates			
47,963 ± 4,572 [32]	42,083 ± 3,809 [6]	51,863 ± 8,784 [16] $F = 0.346$, df = 2, 26, $P = 0.711$	44,729 ± 4,137 [7]

[a] These are the same proteins (including nonenzymatic proteins) as those considered in the analysis of Table XIV. Numerals in brackets are numbers of proteins, and means and standard errors are given.

TABLE XVI. Multiple Regression Analysis of Arcsine-Transformed H_E (H_{arc}) and Subunit Number n and Subunit Size s in Vertebrates and Insects

	Vertebrates	Insects
Number of proteins	37	26
Correlations of the three variables		
r_{ns}	−0.174, $P = 0.303$	−0.012, $P = 0.953$
r_{nH_E}	−0.407, $P = 0.013$	−0.342, $P = 0.088$
r_{sH_E}	0.292, $P = 0.080$	0.510, $P = 0.008$
Multiple correlation coefficients		
$R_{H_E,n,s}$	0.464, $P = 0.016$	0.611, $P = 0.005$
$R^2_{H_E,n,s}$	0.216	0.373
R^2 change[a]: s	0.050, $P = 0.148$	0.256, $P = 0.006$
n	0.131, $P = 0.023$	0.113, $P = 0.054$

Prediction equations

Vertebrates: $H_{arc} = 15.5425 - 1.4512n + 0.0000441s$

Insects: $H_{arc} = 17.6045 - 2.5658n + 0.0001432s$

[a] R^2 change is the increment in determination due to the addition of the stated variable over determination by the other variable alone.

units, standardization is necessary in order to compare them and determine their relative influence on H_E.

The standardized partial regression coefficients (Sokal and Rohlf, 1981) for the vertebrate data are, for subunit number (keeping size constant), −0.3667, and for subunit size (keeping number constant), 0.2281. Thus, subunit number decreases H_E relatively more than subunit size increases it. For the insect analysis, these two coefficients are −0.3355 and 0.5060, respectively, meaning that in this taxon, subunit size has more influence than subunit number.

In this analysis, subunit size accounts for about 5% of the variation in H_E in vertebrates, and subunit number about 13%. For the insects, these values are about 26% and 11%, respectively. Perhaps a reasonable conclusion from this analysis is that both structural factors are important in influencing H_E, and perhaps to rather similar extents, but further data are required before any better-defined inferences can be drawn.

Enzyme Function and Heterozygosity

Associations have been reported between heterozygosity and such aspects of an enzyme's function as whether or not it is a "glucose-metabolizing" enzyme (Gillespie and Kojima, 1968) or a "regulatory" enzyme (Johnson, 1974). One practical problem with such hypotheses is that it is frequently difficult to decide objectively into which category a particular enzyme belongs. Here, we have classified enzymes according to IUBNC (1984) into one of five categories (oxidoreductases, transferases, hydrolases, lyases, and isomerases—see Table I). One-way ANOVAs of vertebrate and, separately, invertebrate arcsine-transformed H_E data (using the values in the Totals columns of Tables VIII and IX) showed nonsignificant amounts of variance associated with function ($n = 39$, $F = 1.27$, df $= 4, 34$, $P = 0.30$ and $n = 38$, $F = 2.51$, df $= 4, 33$, $P = 0.06$, respectively). The invertebrate analysis borders on significance, due primarily to the high average heterozygosity of the four enzymes classified as isomerases (PGI, MPI, TPI, PGM). Prior to 1983, phosphoglucomutase was classified as a kinase (EC 2.7.5.1) rather than an isomerase (EC 5.4.2.2); the invertebrate analysis using such a classification no longer borders on significance ($n = 39$, $F = 0.92$, df $= 4, 34$, $P = 0.47$ and $n = 38$, $F = 1.48$, df $= 4, 33$, $P = 0.23$, respectively, for vertebrates and invertebrates).

Joint Consideration of Taxon, Enzyme Function, Enzyme Quaternary Structure, and Enzyme Subunit Size

In order to estimate the relative contributions of these factors, when considered together, to variation in enzyme heterozygosity H_E, multiple regression analyses (Table XVII) were carried out on arcsine transformations of the mean values given in Tables VIII and IX. Vertebrates were analyzed separately from invertebrates. Five taxonomic categories (T) were used in the vertebrate analysis (mammals, birds, amphibia, reptiles, and fish) and three in the invertebrate analysis (crustaceans, mollusks, and insects). A mean heterozygosity value for a particular enzyme (E) was included in the analyses if that enzyme was scored for at least 10 species in that taxon. Enzyme function (F) comprised five categories and quaternary structure (Q) four categories; these, along with subunit sizes, are given in Table I. Thirty five enzymes were used for the vertebrate analysis and 24 enzymes for the invertebrate analysis. For both vertebrates and invertebrates, taxon and enzyme effects together account for over 70% of variation in H_E (72.5% for vertebrates, 75.9% for invertebrates).

The first analysis of vertebrate data showed that 56.9% of the variation in H_E is explained by enzyme-specific effects and 11.3% by taxon effects. These values are similar to those derived from the smaller but orthogonal data set described earlier. The second analysis considers the joint effects of function, quaternary structure, subunit size, and taxon. These account for, respectively, 2.9%, 12.3%, 7.7%, and 13.8% of the variation in H_E, and all are

TABLE XVII. Multiple Regression Analysis of Variation in Arcsine-Transformed H_E in Relation to Taxon, Enzyme Function, Quaternary Structure, and Subunit Size

Variable	R^2 change[a]	P
Vertebrates		
Enzyme	0.569	<0.001
Taxon	0.114	<0.001
Overall $R^2 = 0.725$		
Taxon	0.138	<0.001
Enzyme function	0.029	0.017
Enzyme quaternary structure	0.123	<0.001
Enzyme subunit size	0.077	0.013
Overall $R^2 = 0.423$		
Invertebrates		
Enzyme	0.542	0.001
Taxon	0.178	<0.001
Overall $R^2 = 0.759$		
Taxon	0.218	<0.001
Enzyme function	0.068	0.205
Enzyme quaternary structure	0.046	0.257
Enzyme subunit size	0.071	0.014
Overall $R^2 = 0.483$		

[a] R^2 change is the increment in the coefficient of determination obtained after all other variables are put in the equation.

statistically significant. These values relate to the change in R^2 obtained after all other variables are put into the equation. Thus, for example, for enzymes of given function, subunit size, and taxon, quaternary structure explains an additional 12.3% of the residual variation. The reduction in R^2 from 0.725 in the first to 0.423 in the second analysis implies that unidentified enzymic factors also contribute significantly to variation in H_E.

The invertebrate analyses showed that 54.2% of the variation in H_E is explained by enzyme-specific effects and 17.8% by taxon effects. Again these values are quite similar to those described earlier. When the joint effects of function, quaternary structure, subunit size, and taxon are considered, only the effects of taxon and subunit size are significant, accounting for 21.8% and 7.1%, respectively, of the variation in H_E.

DISCUSSION

The average heterozygosity of vertebrates is less than that of invertebrates, a finding already extensively reported in the electrophoresis literature (e.g., Selander and Kaufman, 1973; Ward, 1977; Nevo *et al.*, 1984). Within the vertebrates, amphibia have the highest heterozygosity, and fish the lowest. Other analyses have described a positive relationship between heterozygosity and population size N (Soulé, 1976; Nevo *et al.*, 1984) and between heterozygosity and the product of N and generation time g (Nei and Graur, 1984). Thus, both N and Ng are likely to be higher in amphibia than in other vertebrates. The high heterozygosity of amphibians results partly from extensive genetic population structuring, but even local populations of amphibia have heterozygosity values $\bar{H}_S$ which significantly exceed corresponding values from all other vertebrates bar reptiles. The extensive population structuring of amphibia gives rise to the highest $\bar{G}_{ST}$ and $\bar{D}_m$ values of any vertebrate or invertebrate taxon considered here. This result is not unexpected; gene flow among amphibian populations is expected to be low given their relatively low mobility and the fact that most are restricted to the vicinity of patchily distributed ponds and lakes. The magnitude of G_{ST} varies greatly among amphibians, but on average about 30% of total genetic variation in this class results from population differentiation, compared with less than 10% for the most mobile vertebrate phylum, the birds. Thus, the data are consistent with theories that relate variation in G_{ST} to variation in rates of gene flow or, more exactly, to variation in Nm, where m is the migration rate per generation. We have not attempted here further to partition variation in G_{ST}. An example of such partitioning is given by the analysis of Gyllensten (1985), who showed that freshwater fishes had higher G_{ST} values than marine fishes, and related this to the lack of geographic barriers in the latter group.

Among the invertebrates, insects and mollusks show similar levels of variation, both higher than the crustaceans, with population structuring (higher G_{ST} values) more evident in mollusks than insects. Again, this result is not unexpected given that most insects are highly mobile whereas most mollusks are relatively sedentary, although some species of mollusks, sedentary when juvenile and adult, enjoy high levels of gene flow through planktonic dispersal of eggs and larvae. Indeed, the distribution of G_{ST} among mollusks indicates that an appreciable proportion show little population differentiation (about one third have $G_{ST} < 0.1$, compared with about one sixth of the amphibians, which have a similar mean value of G_{ST}), while some show very extensive differentiation.

A consideration of the various analyses of variance and multiple regression analyses shows that for both invertebrates and vertebrates, taxon effects and enzyme effects together account for between 70 and 75% of variation in mean enzyme heterozygosity H_E, leaving rather little variation to be explained by other factors. Subunit number and subunit size each explains about 10–20% of the variance, although these figures vary somewhat depending on the taxa and enzyme category. For example, about 50% of the variation in H_E in dimeric enzymes of amphibia and insects can be ascribed to variation in subunit molecular weight, although this figure appears to be substantially less for monomeric and tetrameric enzymes in these taxa. It should be stressed here that it is variation in mean heterozygosity among enzymes within particular taxa that is being assessed, residual variation (between species, within enzyme within taxa) not being accounted for.

Thus, variation in average heterozygosity among enzymes is related, at least partially, to variation in quaternary structure. Monomers are on average more variable than multimers, and there is some evidence that dimers are more variable than tetramers. There are several possible causes of this relationship. As heterozygosity is known to be related to subunit size, it may be that multimeric enzymes have smaller subunits than monomers. However, there are no significant or consistent differences in subunit size among the different quaternary classes considered here and so this potentially confounding factor does not appear important.

What possible factors concerning quaternary structure *per se* could constrain the variability of multimeric enzymes? An obvious factor is the requirement for multimers to form intersubunit contacts, contacts which are evolutionarily conserved over long periods of time. Homologous multimeric enzymes from very different species commonly form stable hybrid molecules under experimental conditions, implying that the amino acid residues involved in these contacts change little over long periods of time. Such results have been reported for aldolase [chicken, lobster, wheat (Swain and Lebherz, 1986)], glyceraldehyde 3-phosphate dehydrogenase [yeast and rabbit (Spotorno and Hollaway, 1970; Osborne and Hollaway, 1974); *Ascaris* and rabbit (Kochman *et al.*, 1974)], superoxide dismutase [eel and snail, and *Neurospora* and salmon, among others (Tegelstrom, 1975)], and (for *Drosophila* species) acid phosphatase (MacIntyre and Dean, 1978). Subunit contact sites of vertebrate hemoglobins are highly conserved, although invertebrates have subunit contact sites which differ not only from those of vertebrates, but also from one another (Perutz, 1990).

The number or proportion of residues concerned with making intersubunit contacts requires a knowledge of the amino acid sequence and the 3D structure of the enzyme molecule, information still lacking for most types of enzymes. With respect to vertebrate hemoglobins, about 30 of the approximately 140 amino acid residues in each of the α and β chains are involved in packing ($\alpha_1\beta_1$ and $\alpha_2\beta_2$) and sliding ($\alpha_1\beta_2$ and $\alpha_2\beta_1$) contacts. These residues tend to be strongly conserved (Dickerson and Geis, 1983). As many as 70 of the 331 amino acids per subunit of porcine cytoplasmic MDH may be involved in subunit–subunit interactions, with 13 of these making intersubunit hydrogen bonds (Birktoft *et al.*, 1989). Stabilization of glyceraldehyde-3-phosphate dehydrogenase is primarily through hydrophobic interactions of the S-loop amino acids, which make up 24 of the approximately 310 residues per polypeptide chain. The S-loop sequence is highly conserved, with 19 of the 24 residues being identical in humans and other vertebrates, lobster, *Drosophila melanogaster,* and yeast (Banas *et al.,* 1987).

Clearly, the number and proportion of amino acids involved in making intersubunit contacts can be considerable, but of course they are not the only conserved residues. Zuckerkandl (1976) classified contact amino acids into two groups: (1) those with interactions with other polypeptide chains or different parts of the same chain and (2) those with interactions with invariant molecules such as prosthetic groups, substrates, and cofactors. Active-site residues are known to be conserved. For example, the sequence of 17 amino acids forming the active thiol group of glyceraldehyde-3-phosphate dehydrogenase is identical in 10 different species, including mammals, birds, fish, insects, and yeast (Dixon and Webb, 1979). The two amino acid groups of Zuckerkandl (1976) are not necessarily mutually exclusive. An example is Arg-161 in porcine sMDH, found in the subunit interface but also implicated in substrate binding (Birktoft and Banaszak, 1983). In a minority of enzymes, such as aspartate aminotransferase, glutathione reductase, aspartate transcarbamylase, catalase, and glutamine synthetase (Ford *et al.,* 1980; Thieme *et al.,* 1981; Krause *et al.,* 1985; Fita and Rossmann, 1985; Almassy *et al.,* 1986), the active site is actually formed from the intimate association of two subunits.

In general, the need to maintain a multimeric quaternary structure will add to the constraints already imposed by the need to maintain the integrity of active sites and secondary and tertiary structures, leading to the expectation that multimers should show less variability than monomers. Indeed, as Koehn and Eanes (1979) point out, the simple aggregation of monomers into multimers is in itself expected to reduce variability, as the numbers of exposed amino acids will be lessened, and it is primarily substitutions in these regions that are detected by electrophoresis. It has been estimated that about 14% of the surface area per subunit is lost in the formation of dimers (Teller, 1976; Chothia and Janin, 1976), and exterior residues are on average more variable than interior residues (Zuckerkandl and Pauling, 1965; Go and Miyazawa, 1980).

The expectation that multimers should be, on average, less variable than monomers is fulfilled, but the corresponding relationship between dimers and tetramers is uncertain. It is not clear whether tetramers have more residues involved in intersubunit contacts than do dimers, although the former presumably do suffer a proportionately greater loss in surface area per subunit than dimers, and this is expected to be reflected in decreased variability. There is some evidence that tetramers are indeed less variable on average than dimers.

Subunit size is also expected to influence protein variability. *A priori,* one would expect, other factors (such as numbers of residues in active sites and subunit contact sites) being constant, that proteins of high subunit molecular weight should be more variable than proteins of small subunit molecular weight, as they will have a larger surface area on average and more amino acid sites free to vary. Koehn and Eanes (1979), in an earlier survey, estimated that size variation explained between 60 and 80% of the total variance in mean enzyme heterozygosity. This seems an exaggerrated estimate compared with our figures of between 0% and 50% for the better-sampled categories, and their conclusion that little unexplained variance remained to be attributed to variation in quaternary structure is not supported by our analyses.

An interesting pair of enzymes to consider when discussing the joint effects of subunit number and subunit size is MDH and ME. While these two enzymes might be expected to have similar active-site constraints on variability (both utilizing malate as substrate and both having similar cofactors, NAD and NADP, respectively), MDH is a dimer with a subunit weight of around 30 daltons, whereas ME is a tetramer with a subunit size of around 60 daltons. In most taxa, ME is appreciably more variable than MDH, and this remains the case when cytoplasmic and mitochondrial isozymes of both enzymes are considered separately (Ward and Skibinski, 1988). Thus, for these two enzymes, subunit size seems to have more influence than subunit number in predicting resultant levels of variability. However, the multiple regression analyses of various datasets presented earlier show that when all enzymes are considered jointly, both these structural factors do influence heterozygosity, and to broadly similar extents.

But other factors besides size and subunit number are important in determining average levels of protein variability. For example, for reasons that are not clear, the amino acid sequences of monomeric myoglobins from different species are *more* conserved than those of the closely related but tetrameric hemoglobins (Dickerson and Geis, 1983), which have similar subunit sizes to myoglobin. One additional factor influencing variability is subcellular location, with mitochondrial isozymes being less variable than cytoplasmic isozymes (Ward and Skibinski, 1988), while the present analyses indicate that enzyme function *per se* and whether or not the subunits of multimeric enzymes hybridize with the products of other homologous loci have little effect. There are certainly taxon-specific effects, in so far as some taxa show higher heterozygosity than other taxa, although these effects are usually enzyme independent. An exception to this generalization lies in the comparison of G3PDH and MDH in vertebrates and invertebrates. In vertebrates, G3PDH is more variable than MDH (H_E of 0.062 and 0.029, respectively), while in invertebrates, G3PDH is less variable than MDH (H_E of

0.034 and 0.063, respectively). These differences are all the more notable as these two enzymes have very similar structures (both are dimers, with subunit molecular weights around 30,000). However, other pairs of enzymes with similar structural constraints (e.g., *l*-LDH and GAPDH, and ALAT and AAT) do have very similar mean H_E within taxa.

The results of our analyses indicate that a substantial proportion of the variation in mean heterozygosity between groups can be accounted for statistically by enzyme- or taxon-specific effects. It appears that this finding can be consistent with both neutral and selection theories by assuming that the degree of constraint varies in an enzyme- or taxon-specific way. High constraint will leave fewer amino acid sites free to vary and accept neutral or selectively maintained polymorphisms, resulting in low heterozygosity. Conversely, low constraint under both neutral and selection theories will tend to give high heterozygosity.

In neutral theory, differences in constraint will result in differences in neutral mutation rate between enzymes. Thus, variation in heterozygosity between enzymes will be attributed to differences in neutral mutation rate. Effective population size will not be important because on average in the analysis this variable will be roughly the same for each enzyme. By contrast, the differences in heterozygosity between taxa will be attributed to variation in effective population size. Average neutral mutation rate will be similar for different taxa, as roughly the same set of enzymes would contribute to each taxon.

It is not clear what the analogous parameters might be in selection theory. To explain differences in heterozygosity between enzymes, it would be necessary to search for a parameter which related not to external ecological circumstances of species, but rather to internal differences in function or other properties of enzymes common to all species. It is not satisfactory for this parameter to be mutation rate, because this is excluded from equilibrium models of balancing selection. To explain the differences in heterozygosity between major taxa, selection theory would seek a parameter relating to the biological differences distinguishing these taxa. For example, the parameter might relate to physiological or ecological differences between birds, amphibians, and the other taxa. Selection theory might well account for heterozygosity variation by demonstrating locus by locus and species by species the precise selective factors and coefficients responsible for maintaining allele frequencies at their observed polymorphic values. Alone, however, this approach would not provide an explanation for the global differences in mean heterozygosity between enzymes and taxa.

Thus, whatever the failings of neutral theory, this is an appealing theory in that it appears able to account for the observed variation using only two parameters, mutation rate and effective population size. We would like proponents of selection theory to suggest analogous parameters which can account for the observed variation.

In due course, more refined estimates of heterozygosity based on nucleotide sequence diversity will become available, but it will be many years before they will be able to rival, in terms of numbers of cases (species and numbers of loci), the heterozygosity estimates already available from allozyme analyses. A few DNA diversity estimates are available for some proteins, especially in *Drosophila* (Kreitman, 1983; Aquadro *et al.,* 1988; Riley *et al.,* 1989), but such data remain scanty. The striking variation in allozyme heterozygosity among proteins revealed in the present chapter may not be reflected in corresponding variation in nucleotide sequence diversity, even if attention is restricted to expressed regions of the gene. For example, allozymically silent DNA substitutions are prevalent at the DNA level, and nucleotide diversity is not dependent on length of DNA (and there is therefore no *a priori* reason to expect large-subunit proteins to have higher nucleotide diversity than small-subunit enzymes). Aldolase, a tetrameric enzyme, has low average heterozygosity and, in common with many other species, the genes coding for this enzyme are allozymically monomorphic in mole rats of the *Spalax ehrenbergi* superspecies. However, at the DNA level (and including introns and flanking regions), mole-rat aldolase is highly polymorphic (Nevo *et al.,* 1990). The contrast between allozyme variability and DNA sequence variation in *D. melanogaster* and *D. simulans* has been described by Aquadro *et al.* (1988). Allozyme variability is higher in *D. melanogaster,* but DNA variation is higher in *D. simulans.* When appropriate information does become available, it will be very interesting to see how well the allozyme estimates presented and analyzed here correspond to sequence diversity estimates or to other estimates of DNA variability.

SUMMARY

Average heterozygosities (and standard errors) per protein are presented for a variety of taxa: mammals (172 species), birds (80), fish (195), amphibia (116), reptiles (85), insects (170), mollusks (105), crustacea (80), and "other invertebrates" (15). These data were analyzed with respect to the extent of intraspecies subpopulation differentiation in these taxa, and with respect to hypotheses relating variation in heterozygosity to enzyme structure and function. Vertebrates were on average less variable than invertebrates. Within the vertebrates, average heterozygosities per taxon were in the (descending) order amphibia > mammals = birds = reptiles > fish. The corresponding order for invertebrates was mollusks = insects = "other invertebrates" > crustaceans. Values of G_{ST} were greatest for amphibia and mollusks, and least for birds, insects, and "other invertebrates." For vertebrates, about 21% of variance in average protein heterozygosity was attributable to taxon effects and 52% to protein effects. These figures were 34% and 41%, respectively, for invertebrates. The 15 most variable proteins in vertebrates and invertebrates are listed, but the point is made that even those with high average heterozygosity (such as MPI and EST) are effectively monomorphic ($H_j < 0.05$) in 40–60% of species. A method of weighting heterozygosity estimates is suggested which allows the average heterozygosities of species assayed for different suites of enzymes to be more properly compared. A literature survey confirmed that subunit numbers and subunit sizes are highly conserved in most proteins. Several analyses of variance and multiple regressions indicated that variation in subunit number and subunit size each explains about 10–20% of variance in average protein or enzyme heterozygosity. The hypothesis that multimeric enzymes forming interlocus hybrids are less variable than multimeric enzymes not forming such hybrids was not supported by a consideration of heterozygosity variation at the duplicate GPI and MDH loci in fish in comparison with the single loci found for these enzymes in other vertebrate groups. No consistent relationships were found between enzyme heterozygosity and enzyme function.

ACKNOWLEDGMENTS

We gratefully acknowledge the help given by Jonathan Lees, Cath Fisher, and Christine Beynon in setting up the data base, and Loughborough University of Technology and especially the Natural Environment Research Council (grant number GR3/6976) for financial support. We also thank the following for providing unpublished data for this analysis: C. Anderson, A. J. Baker, T. Child, R. A. Galleguillos, H. J. Geiger, D. Hedgecock, E. Nevo, J. L. Patton, K. G. Ross, P. W. Sattler, and G. P. Wallis.

REFERENCES

Afolayan, A., and Daini, O. A., 1986, Isolation and properties of creatine kinase from the breast muscle of tropical fruit bat, *Eidolon helvum* (Kerr), *Comp. Biochem. Physiol.* **85B:**463–468.

Ahmad, M., Skibinski, D. O. F., and Beardmore, J. A., 1977, An estimate of the amount of genetic variation in the common mussel *Mytilus edulis, Biochem. Genet.* **15:**833–846.

Aikawa, T., Umemori-Aikawa, Y., and Fisher, J. R., 1977, Purification and properties of the adenosine deaminase from the midgut gland of a marine bivalved mollusc *Atrina* spp., *Comp. Biochem. Physiol.* **58B:**357–364.

Almassy, R. J., Janson, C. A., Hamlin, R., Xuong, N.-H., and Eisenberg, D., 1986, Novel subunit–subunit interactions in the structure of glutamine synthetase, *Nature* **323:**304–309.

Aquadro, C. F., Lado, K. M., and Noon, W. A., 1988, The *rosy* region of *Drosophila melanogaster* and *Drosophila simulans.* I. Contrasting levels of naturally occurring DNA restriction map variation and divergence, *Genetics* **119:**875–888.

Atrian, S., and Gonzalez-Duarte, R., 1985, Purification and molecular characterization of alcohol dehydrogenase from *Drosophila hydei:* Conservation in the biochemical features of the enzyme in several species of *Drosophila, Biochem. Genet.* **23:**891–911.

Avise, J. C., and Kitto, G. B., 1973, Phosphoglucose isomerase gene duplication in the bony fishes: An evolutionary history, *Biochem. Genet.* **8:**113–132.

Banas, T., Malarska, A., Krotkiewska, B., and Marcinkowska, A., 1987, Glyceraldehyde-3-phosphate dehydrogenase. Investigation on the regions responsible for self-assembly of subunits, *Comp. Biochem. Physiol.* **87B:**391–401.

Barbosa, V. M., and Nakano, M., 1987, Muscle D-glyceraldehyde-3-phosphate dehydrogenase from *Anas* sp.—I. Purification and properties of the enzyme, *Comp. Biochem. Physiol.* **88B:**563–568.

Barra, D., Bossa, F., Doonan, S., Fahmy, H. M. A., Hughes, G. J., Martini, F., Petruzzelli, R., and Wittman-Liebold, B., 1980, The cytosolic and mitochondrial aspartate aminotransferases from pig heart. A comparison of their primary structures, predicted secondary structures and some physical properties, *Eur. J. Biochem.* **108:**405–414.

Basaglia, F., 1989, Some aspects of lactate dehydrogenase, malate dehydrogenase and glucosephosphate isomerase in fish, *Comp. Biochem. Physiol.* **92B:**213–226.

Beckman, L., and Johnson, F. M., 1964, Variations in larval alkaline phosphatase controlled by *Aph* alleles in *Drosophila melanogaster, Genetics* **49:**829–835.

Beeckmans, S., and Kanarek, L., 1977, A new purification procedure for fumarase based on affinity chromatography. Isolation and characterization of pig-liver fumarase, *Eur. J. Biochem.* **78:**437–444.

Bird, T. G., Salin, M. L., Boyle, J. A., and Heitz, J. R., 1986, Superoxide dismutase in the housefly, *Musca domestica* (L.), *Arch. Insect Biochem. Physiol.* **3:**31–43.

Birktoft, J. J., and Banaszak, L. J., 1983, The presence of a histidine–aspartic acid pair in the active site of 2-hydroxyacid dehydrogenase. X-ray refinement of cytoplasmic malate dehydrogenase, *J. Biol. Chem.* **258:**472–482.

Birktoft, J. J., Rhodes, G., and Banaszak, L. J., 1989, Refined crystal structure of cytoplasmic malate dehydrogenase at 2.5-A resolution, *Biochemistry* **28:**6065–6081.

Borack, L. I., and Sofer, W., 1971, β-L-Hydroxyacid dehydrogenases, *J. Biol. Chem.* **246:**5345–5350.

Buczylko, J., Hargrave, P. A., and Kochman, M., 1980, Fructose-bisphosphate aldolase from *Helix pomatia*—II. Chemical and physical properties, *Comp. Biochem. Physiol.* **67B:**233–238.

Bulnheim, H.-P., and Scholl, A., 1981, Genetic variation between geographic populations of the amphipods *Gammarus zaddachi* and *G. salinus, Mar. Biol.* **64:**105–115.

Buroker, N. E., Hershberger, W. K., and Chew, K. K., 1975, Genetic variation in the Pacific Oyster, *Crassostrea gigas, J. Fish. Res. Board Can.* **32:**2471–2477.

Carlier, M.-F., and Pantaloni, D., 1973, NADP-linked isocitrate dehydrogenase from beef liver. Purification, quaternary structure and catalytic activity, *Eur. J. Biochem.* **37:**341–354.

Carvajal, N., Gonzalez, R., and Moran, A., 1986, Properties of pyruvate kinase from the heart of *Concholepas concholepas, Comp. Biochem. Physiol.* **85B:**577–580.

Catmull, J., Yellowlees, D., and Miller, D. J., 1987, NADP$^+$-dependent glutamate dehydrogenase from *Acropora formosa:* Purification and properties, *Mar. Biol.* **95:**559–564.

Cederbaum, S. D., and Yoshida, A., 1976, Glucose 6-phosphate dehydrogenase in rainbow trout, *Biochem. Genet.* **14:**245–258.

Chothia, C., and Janin, J., 1976, The nature of the accessible and buried surfaces in proteins, *J. Mol. Biol.* **105:**1–14.

Cini, J. K., and Gracy, R. W., 1986, Molecular basis of the isozymes of bovine glucose-6-phosphate isomerase, *Arch. Biochem. Biophys.* **249:**500–505.

Cobbs, G., 1976, Polymorphism for dimerizing ability at the esterase-5 locus in *Drosophila pseudoobscura, Genetics* **82:**53–62.

Colowick, P., 1975, The hexokinases, in: *The Enzymes* (P. D. Boyer, ed.), Vol. IX, pp. 1–48, Academic Press, Orlando, Florida.

Cross, J. F., Ward, R. D., and Abreu-Grobois, A., 1979, Duplicate loci and allelic variation for mitochondrial malic enzyme in the Atlantic salmon (*Salmo salar* L.), *Comp. Biochem. Physiol.* **62B:**403–406.

Daddona, P. E., and Kelley, W. N., 1979, Human adenosine deaminase. Stoichiometry of the adenosine deaminase-binding protein complex, *Biochim. Biophys. Acta* **580:**302–311.

Dando, P. R., 1980, Duplication of the glucose phosphate isomerase (E.C. 5.3.1.9) locus in vertebrates, *Comp. Biochem. Physiol.* **66B:**373–378.

Darnall, D. W., and Klotz, I. M., 1975, Subunit constitution of proteins: A table, *Arch. Biochem. Biophys.* **166:**651–682.

Dell'Agata, M., Pannunzio, G., Annicchiarico, M., Coscarella, A., and Ferracin, A., 1990, Unusual steady-state kinetic properties of a chilopod enzyme: L (+) lactate dehydrogenase purified from *Scolopendra cingulata, Comp. Biochem. Physiol.* **96B:**439–444.

Dickerson, R. E., and Geis, I., 1983, *Hemoglobin: Structure, Function, Evolution, and Pathology,* Benjamin/Cummings, Menlo Park, California.

Dixon, M., and Webb, E. C., 1979, *Enzymes,* 3rd ed., Longman, London.

Doane, W. W., Abraham, I., Kolar, M. M., Martenson, R. E., and Deibler, G. E., 1975, Purified *Drosophila* α-amylase isozymes: Genetical, biochemical, and molecular characterization, in: *Isozymes IV. Genetics and Evolution* (C. L. Markert, ed.), pp. 585–607, Academic Press, New York.

Ellington, W. R., and Long, G. L., 1978, Purification and characterization of a highly unusual tetrameric D-lactate dehydrogenase from the muscle of the giant barnacle, *Balanus nubilus* Darwin, *Arch. Biochem. Biophys.* **186:**265–274.

Ferris, S. D., and Whitt, G. S., 1978, Genetic and molecular analysis of nonrandom dimer assembly of the creatine kinase isozymes of fishes, *Biochem. Genet.* **16:**811–829.

Fisher, S. E., Shaklee, J. B., Ferris, S. D., and Whitt, G. S., 1980, Evolution of five multilocus isozyme systems in the chordates, *Genetica* **52/53:**73–85.

Fita, I., and Rossman, M. G., 1985, The active center of catalase, *J. Mol. Biol.* **185:**21–37.

Ford, G. C., Eichele, G., and Jansonius, J. N., 1980, Three-dimensional structure of a pyridoxal-phosphate-dependent enzyme, mitochondrial aspartate aminotransferase, *Proc. Natl. Acad. Sci. USA* **77:**2559–2563.

Fosset, M., Chappelet-Tordo, D., and Lazdunski, M., 1974, Intestinal alkaline phosphatase. Physical properties and quaternary structure, *Biochemistry* **13:**1783–1788.

Frydenberg, O., and Simonsen, V., 1973, Genetics of *Zoarces* populations. V. Amount of protein polymorphism and degree of genic heterozygosity, *Hereditas* **75:**221–231.

Fucci, L., Gaudio, L., Rao, R., Spano, A., and Carfagna, M., 1979, Properties of the two common electrophoretic variants of phosphoglucomutase in *Drosophila melanogaster*, *Biochem. Genet.* **17**:825–836.

Fujio, Y., Yamanaka, R., and Smith, P. J., 1983, Genetic variation in marine molluscs, *Bull. Jpn. Soc. Sci. Fish.* **49**:1809–1817.

Fushitani, K., Matsuura, M. S. A., and Riggs, A. F., 1988, The amino acid sequences of chains *a*, *b* and *c* that form the trimeric subunit of the extracellular haemoglobin from *Lumbricus terrestris*, *J. Biol. Chem.* **263**:6502–6517.

Gavilanes, F. G., Gavilanes, J. G., and Martin-Dudoignon, R., 1981, Triosephosphate isomerase from the insect *Ceratitis capitata* molecular and enzymic properties, *Comp. Biochem. Physiol.* **70B**:257–262.

Geer, B. W., Krochko, D., Oliver, M. J., Walker, V. K., and Williamson, J. H., 1980, A comparative study of the NADP-malic enzymes from *Drosophila* and chick liver, *Comp. Biochem. Physiol.* **65B**:25–34.

Gillespie, J. H., and Kojima, K., 1968, The degree of polymorphism in enzymes involved in energy production compared to that in nonspecific enzymes in two *Drosophila ananassae* populations, *Proc. Natl. Acad. Sci. USA* **61**:582–585.

Gillespie, J. H., and Langley, C. H., 1974, A general model to account for enzyme variation in natural populations, *Genetics* **76**:837–848.

Gleason, F. H., Price, J. S., Mann, R. A., and Stuart, T. D., 1971, Lactate dehydrogenases from crustaceans and arachnids, *Comp. Biochem. Physiol.* **40B**:387–394.

Gō, M., and Miyazawa, S., 1980, Relationship between mutability, polarity and exteriority of amino acid residues in protein evolution, *Int. J. Pept. Protein Res.* **15**:211–224.

Gonzalez-Villasenor, L. I., and Powers, D. A., 1985, A multilocus system for studying tissue and subcellular specialization. The three NADP-dependent isocitrate dehydrogenase isozymes of the fish *Fundulus heteroclitus*, *J. Biol. Chem.* **260**:9106–9113.

Gooding, R. H., and Rolseth, B. M., 1978, Genetics of *Glossina morsitans morsitans* (Diptera: Glossinidae). II. Electrophoretic banding patterns of midgut alkaline phosphatase, *Can. Entomol.* **110**:1241–1246.

Gooding, R. H., and Rolseth, B. M., 1979, Genetics of *Glossina morsitans morsitans* (Diptera; Glossinidae) IV. Electrophoretic banding patterns of octanol dehydrogenase and arginine phosphokinase, *Can. Entomol.* **111**:1307–1310.

Guha, A., Lai, C. Y., and Horeciter, B. L., 1971, Lobster muscle aldolase: Isolation, properties and primary structure at the substrate-binding site, *Arch. Biochem. Biophys.* **147**:692–706.

Gyllensten, U., 1985, The genetic structure of fish: Differences in the intraspecific distribution of biochemical genetic variation between marine, anadromous, and freshwater species, *J. Fish Biol.* **26**:691–699.

Harper, R. A., and Armstrong, F. B., 1972, Alkaline phosphatase of *Drosophila melanogaster*. I. Partial purification and characterization, *Biochem. Genet.* **6**:75–82.

Harris, H., Hopkinson, D. A., and Edwards, Y. H., 1977, Polymorphism and the subunit structure of enzymes: A contribution to the neutralist–selectionist controversy, *Proc. Natl. Acad. Sci. USA* **74**:698–701.

Harrison, R. G., 1979, Speciation in North American field crickets: Evidence from electrophoretic comparisons, *Evolution* **33**:1009–1023.

Head, E. J. H., 1980, NADP-dependent isocitrate dehydrogenase from the mussel *Mytilus edulis* L. I. Purification and characterisation, *Eur. J. Biochem.* **111**:575–579.

Hebert, P. D. N., Muncaster, B. W., and Mackie, G. L., 1989, Ecological and genetic studies on *Dreissena polymorpha* (Pallas): A new mollusc in the Great Lakes, *Can. J. Fish. Aquat. Sci.* **46**:1587–1591.

Hedgecock, D., and Ayala, F. J., 1974, Evolutionary divergence in the genus *Taricha* (Salamandridae), *Copeia* **1974**:738–746.

Hill, A., Johnson, M. S., and Merrifield, H., 1983, An electrophoretic and morphological examination of *Bothriembryon kendricki* (Pulmonata: Bulimulidae), a new species previously considered conspecific with *B. bulla* (Menke), *Aust. J. Zool.* **31**:227–242.

Hopkinson, D. A., Mestriner, M. A., Cortner, J., and Harris, H., 1973, Esterase D: A new human polymorphism, *Ann. Hum. Genet.* **37**:119–137.

Hopkinson, D. A., Edwards, Y. H., and Harris, H., 1976, The distributions of subunit numbers and subunit sizes of enzymes: A study of the products of 100 human gene loci, *Ann. Hum. Genet.* **39**:383–411.

Hori, S. H., and Tanda, S., 1980, Purification and properties of wild-type and mutant glucose 6-phosphate dehydrogenases and of 6-phosphogluconate dehydrogenase from *Drosophila melanogaster, Jpn. J. Genet.* **55**:211–223.

Hori, S. H., Yonezawa, S., Mochizuki, Y., Sado, Y., and Kamada, T., 1975, Evolutionary aspects of animal glucose-6-phosphate dehydrogenase isozymes in: *Isozymes IV. Genetics and Evolution* (C. L. Markert, ed.), pp. 839–852, Academic Press, New York.

Illingworth, J. A., and Tipton, K. F., 1970, Purification and properties of the nicotinamide adenine dinucleotide phosphate-dependent isocitrate dehydrogenase from pig liver cytoplasm, *Biochem. J.* **118**:253–258.

Ingolia, D. E., Yeung, C.-Y., Orengo, I. F., Harrison, M. L., Frayne, E. G., Rudolph, F. B., and Kellems, R. E., 1985, Purification and characterization of adenosine deaminase from a genetically enriched mouse cell line, *J. Biol. Chem.* **260**:13261–13267.

Inkerman, P. A., Winzor, D. J., and Zerner, B., 1975, Carboxylesterases (EC 3.1.1). The molecular sizes of chicken and pig liver carboxylesterases, *Can. J. Biochem.* **53**:547–560.

IUBNC (International Union of Biochemistry. Nomenclature Committee), 1984, *Enzyme Nomenclature,* Academic Press, Orlando, Florida.

Janska, H., Kubicz, A., Szalewicz, A., and Harazna, J., 1988, The high molecular weight and the low molecular weight acid phosphatases of the frog liver and their phosphotyrosine activity, *Comp. Biochem. Physiol.* **90B**:173–178.

Janska, H., Kubicz, A., and Szalewicz, A., 1989, The lower molecular weight acid phosphatase from the frog liver: Isolation of homogeneous AcPase III and IV representing glycoforms with different bioactivity, *Comp. Biochem. Physiol.* **92B**:341–346.

Jaussi, R., Cotton, B., Juretic, N., Christen, P., and Schumperli, D., 1985, The primary structure of the precursor of chicken mitochondrial aspartate aminotransferase. Cloning and sequence analysis of cDNA, *J. Biol. Chem.* **260**:16060–16063.

Johnson, G. B., 1974, Enzyme polymorphism and metabolism, *Science* **184**:28–37.

Jornvall, H., Persson, M., and Jeffery, J., 1981, Alcohol and polyol dehydrogenases are both divided into two protein types, and structural properties cross-relate the different enzyme activities within each type, *Proc. Natl. Acad. Sci. USA.* **78**:4226–4230.

Karn, R. C., and Malacinski, G. M., 1978, The comparative biochemistry, physiology, and genetics of animal α-amylases, *Adv. Comp. Physiol. Biochem.* **7**:1–103.

Keung, W.-M., Ho, Y.-W., Fong, W.-P., and Lee, C.-Y., 1989, Isolation and characterization of shrew (*Suncus marinus*) liver alcohol dehydrogenases, *Comp. Biochem. Physiol.* **93B**:169–173.

Knight, A. J., and Ward, R. D., 1986, Purine nucleoside phosphorylase polymorphism in the genus *Littorina* (Prosobranchia: Mollusca), *Biochem. Genet.* **24**:405–413.

Kochman, M., Golebiowska, J., Baranowski, T., Dedman, J. R., Fodge, D. W., and Harris, B. G., 1974, Hybridization of glyceraldehyde-3-phosphate dehydrogenase, *FEBS Lett.* **41**:104–107.

Koehn, R. K., and Eanes, W. F., 1977, Subunit size and genetic variation of enzymes in natural populations of *Drosophila, Theor. Popul. Biol.* **11**:330–341.

Koehn, R. K., and Eanes, W. F., 1978, Molecular structure and protein variation within and among populations. *Evol. Biol.* **11**:39–100.

Koehn, R. K., and Eanes, W. F., 1979, Molecular structure, polypeptide size, and genetic variation of enzymes, in: *Isozymes: Current Topics in Biological and Medical Research,* (M. C. Rattazzi, J. G. Scandalios, and G. S. Whitt, eds.), Vol. 3, pp. 185–211, Liss, New York.

Kolatkar, P. R., Meador, W. E., Stanfield, R. L., and Hackert, M. L., 1988, Novel subunit structure observed for noncooperative haemoglobin from *Urechis caupo, J. Biol. Chem.* **263**:3462–3465.

Komuniecki, R. W., and Roberts, L. S., 1977, Hexokinase from the rat tapeworm, *Hymenolepis diminuta, Comp. Biochem. Physiol.* **76B**:45–49.

Krause, K. L., Volz, K. W., and Lipscomb, W. N., 1985, Structure at 2.9-Å resolution of aspartate carbamoyltransferase complexed with the bisubstrate analogue *N*-(phosphonacetyl)-L-aspartate, *Proc. Natl. Acad. Sci. USA* **82**:1643–1647.

Kreitman, M., 1983, Nucleotide polymorphism at the alcohol dehydrogenase locus of *Drosophila melanogaster, Nature* **304**:412–417.

Lavery, S., and Shaklee, J. B., 1989, Population genetics of two tropical sharks, *Carcharhinus tilstoni* and *C. sorrah,* in Northern Australia, *Aust. J. Mar. Freshwater Res.* **40**:541–547.

Lee, C.-Y., Langley, C. H., and Burkhart, J., 1978, Purification and molecular weight determination of glucose-6-phosphate dehydrogenase and malic enzyme from mouse and *Drosophila, Analyt. Biochem.* **86**:697–706.

Lee, Y. M., Ayala, F. J., and Misra, H. P., 1981, Purification and properties of superoxide dismutase from *Drosophila melanogaster, J. Biol. Chem.* **256**:8506–8509.

Leigh Brown, A. J., and Langley, C. H., 1979, Correlation between heterozygosity and subunit molecular weight, *Nature* **277**:649–651.

Leigh Brown, A. J., and Voelker, R. A., 1980, Genetic and biochemical studies on glutamate-pyruvate transaminase from *Drosophila melanogaster, Biochem. Genet.* **18**:303–309.

Ling, I. T., Cooksley, S., Bates, P. A., Hempelmann, E., and Wilson, R. J. M., 1986, Antibodies to the glutamate dehydrogenase of *Plasmodium falciparum, Parasitology* **92**:313–324.

Long, G. L., and Kaplan, N. O., 1968, D-lactate specific pyridine nucleotide lactate dehydrogenase in animals, *Science* **162**:685–686.

Long, G. L., and Kaplan, N. O., 1973, Diphosphopyridine nucleotide-linked D-lactate dehydrogenases from the horseshoe crab, *Limulus polyphemus* and the seaworm, *Nereis virens.* I. Physical and chemical properties, *Arch. Biochem. Biophys.* **154**:696–710.

Ma, P. F., and Fisher, J. R., 1968, Multiple adenosine deaminases in the amphibia and their possible phylogenetic significance, *Comp. Biochem. Physiol.* **27B**:687–697.

MacFarlane, N., Mathews, B., and Dalziel, K., 1977, The purification and properties of NADP-dependent isocitrate dehydrogenase from ox-heart mitochondria, *Eur. J. Biochem.* **74**:553–559.

MacIntyre, R. J., 1971, Evolution of acid phosphatase-1 in the genus *Drosophila* as estimated by subunit hybridization. 1. Methodology, *Genetics* **68**:483–508.

MacIntyre, R. J., and Dean, M. R., 1978, Evolution of acid phosphatase-1 in the genus *Drosophila* as estimated by subunit hybridization. Interspecific tests, *J. Mol. Evol.* **12**:143–171.

Manning, A. M., Trotman, C. N. A., and Tate, W. P., 1990, Evolution of a polymeric globin in the brine shrimp *Artemia, Nature* **348**:653–656.

Markley, H. G., Faillace, L. A., and Mezey, E., 1973, Xanthine oxidase activity in rat brain, *Biochim. Biophys. Acta* **309**:23–31.

May, B., and Holbrook, F. R., 1978, Absence of genetic variability in the green peach aphid, *Myzus persicae* (Hemiptera: Aphididae), *Ann. Entomol. Soc. Am.* **71**:809–812.

Mayzaud, O., 1985, Purification and kinetic properties of the α-amylase from the copepod *Acartia clausi* (Giesbrecht, 1889), *Comp. Biochem. Physiol.* **82B**:725–730.

McCord, J. M., 1979, Superoxide dismutases: Occurrence, structure, function, and evolution, in: *Isozymes: Current Topics in Biological and Medical Research,* (M. C. Rattazzi, J. G. Scandalios, and G. S. Whitt, eds.), Vol. 3, pp. 1–22, Liss, New York.

McCormick, A., Paynter, K. T., Brodey, M. M., and Bishop, S. H., 1986, Aspartate aminotransferases from ribbed mussel gill tissue: Reactivity with B-L-cysteinesulfinic acid and other properties, *Comp. Biochem. Physiol.* **84B**:163–166.

Miake, F., Torikata, T., Koga, K., and Hayashi, K., 1977, Isolation and characterization of $NADP^+$ specific isocitrate dehydrogenase from the pupa of *Bomyx mori, J. Biochem.* **82**:449–454.

Mitchell, J. M., Daron, H. H., and Frandsen, J. C., 1982, Purification and characterization of aldolase from the parasitic protozoan *Eimeria stiedai* (Coccidia), *Comp. Biochem. Physiol.* **73B**:221–229.

Mochizuki, Y., and Hori, S. H., 1976, Hexose 6-phosphate dehydrogenase in starfishes, *Comp. Biochem. Physiol.* **54B**:489–494.

Mochizuki, Y., and Hori, S. H., 1977, Purification and properties of hexokinase from the starfish, *Asterias amurensis, J. Biochem.* **81**:1849–1856.

Moens, L., and Kondo, M., 1978, Evidence for a dimeric form of *Artemia salina* extracellular haemoglobins with high molecular weight subunits, *Eur. J. Biochem.* **82**:65–72.

Morris, S. R., 1979, Genetic variation in the genus *Littorina,* Ph.D. Thesis, University of Wales, Swansea, United Kingdom.

Morrison, J. F., 1975, Arginine kinase and the invertebrate guanidino kinases, in: *The Enzymes* (P. D. Boyer, ed.), Vol. VIII, pp. 457–486, Academic Press, Orlando, Florida.

Moser, D., Johnson, L., and Lee, C.-Y., 1980, Multiple forms of *Drosophila* hexokinase. Purification, biochemical and immunological characterization, *J. Biol. Chem.* **255**:4673–4679.

Mulley, J. L., and Latter, B. D. H., 1980, Genetic variation and evolutionary relationships within a group of thirteen species of Penaeid prawns, *Evolution* **34**:904–916.

Murphy, R. W., Sites, J. W., Jr., Buth, D. G., and Haufler, C. H., 1990, Proteins I: Isozyme electrophoresis, in: *Molecular Systematics* (D. M. Hillis and C. Moritz, eds.), pp. 45–126, Sinauer Associates, Sunderland, Massachusetts.

Nadler, S. A., 1986, Biochemical polymorphism in *Parascaris equorum, Toxocara canis* and *Toxocara cati, Mol. Biochem. Parasitol.* **18**:45–54.

Nahmias, J. A., and Bewley, G. C., 1984, Characterization of catalase purified from *Drosophila melanogaster* by hydrophobic interaction chromatography, *Comp. Biochem. Physiol.* **77B**:355–364.

Nakagawa, T., and Nagayama, F., 1989*a,* Enzymatic properties of glyceraldehyde-3-phosphate dehydrogenase from fish muscle, *Comp. Biochem. Physiol.* **93B**:379–384.

Nakagawa, T., and Nagayama, F., 1989*b,* Enzymatic properties of fish muscle aldolase, *Comp. Biochem. Physiol.* **92B**:405–410.

Nei, M., 1973, Analysis of gene diversity in subdivided populations, *Proc. Natl. Acad. Sci. USA* **70**:3321–3323.

Nei, M., and Graur, D., 1984, Extent of protein polymorphism and the neutral mutation theory, *Evol. Biol.* **17**:73–118.

Nei, M., Fuerst, P. A., and Chakraborty, R., 1978, Subunit molecular weight and genetic variability of proteins in natural populations, *Proc. Natl. Acad. Sci. USA* **75**:3359–3362.

Nevaldine, B. H., Basel, A. R., and Hsu, R. Y., 1974, Mechanism of pigeon liver malic enzyme subunit structure, *Biochim. Biophys. Acta* **336**:283–293.

Nevo, E., Beiles, A., and Ben-Shlomo, R., 1984, The evolutionary significance of genetic diversity: Ecological, demographic and life history correlates, in: *Evolutionary Dynamics of Genetic Diversity* (G. S. Mani, ed.), pp. 13–213, Springer-Verlag, Berlin.

Nevo, E., Joh, K., Hori, K., and Beiles, A., 1990, Aldolase DNA polymorphism in subterranean mole-rats: Genetic differentiation and environmental correlates, *Heredity* **65**:307–320.

Noda, L., 1975, Adenylate kinase, in: *The Enzymes* (P. D. Boyer, ed.), Vol. VIII, pp. 279–305, Academic Press, Orlando, Florida.

Norton, I. L., Pfuderer, P., Stringer, C. D., and Hartman, F. C., 1970, Isolation and characterization of rabbit muscle triosephosphate isomerase, *Biochemistry* **9**:4952–4958.

Ohkubo, M., Sakiyama, S., and Fujimura, S., 1983, Purification and characterization of N^1-methylnicotinamide oxidases I and II separated from rat liver, *Arch. Biochem. Biophys.* **221**:534–542.

Ohnishi K.-I., and Hori, S. H., 1977, A comparative study of invertebrate glucose 6-phosphate dehydrogenases, *Jpn. J. Genet.* **52**:95–106.

Ohta, T., and Kimura, M., 1973, A model of mutation appropriate to estimate the number of electrophoretically detectable alleles in a finite population, *Genet. Res.* **22**:201–204.

Okada, N., Azuma, M., and Eguchi, M., 1989, Alkaline phosphatase isozymes in the midgut of silkworm: Purification of high pH-stable microvillus and labile cytosolic enzymes, *J. Comp. Physiol. B.* **159**:123–130.

Onoufriou, A., and Alahiotis, S. N., 1982, *Drosophila* lactate dehydrogenase: Molecular and genetic aspects, *Biochem. Genet.* **20**:1195–1209.

Osborne, H. H., and Hollaway, M. R., 1974, The hybridization of glyceraldehyde 3-phosphate dehydrogenases from rabbit muscle and yeast. Kinetics and thermodynamics of the reaction and isolation of the hybrid, *Biochem. J.* **143**:651–662.

Panara, F., 1985, Isolation and partial characterization of high and low molecular weight acid phosphatases from chicken liver, *Int. J. Biochem.* **17**:1213–1217.

Panara, F., and Pascolini, R., 1989, Acid phosphatases from liver of *Cyprinus carpio, Comp. Biochem. Physiol.* **92B**:751–754.

Panara, F., Angiolillo, A., and Pascolini, R., 1989, Acid phosphatases from liver of *Rana esculenta.* Subcellular localization and partial characterization of multiple forms, *Comp. Biochem. Physiol.* **93B**:877–882.

Payne, D. M., Powley, D. G., and Harris, B. G., 1979, Purification, characterization, and the presumptive role of fumarase in the energy metabolism of *Ascaris suum, J. Parasitol.* **65**:833–841.

Perutz, M. F., 1990, Molecular inventiveness, *Nature* **348**:583–584.

Peters, T., 1975, Serum albumin, in: *The Plasma Proteins* (F. W. Putman, ed.), Vol. I, pp. 133–181, Academic Press, New York.

Putnam, F. W., 1975, Transferrin, in: *The Plasma Proteins* (F. W. Putman, ed.), Vol. I, pp. 266–316, Academic Press, New York.

Rajagopalan, K. V., and Handler, P., 1967, Purification and properties of chicken liver xanthine dehydrogenase, *J. Biol. Chem.* **242**:4097–4107.

Ray, W. J., and Peck, E. J., 1975, Phosphomutases, in: *The Enzymes* (P. D. Boyer, ed.), Vol. VI, pp. 408–478, Academic Press, Orlando, Florida.

Rehse, P. H., and Davidson, W. S., 1986, Purification and properties of a C-type isozyme of lactate dehydrogenase from the liver of the Atlantic cod (*Gadus morhua*), *Comp. Biochem. Physiol.* **84B**:145–150.

Riggs, A. F., 1991, Aspects of the origin and evolution of non-vertebrate hemoglobins. *Am. Zool.* **31**:535–545.

Riley, M. A., Hallas, M. E., and Lewontin, R. C., 1989, Distinguishing the forces controlling genetic variation at the Xdh locus in *Drosophila pseudoobscura, Genetics* **123**:359–369.

Royer, W. E., Love, W. E., and Fenderson, F. F., 1985, Cooperative dimeric and tetrameric clam haemoglobins are novel assemblages of myoglobin folds, *Nature* **316:**277–280.

Ruth, R. C., and Wold, F., 1976, The subunit structure of glycolytic enzymes, *Comp. Biochem. Physiol.* **54B:**1–6.

Ryden, L., Ofverstedt L.-G., Beinert, H., Emptage, M. H., and Kennedy, M. C., 1984, Molecular weight of beef heart aconitase and stoichiometry of the components of its iron–sulfur cluster, *J. Biol. Chem.* **259:**3141–3144.

Sanchez, J. L., Abad, M., Martin, L. O. G., and Galarza, A., 1983, Hepatopancreas hexokinase isozymes from the mussel *Mytilus galloprovincialis, Comp. Biochem. Physiol.* **74B:** 807–814.

Scheid, M. J., and Awapara, J., 1972, Stereospecificity of some invertebrate lactic dehydrogenases, *Comp. Biochem. Physiol.* **43B:**619–626.

Scholze, H., 1983, Studies on aconitase species from *Saccharomyces cerevisiae,* porcine and bovine heart, obtained by a modified isolation method, *Biochim. Biophys. Acta* **746:**133–137.

Searle, J. B., 1986, A trimeric esterase in the common shrew (*Sorex araneus*), *J. Hered.* **77:**121–122.

Selander, R. K., 1976, Genic variation in natural populations, in: *Molecular Evolution* (F. J. Ayala, ed.), pp. 21–45, Sinauer Associates, Sunderland, Massachusetts.

Selander, R. K., and Kaufman, D. W., 1973, Genic variability and strategies of adaptation in animals, *Proc. Natl. Acad. Sci. USA* **70:**1875–1877.

Selander, R. K., and Yang, S. Y., 1970, Horseshoe crab lactate dehydrogenases: Evidence for dimeric structure, *Science* **169:**179–180.

Shaklee, J. B., 1983, Mannosephosphate isomerase in the Hawaiian spiny lobster *Panulirus marginatus:* A polymorphic, sex-linked locus useful in investigating embryonic and larval sex ratios, *Mar. Biol.* **73:**193–201.

Shows, T. B., Chapman, V. M., and Ruddle, F. H., 1970, Mitochondrial malate dehydrogenase and malic enzyme: Mendelian inherited electrophoretic variants in the mouse, *Biochem. Genet.* **4:**707–718.

Siegismund, H. R., 1985, Genetic studies of *Gammarus.* III. Inheritance of electrophoretic variants of the enzymes mannose phosphate isomerase and glucose phosphate isomerase in *Gammarus oceanicus, Hereditas* **102:**25–31.

Skibinski, D. O. F., and Ward, R. D., 1981, Relationship between allozyme heterozygosity and rates of divergence, *Genet. Res.* **38:**71–92.

Smith, E. L., Austen, B. M., Blumenthal, K. M., and Nyc, J. F., 1975, Glutamate dehydrogenases, in: *The Enzymes* (P. D. Boyer, ed.), Vol. XI, pp. 293–367, Academic Press, Orlando, Florida.

Smith, J. K., and Zimmerman, E. G., 1976, Biochemical genetics and evolution of North American blackbirds, (family Icteridae), *Comp. Biochem. Physiol.* **53B:**319–324.

Sokal, R. R., and Rohlf, F. J., 1981, *Biometry,* 2nd ed., W. H. Freeman, New York.

Soulé, M., 1976, Allozyme variation: Its determinants in space and time, in: *Molecular Evolution* (F. J. Ayala, ed.), pp. 60–77, Sinauer Associates, Sunderland, Massachusetts.

Spotorno, G. L. M., and Hollaway, M. R., 1970, Hybrid molecules of yeast and rabbit GPD containing native and modified subunits, *Nature* **226:**756–757.

Stoneking, M., May, B., and Wright, J. E., 1979, Genetic variation, inheritance, and quaternary structure of malic enzyme in brook trout (*Salvelinus fontinalis*), *Biochem. Genet.* **17:**599–619.

Storey, K. B., 1976, Purification and properties of squid mantle adenylate kinase. Role of NADH in control of the enzyme, *J. Biol. Chem.* **251:**7810–7815.

Sullivan, D. T., Carroll, W. T., Kanik-Ennulat, C. L., Hitti, Y. S., Lovett, J. A., and Von Kalm, L., 1985, Glyceraldehyde-3-phosphate dehydrogenase from *Drosophila melanogaster*. Identification of two isozymic forms encoded by separate genes, *J. Biol. Chem.* **260**:4345–4350.

Swain, M. S., and Lebherz, H. G., 1986, Hybridization between fructose diphosphate aldolase subunits derived from diverse biological systems: Anomalous hybridization behavior of some aldolase subunit types, *Arch. Biochem. Biophys.* **244**:35–41.

Tegelstrom, H., 1975, Interspecific hybridisation *in vitro* of superoxide dismutase from various species, *Hereditas* **81**:185–198.

Teller, D. C., 1976, Accessible area, packing volumes and interaction surfaces of globular proteins, *Nature* **260**:729–731.

Thieme, R., Pai, E. F., Schirmer, R. H., and Schulz, G. E., 1981, Three dimensional structure of glutathione reductase at 2Å resolution, *J. Mol. Biol.* **152**:763–782.

Tracey, M. L., Nelson, K., Hedgecock, D., Shleser, R. A., and Pressick, M. L., 1975, Biochemical genetics of lobsters: Genetic variation and the structure of American lobster (*Homarus americanus*) populations, *J. Fish. Res. Board Can.* **32**:2091–2101.

Turner, A. C., Lushbaugh, W. B., and Hutchinson, W. F., 1986, *Dirofilaria immitis:* Comparison of cytosolic and mitochondrial glutamate dehydrogenases, *Exp. Parasitol.* **61**:176–183.

Umemori-Aikawa, Y., and Aikawa, T., 1974, Adenosine deaminase from the clam, *Tapes philippinarum:* Partial purification and detection of activity on disc electrophoresis, *Comp. Biochem. Physiol.* **49B**:353–359.

Unakami, S., Komoda, T., Watanabe, M., Tanimoto, Y., Sakagishi, Y., and Ikezawa, H., 1987, Molecular nature of three liver alkaline phosphatases detected by drug administration *in vivo:* Differences between soluble and membranous enzymes, *Comp. Biochem. Physiol.* **88B**:111–118.

Unemura, S., Nishino, T., Murakami, K., and Tsushima, K., 1982, Trimeric purine nucleoside phosphorylase from chicken liver having a proteolytic nick on each subunit and its kinetic properties, *J. Biol. Chem.* **257**:13374–13378.

Vieira, M. M., Veiga, L. A., and Nakano, M., 1983, Muscle D-glyceraldehyde-3-phosphate dehydrogenase from *Caiman* sp.—I. Purification and properties of the enzyme, *Comp. Biochem. Physiol.* **74B**:781–790.

Walsh, P. J., 1981, Purification and characterization of glutamate dehydrogenases from three species of sea anemonies: Adaptation to temperature within and among species from different thermal environments, *Mar. Biol. Lett.* **2**:289–299.

Ward, R. D., 1977, Relationship between enzyme heterozygosity and quaternary structure, *Biochem. Genet.* **15**:123–135.

Ward, R. D., 1978, Subunit size of enzymes and genetic heterozygosity in vertebrates, *Biochem. Genet.* **16**:799–810.

Ward, R. D., and Beardmore, J. A., 1977, Protein variation in the plaice, *Pleuronectes platessa* L., *Genet. Res.* **30**:45–62.

Ward, R. D., and Skibinski, D. O. F., 1985, Observed relationships between protein heterozygosity and protein genetic distance and comparisons with neutral expectations, *Genet. Res.* **45**:315–340.

Ward, R. D., and Skibinski, D. O. F., 1988, Evidence that mitochondrial isozymes are genetically less variable than cytoplasmic isozymes, *Genet. Res.* **51**:121–127.

Ward, R. D., and Warwick, T., 1980, Genetic differentiation in the molluscan species *Littorina rudis* and *Littorina arcana* (Prosobranchia: Littorinidae), *Biol. J. Linn. Soc.* **14**:417–428.

Weeda, E., 1981, Some properties of mitochondrial NAD^+ linked malic enzyme and malate dehydrogenase from the flight muscles of *Leptinotarsa decemlineata, Insect Biochem.* **11**:679–684.

Weisman, M. I., Caiolfa, V. R., and Parola, A. H., 1988, Adenosine deaminase-complexing protein from bovine kidney. Isolation of two distinct subunits, *J. Biol. Chem.* **263:**5266–5270.

Weller, D. L., 1982, Isoelectric point and molecular weight of a diaphorase of *Entamoeba invadens, J. Parasitol.* **68:**343–345.

Widmer, G., Dvorak, J. A., and Miles, M. A., 1986, A biochemical comparison of glucosephosphate isomerase isozymes from *Trypanosoma cruzi, Biochim. Biophys. Acta* **873:**119–126.

Williamson, J. H., and Bentley, M. M., 1983, Comparative properties of three forms of glucose-6-phosphate dehydrogenase in *Drosophila melanogaster, Biochem. Genet.* **21:**1153–1166.

Williamson, J. H., Krochko, D., and Bentley, M. M., 1980*a,* Properties of *Drosophila* NADP$^+$-isocitrate dehydrogenase purified on Procion Brilliant Blue–sepharose-4B, *Comp. Biochem. Physiol.* **65B:**339–343.

Williamson, J. H., Krochko, D., and Geer, B. W., 1980*b,* 6-Phosphogluconate dehydrogenase from *Drosophila melanogaster.* 1. Purification and properties of the A isozyme, *Biochem. Genet.* **18:**87–101.

Winberg, J.-O., Hovik, R., McKinley-McKee, J. S., Juan, E., and Gonzalez-Duarte, R., 1986, Biochemical properties of alcohol dehydrogenase from *Drosophila lebanonensis, Biochem. J.* **235:**481–490.

Yoshihara, S., and Tatsumi, K., 1985, Guinea pig liver aldehyde oxidase as a sulfoxide reductase: Its purification and characterization, *Arch. Biochem. Biophys.* **242:**213–224.

Zink, R. M., 1982, Patterns of genic and morphologic variation among sparrows in the genera *Zonotrichia, Nelospiza, Juneo* and *Passerella, Auk* **99:**632–649.

Zouros, E., 1976, Hybrid molecules and the superiority of the heterozygote, *Nature* **262:**227–229.

Zuckerkandl, E., 1976, Evolutionary processes and evolutionary noise at the molecular level. I. Functional density in proteins, *J. Mol. Evol.* **7:**167–183.

Zuckerkandl, E., and Pauling, L., 1965, Evolutionary divergence and convergence in proteins, in: *Evolving Genes and Proteins* (V. Bryson and H. J. Vogel, eds.), pp. 97–196, Academic Press, New York.

DATA BASE REFERENCES

Abban, E. K., and Skibinski, D. O. F., 1988, Protein variation in *Schilbe mystus* (L.) and *Eutropius niloticus* (Ruppel) (Pisces siluriformes) in the volta basin of Ghana, West Africa, *Aquat. Fish. Manag.* **19:**25–37.

Achituv, Y., and Mizrahi, L., 1987, Allozyme differences between tidal levels in *Tetraclita squamosa* Pilsbry from the Red Sea, *J. Exp. Mar. Biol. Ecol.* **108:**181–190.

Adams, M., Baverstock, P. R., Saunders, D. A., Schodde, R., and Smith, G. T., 1984, Biochemical systematics of the Australian cockatoos (Psittaciformes: Cacatuinae), *Aust. J. Zool.* **32:**363–378.

Adams, M., Baverstock, P. R., Watts, C. H. S., and Reardon, T., 1987, Electrophoretic resolution of species boundaries in Australian microchiroptera. I. *Eptesicus* (Chiroptera: Vespertilionidae), *Aust. J. Biol. Sci.* **40:**143–162.

Adams, S. E., Smith, M. H., and Baccus, R., 1980, Biochemical variation in the American alligator, *Herpetologica* **36:**289–296.

Adest, G. A., 1977, Genetic relationships in the genus *Uma* (Iguanidae), *Copeia* **1977:**47–52.

Agatsuma, T., and Habe, S., 1986, Genetic variability and differentiation of natural populations in three Japanese king flukes *Paragonimus chirai, Paragonimus iloktsnensis* and *Paragonimus sadoensis* (Digenea, Troglotrematidae), *J. Parasitol.* **72**:417–433.

Andersson, L., Ryman, N., Rosenberg, R., and Stahl, G., 1981, Genetic variability in Atlantic herring (*Clupea harengus harengus*): Description of protein loci and population data, *Hereditas* **95**:69–78.

Andoh, T., and Goto, A., 1988, Biochemical evidence for reproductive isolation between the sympatric populations of *Cottus amblystomopsis* and *C. nozawae, Jpn. J. Ichthyol.* **35**:176–183.

Ankney, C. D., Dennis, D. G., Wishard, L. N., and Seeb, J. E., 1986, Low genic variation between black ducks and mallards, *Auk* **103**:701–709.

Apfelbaum, L. I., and Blauer, A., 1984, Genetic similarity between species of *Akodon* (Rodentia, Cricetidae), *J. Exp. Zool.* **229**:1–6.

Arduino, P., and Bullini, L., 1985, Reproductive isolation and genetic divergence between the small ermine moths *Yponomeuta padellus* and *Y. malinellus* (Lepidoptera: Yponomeutidae), *Memorie, Accademia Nazionale Dei Lincei, Roma,* **18**:33–61.

Ashton, R. E., Jr., Braswell, A. L., and Guttman, S. I., 1980, Electrophoretic analysis of three species of *Necturus* (Amphibia: Proteidae), and the taxonomic status of *Necturus lewisi* (Brimley), *Brimleyana* **4**:43–46.

Attard, J., and Pasteur, N., 1984, Genetic variability and differentiation in five species of crayfish (Astacidae), *Biochem. Syst. Ecol.* **12**:108–118.

Attard, J., and Vianet, R., 1985, Variabilité génétique et morphologique de cinq populations de l'écrevisse européenne *Austropotamobius pallipes* (Lereboullet, 1858) (Crustacea, Decapoda), *Can. J. Zool.* **63**:2933–2939.

Augustyn, C. J., and Grant, W. S., 1988, Biochemical and morphological systematics of *Loligo vulgaris vulgaris* Lamarck and *Loligo vulgaris reynaudii* D'Orbigny Nov. Comb. (Cephalopoda: Myopsida), *Malacologia* **29**:215–233.

Avise, J. C., and Selander, R. K., 1972, Evolutionary genetics of cave dwelling fishes of the genus *Astyanax, Evolution* **26**:1–19.

Avise, J. C., Smith, M. H., and Selander, R. K., 1974, Biochemical polymorphism and systematics in the genus *Peromyscus* VI. The *boylii* group, *J. Mammal.* **55**:751–763.

Ayala, F. J., 1975, Genetic differentiation during speciation, *Evol. Biol.* **8**:1–78.

Ayala, F. J., Tracey, M. L., Barr, L. G., McDonald, J. F., and Pérez-Salas, S., 1974, Genetic variation in natural populations of five *Drosophila* species and the hypothesis of the selective neutrality of protein polymorphisms, *Genetics* **77**:343–384.

Ayala, F. J., Valentine, J. W., Barr, L. G., and Zumwalt, G. S., 1984, Genetic variability in a temperate intertidal phoronid, *Phoronopsis viridis, Biochem. Genet.* **11**:413–427.

Baker, A. J., and Moeed, A., 1987, Rapid genetic differentiation and founder effect in colonizing populations of common mynas (*Acridotheres tristis*), *Evolution* **41**:539–546.

Baker, A. J., Honeycutt, R. L., Arnold, M. L., Sarich, V. M., and Genoways, H. H., 1981, Electrophoretic and immunological studies on the relationship of the Brachyphyllinae and Glossophaginae, *J. Mammal.* **62**:665–672.

Balletto, E., Cherchi, M. A., Salvido, S., Lattes, A., Malacrida, A., Gasperi, G., and Doria, G., 1986, Area effect in southwestern European greenfrogs (*Rana esculenta*) (Amphibia: Ranidae), *Boll. Zool.* **53**:97–110.

Barrowclough, G. F., 1980, Genetic and phenotypic differentiation in a wood warbler (genus *Dendroica*) hybrid zone, *Auk* **97**:655–658.

Barrowclough, G. F., and Corbin, K. W., 1978, Genetic variation and differentiation in the Parulidae, *Auk* **95**:691–702.

Beck, M. L., and Price, J. O., 1981, Genetic variation in the terrestrial isopod *Armadillium vulgare, J. Hered.* **72**:15–18.

Beckwitt, R., 1983, Genetic structure of *Genyonemus lineatus,* (Sciaenidae) and *Paralabrax clathratus* (Serrandiae) in southern California, *Copeia* **1983**:691–696.

Bell, L. J., Moyer, J. T., and Numach, K., 1982, Morphological and genetic variation in Japanese populations of the anenome fish (*Anaphyrion clarkii*), *Mar. Biol.* **72**:99–108.

Bellemin, J., Adest, G., and Gorman, G. C., 1978, Genetic uniformity in northern populations of *Thamnophis sirtalis* (Serpentes: Colubridae), *Copeia* **1978**:150–151.

Benharrat, K., Quignard, J.-P., and Pasteur, N., 1981, The black gobies (*Gobius niger* Linne, 1758) of the French Mediterranean coast: Variation in the enzyme polymorphism of lagoon and marine populations, *Cybium* **5**:29–34.

Bentz, B. J., and Stock, M. W., 1986, Phenetic and phylogenetic relationships among the species of *Dendroctonus* bark beetles (Coleoptera: Scolytidae), *Ann. Entomol. Soc. Am.* **79**:527–534.

Berglund, A., and Lagercrantz, V., 1983, Genetic differentiation in populations of two *Paleamon* prawn species at the Atlantic coast: Does gene flow prevent local adaptation?, *Mar. Biol.* **77**:49–58.

Berlocher, S. H., and Bush, G. L., 1982, An electrophoretic analysis of *Rhagoletis* (Diptera: Tephritidae) phylogeny, *Syst. Zool.* **31**:136–155.

Best, T. L., Sullivan, R. M., Cook, J. A., and Yates, T. L., 1986, Chromosomal, genic and morphological variation in the agile kangaroo rat, *Dipodomys agilis* (Rodentia: Heteromyidae), *Syst. Zool.* **35**:311–324.

Bezy, R. L., Gorman, G. C., Kim, Y. J., and Wright, J. W., 1977, Chromosomal and genetic divergence in the fossorial lizards of the family Anniellidae, *Syst. Zool.* **26**:57–71.

Blanc, F., and Cariou, M. L., 1980, High genetic variability of lizards of the sand-dwelling lacertid genus *Acanthodactylus, Genetica* **54**:141–147.

Blanc, F., and Cariou, M. L., 1987, Geographical variability of allozyme variation and genetic divergence between Tunisian lizards of the sand-dwelling lacertid genus *Acanthodactylus, Genetica* **72**:13–25.

Blanc, F., Ledeme, P., and Blanc, C.-P., 1986, Variation géographique de la diversité génétique chez la perdrix grise (*Perdix perdix*), *Gibier Faune Sauvage* **3**:5–41.

Bodaly, R. A., Ward, R. D., and Mills, C. A., 1989, A genetic stock study of perch, *Perca fluviatilis* L., in Windermere, *Fish. Biol.* **34**:965–967.

Bohlin, R. G., and Zimmerman, E. G., 1982, Genic differentiation of two chromosome races of the *Geomys bursarius* complex, *J. Mammal.* **63**:218–228.

Braun, M. S., and Robbins, M. B., 1986, Extensive protein similarity of the hybridizing chickadees *Parus atricapillus* and *Parus carolinensis, Auk* **103**:667–675.

Brittnacher, J. G., Sims, S. R., and Ayala, F. J., 1978, Genetic differentiation between species of the genus *Speyeria* (Lepidoptera: Nymphalidae), *Evolution* **32**:199–210.

Britton, J., and Thaler, L., 1978, Evidence for the presence of two sympatric species of mice (genus *Mus* L.) in southern France based on biochemical genetics, *Biochem. Genet.* **16**:213–225.

Britton-Davidian, J., Nadeau, J. H., Croset, H., and Thaler, L., 1989, Genic differentiation and origin of Robertsonian populations of the house mouse (*Mus musculus domesticus* Rutty), *Genet. Res.* **53**:29–44.

Brown, K., 1981, Low genetic variability and high similarities in the crayfish genera *Cambarus* and *Procambarus, Am. Midl. Nat.* **105**:225–232.

Browne, R. A., 1977, Genetic variation in island and mainland populations of *Peromyscus leucopus, Am. Midl. Nat.* **97**:1–9.

Bucklin, A., and Hedgecock, D., 1982, Biochemical genetic evidence for a third species of *Metridium* (Colenterata: Actiniara), *Mar. Biol.* **66**:1–8.

Buroker, N. E., 1982, Allozyme variation in three non-sibling *Ostrea* species, *J. Shellfish Res.* **2**:157–164.

Buroker, N. E., 1983, Population genetics of the American oyster *Crassostrea virginica* along the Atlantic coast and the Gulf of Mexico, *Mar. Biol.* **75**:99–112.

Buroker, N. E., 1983, Systematic status of two oyster populations of the genus *Triostrea* from New Zealand and Chile, *Mar. Biol.* **77**:191–200.

Buroker, N. E., Hershberger, W. K., and Chew, K. K., 1979, Population genetics of the family Ostreidae. I. Intraspecific studies of *Crassostrea gigas* and *Saccostrea commercialis, Mar. Biol.* **54**:157–169.

Buroker, N. E., Hershberger, W. K., and Chew, K. K., 1979, Population genetics of the family Ostreidae. II. Interspecific studies of the genera *Crassostrea* and *Saccostrea, Mar. Biol.* **54**:171–184.

Buth, D. G., 1979, Biochemical systematics of the cyprinid genus *Notropis.* I. The subgenus *luxilus, Biochem. Syst. Ecol.* **1977**:69–79.

Buth, D. G., 1980, Evolutionary genetics and systematic relationships in the catostomid genus, *Hypentelium, Copeia* **1980**:280–290.

Buth, D. G., and Burr, B. M., 1978, Isozyme variability in the cyprinid genus *Campostoma, Copeia* **1978**:298–311.

Buth, D. G., and Crabtree, C. B., 1982, Genetic variability and population structure of *Catostomus santaanae,* in the Santa Clara drainage, *Copeia* **1982**:439–444.

Buth, D. G., Burr, J. R., and Schenk, J. R., 1980, Electrophoretic evidence for relationships and differentiation among members of the percid subgenus *Microperca, Biochem. Syst. Ecol.* **8**:297–304.

Butlin, R. K., and Hewitt, G. M., 1987, Genetic divergence in the *Chorthippus parallelus* species group (Orthoptera: Acrididae), *Biol. J. Linn. Soc.* **31**:301–310.

Cabrera, V. M., 1983, Genetic distance and evolutionary relationships in the *Drosophila obscura* group, *Evolution* **37**:675–689.

Caccone, A., Allegrucci, G., Cesaroni, D., Cobolli Sbordoni, M., De Matthaeis, E., LaRosa, G., and Sbordoni, V., 1986, Genetic variability and divergence between cave dwelling populations of *Typhlocirolana* from Majorca and Sicily, *Biochem. Syst. Ecol.* **14**:215–221.

Calhoun, S. W., Greenbaum, I. F., and Fuxa, K. P., 1988, Biochemical and karyotypic variation in *Peromyscus maniculatus* from Western North America, *J. Mammal.* **69**:34–45.

Campton, D. E., and Utter, F. M., 1987, Genetic structure of anadromous cutthroat trout (*Salmo clarki clarki*) populations in the Puget Sound area: Evidence for restricted gene flow, *Can. J. Fish. Aquat. Sci.* **44**:573–582.

Capparella, A. P., and Lanyon, S. M., 1985, Biochemical and morphometric analysis of the sympatric, neotropical, sibling species, *Mionectes macconnelli* and *M. oleagineus,* in: *Ornithological Monographs #36: Neotropical Ornithology* (P. L. Buckley *et al.*, eds.), pp. 347–355, American Ornithologists Union, Washington, D. C.

Carlson, D. M., Kettler, M. K., Fisher, S. E., and Whitt, G. S., 1982, Low genetic variability in paddlefish populations, *Copeia* **1982**:721–725.

Case, S. M., 1978, Biochemical systematics of members of the genus *Rana* native to western North America, *Syst. Zool.* **27**:299–311.

Case, S. M., 1978, Electrophoretic variation in two species of ranid frogs *Rana boylei* and *R. muscosa, Copeia* **1978**:311–320.

Catalan, J., Poitevin, F., Fons, R., Guerasimov, S., and Croset, H., 1988, Biologie évolutive des populations ouest-européennes de crocidures (Mammalia, Insectivora). III Structure génétique des populations continentales et insulaires de *Crocidura russula* (Hermann, 1780) et de *Crocidura suaevolens* (Pallas, 1811), *Mammalia* **52**:387–400.

Chambers, S. M., 1980, Genetic divergence between populations of *Goniobasis* (Pleuroceridae) occupying different drainage systems, *Malacologia* **20:**63–82.

Chesser, R. K., 1983, Genetic variability within and among populations of the black tailed prairie dog, *Evolution* **37:**320–331.

Cianchi, R., Urbanelli, J., Coluzzi, M., and Bullini, L., 1978, Genetic distance between two sibling species of the *Aedes mariae* complex (Diptera: Culicidae), *Parassitologia* **20:**39–46.

Cianchi, R., Maini, S., and Bullini, L., 1980, Genetic distance between pheremone strains of the European corn borer, *Ostrinia nubilalis:* Different contribution of variable substrate, regulatory and non-regulatory enzymes, *Heredity* **45:**383–388.

Cockley, D. E., Gooch, J. L., and Weston, D. P., 1977, Genic diversity in cave-dwelling crickets (*Ceuthophilus gracilipes*), *Evolution* **31:**313–318.

Cothran, E. G., Zimmerman, E. G., and Nadler, C. F., 1977, Genic differentiation and evolution in the ground squirrel subgenus *Ictidomys* (Genus *Spermophilus*), *J. Mammal.* **58:**610–622.

Crabtree, B. C., and Buth, D. G., 1987, Biochemical systematics of the Catostomid genus *Catostomus:* Assessment of *C. clarki, C. plebius* and *C. discobolus* including the Zuni sucker, *C. d. yarrowi, Copeia* **1987:**843–854.

Crisp, D. J., and Ekartne, K., 1984, Polymorphism in *Pomatoceras, Zool. J. Linn. Soc.* **80:**157–175.

Crouau-Roy, B., 1986, Population studies on Pyrenean troglobitic beetles: Local genetic differentiation and microgeographic variations in natural populations, *Biochem. Syst. Ecol.* **14:**521–526.

Crouau-Roy, B., 1988, Genetic variability and differentiation in a species complex of troglobitic beetles based on isozyme data, *Biochem. Syst. Ecol.* **16:**303–310.

Crouau-Roy, B., 1989, Genetic population structure in a troglobitic beetle (*Speonomus zophosinus*), *Genetica* **78:**13–20.

Crouau-Roy, B., 1989, Population studies on an endemic troglobitic beetle: Geographical patterns of genetic variation, gene flow and genetic structure compared with morphometric data, *Genetics* **121:**571–582.

Daly, J. C., Wilkinson, P., and Shaw, D. D., 1981, Reproductive isolation in relation to allozymic and chromosomal differentiation in the grasshopper *Caledia captiva, Evolution* **35:**1164–1179.

Daugherty, C. H., Allendorf, C. H., Dunlap, W. W., and Knudsen, K. L., 1983, Systematic implications of geographic patterns of genetic variation in the genus *Dicamptodon, Copeia* **1983:**679–691.

Davis, B. J., DeMartini, E. E., and McGee, K., 1981, Gene flow among populations of a teleost (painted greenling, *Oxylebius pictus*) from Puget Sound to southern California, U.S.A., *Mar. Biol.* **65:**17–24.

Day, A. J., and Bayne, B. L., 1988, Allozyme variation in populations of the dog-whelk *Nucella lapilus* (Prosobranchia: Muricacea) from the South West peninsula of England, *Mar. Biol.* **99:**93–100.

De Matthaeis, E., Allegrucci, G., Caccone, A., Cesaroni, D., Cobolli Sbordoni, M., and Sbordoni, V., 1983, Genetic differentiation between *Penaeus kerathurus* and *P. japonicus* (Crustacea: Decapoda), *Mar. Ecol. Prog. Ser.* **12:**191–198.

Dew, R. D., and Kennedy, M. L., 1980, Genetic variation in racoons, *Procyon lotor, J. Mammal.* **61:**697–792.

Dillon, R. T., and Davis, G. M., 1981, The *Goniobasis* of southern Virginia and northwestern North Carolina, U.S.A.: Genetic and shell morphological relationships, *Malacologia* **20:**83–98.

Duggins, C. F., Jr., Karlin, A. A., and Relyea, K. G., 1983, Electrophoretic comparison of *Cyprinodon variegatus* and *Cyprinodon hubbsi,* with comments on the genus *Cyprinodon* (Atheriniformes: Cyprinodontidae), *Northeast Gulf Sci.* **6:**99–108.

Duggins, C. F., Jr., Karlin, A. A., Relyea, K. G., and Yerger, R. W., 1983, Systematics of the genus *Floridichthys, Biochem. Syst. Ecol.* **11:**283–294.

Duncan, R., and Highton, R., 1979, Genetic relationships of the eastern large *Plethodon* of the Ouachita mountains, *Copeia* **1979:**95–110.

Echelle, A. A., Echelle, A. F., and Edds, D. R., 1987, Population structure of four pupfish species (Cyprinodontidae: Cyprinodon) from the Chihuahuan desert region of New Mexico and Texas: Allozyme variation, *Copeia* **1987:**668–681.

Fairbairn, D. J., 1981, Biochemical genetic analysis of population differentiation in Greenland halibut (*Rheinhardtius hippoglossoides*) from the northwest Atlantic, Gulf of St. Lawrence and Bering sea, *Can. J. Fish. Aquat. Sci.* **38:**669–677.

Fairbairn, D. J., 1981, Which witch is which? A study of the stock structure of witch flounders (*Glyptocephalies cynoglossus*) in the Newfoundland region, *Can. J. Fish. Aquat. Sci.* **38:**782–794.

Feder, J. H., 1979, Natural hybridization and genetic divergence between the toads *Bufo boreas* and *Bufo punctatus, Evolution* **33:**1089–1097.

Feder, J. H., Wurst, G. Z., and Wake, D. B., 1975, Genetic variation in western salamanders of the genus *Plethodon* and the status of *Plethodon gordoni, Herpetologica* **34:**64–69.

Ferguson, M. M., Noakes, D. L. G., and Danzmann, R. G., 1981, Morphological and biochemical systematics of chubs, *Nocomis biguttatus* and *Nocomis micropogon,* (Pisces: Cyprinidae), in south Ontario, *Can. J. Zool.* **59:**771–775.

Ferris, S. P., Buth, D. G., and Whitt, G. S., 1982, Substantial genetic differentiation among populations of *Catostomus plebeius, Copeia* **1982:**444–449.

Fevolden, S. E., and Ayala, F. J., 1981, Enzyme polymorphism in Antarctic krill (Euphausiacea); genetic variation between populations and species, *Sarsia* **66:**167–182.

Fevolden, S. E., and Haug, T., 1988, Genetic population structure of Atlantic halibut, *Hippoglossus hippoglossus, Can. J. Fish. Aquat. Sci.* **45:**2–7.

Formas, J. R., Vera, M. I., and Lacrampe, S., 1983, Allozymic and morphological differentiation in the South American frog genus *Eupsophus, Comp. Biochem. Physiol.* **75B:**475–478.

Frydenberg, O., and Simonsen, V., 1973, Genetics of *Zoarces* populations. V. Amount of protein polymorphism and degree of genic heterozygosity, *Hereditas* **75:**221–231.

Fujio, Y., 1977, Natural hybridization between *Platichthys stellatus* and *Kareius bicoloratus, Jpn. J. Genet.* **52:**117–124.

Fujio, Y., Yuzawa, A., Kikuchi, S., and Koganezawa, A., 1986, Genetic study on the population structure of abalone, *Bull. Tohoku Regional Fish. Res. Lab.* **48:**59–65.

Fukao, F., and Okazaki, T., 1987, A study on the divergence of Japanese fishes of the genus *Neoclinus, Jpn. J. Ichthyol.* **34:**309–323.

Fuller, B., and Lester, L. J., 1980, Correlations of allozymic variation with habitat parameters using the grass shrimp, *Palaemonetes pugio, Evolution* **34:**1099–1104.

Gartside, D. F., Rogers, J. S., and Dessauer, H. C., 1977, Speciation with little genic or morphological differentiation in the ribbon snakes *Thamnophis proximus* and *Thamnophis sauritus* (Colubridae), *Copeia* **1977:**697–705.

Gavin, T. A., and May, B., 1988, Taxonomic status and genetic purity of Colombian white-tailed deer, *J. Wildl. Manage.* **52:**1–10.

Geiger, H. J., and Scholl, A., 1982, *Pontia daplidice* in Sudeuropa—eine Gruppe von zwei Arten, *Bull. Soc. Entomol. Suisse* **55:**107–114.

Gemmeke, H., 1981, Genetic differences between field mice (*Apodemus sylvaticus*) from the right and left of the Rhine river, West Germany, *Bonn. Zool. Beitr.* **32:**265–270.

Gharrett, A. J., and Thomason, M. A., 1987, Genetic changes in pink salmon (*Oncorhynchus gorbuscha*) following their introduction into the Great Lakes, *Can. J. Fish. Aquat. Sci.* **44**:787–792.

Gharrett, A. J., Shirley, S. M., and Tromble, G. R., 1987, Genetic relationships among populations of Alaskan chinook salmon (*Oncorhynchus tshawytscha*), *Can. J. Fish. Aquat. Sci.* **44**:765–774.

Gill, A. E., 1976, Genetic divergence of insular populations of deer mice, *Biochem. Genet.* **14**:835–848.

Glover, D. G., Smith, M. H., Ames, L., Joule, J., and Dubach, J. M., 1977, Genetic variation in pika populations, *Can. J. Zool.* **55**:1841–1845.

Gonzalez, A. M., Cabrera, V. M., Larruga, J. M., and Gullon, A., 1982, Genetic distances in the sibling species *Drosophila melanogaster, Drosophila simulans* and *Drosophila mauritiana, Evolution* **36**:517–522.

Goodfellow, W. L., Hocutt, C. H., Morgan, R. P. II, and Stauffer, J. R., Jr., 1984, Biochemical assessment of the taxonomic status of *Rhinichthys bowersi* (Pisces: Cyprinidae), *Copeia* **1984**:652–659.

Gorman, G. C., and Kim, Y. J., 1976, *Anolis* lizards of the Eastern Carribean: A case study in evolution. II. Genetic relationships and genetic variation of the *bimaculatus* group, *Syst. Zool.* **25**:62–77.

Gorman, G. C., and Kim, Y. J., 1977, Genotypic evolution in the face of phenotypic conservativeness: *Abudefduf* (Pomacentridae) from the Atlantic and Pacific sides of Panama, *Copeia* **1977**:694–697.

Gorman, G. C., Soulé, M., Yang, S. Y., and Nevo, E., 1975, Evolutionary genetics of insular Adriatic lizards, *Evolution* **29**:52–71.

Gorman, G. C., Kim, Y. J., and Rubinoff, R., 1976, Genetic relationships of three species of *Bathygobius* from the Atlantic and Pacific sides of Panama, *Copeia* **1976**:361–364.

Gorman, G. C., Kim, Y. S., and Taylor, L. E., 1977, Genetic variation in irradiated and control populations of *Cnemidophorus tigris* (Sauria, Teiidae) from Mercury, Nevada with a discussion of genetic variability in lizards, *Theor. Appl. Genet.* **49**:9–14.

Gorman, G. C., Buth, D. G., Soulé, M., and Yang, S. Y., 1980, The relationships of the *Anolis cristatellus* species group; electrophoretic analysis, *J. Herpetol.* **14**:269–278.

Gould, S. J., and Woodruff, P. S., 1978, Natural history of *Cerion*. VIII: Little Bahama bank—A revision based on genetics, morphometrics and geographic distribution, *Bull. Mus. Comp. Zool.* **148**:371–415.

Graf, J. D., 1982, Biochemical genetics, zoogeography and taxonomy in the family Arvicolidae, (Mammalia: Rodentia), *Rev. Suisse Zool.* **89**:749–788.

Grant, W. S., 1985, Biochemical genetic stock structure of the southern African anchovy, *Engraulis capensis* Gilchrist, *J. Fish Biol.* **27**:23–29.

Grant, W. S., 1986, Biochemical genetic divergence between Atlantic, *Clupea harengus,* and Pacific, *C. pallasi,* herring, *Copeia* **1986**:714–719.

Grant, W. S., and Cherry, M. I., 1985, *Mytilus galloprovincialis* Lmk. in Southern Africa, *J. Exp. Mar. Biol. Ecol.* **90**:179–191.

Grant, W. S., and Ståhl, G., 1988, Evolution of Atlantic and Pacific cod: Loss of genetic variation and gene expression in Pacific cod, *Evolution* **42**:138–146.

Grant, W. S., and Ståhl, G., 1988, Description of electrophoretic loci in Atlantic cod, *Gadus morhua,* and comparison with Pacific cod, *Gadus macrocephalus, Hereditas* **108**:27–36.

Grant, W. S., and Utter, F. M., 1980, Biochemical genetic variation in walleye pollock, *Theragra chalcogramma:* Population structure in the southeastern Bering sea and the Gulf of Alaska, *Can. J. Fish. Aquat. Sci.* **37**:1093–1100.

Grant, W. S., Milner, G. B., Krasnowski, P., and Utter, F. M., 1980, Use of biochemical genetic variants for identification of sockeye salmon (*Oncorhynchus nerka*) stocks in Cook Inlet, Alaska, *Can. J. Fish. Aquat. Sci.* **37**:1236–1247.

Grant, W. S., Bakkala, R., Utter, F. M., Teel, D. J., and Kobayashi, T., 1983, Biochemical genetic population structure of yellowfish sole, *Limandra aspersa* of the North Pacific Ocean and Bering Sea, *Fish. Bull.* **81**:667–676.

Grant, W. S., Teel, D. J., Kobayashi, T., and Schmitt, C., 1984, Biochemical population genetics of Pacific halibut (*Hippoglossus stenolepis*) and comparison with Atlantic halibut (*Hippoglossus hippoglossus*), *Can. J. Fish. Aquat. Sci.* **41**:1083–1088.

Grant, W. S., Zhang, C. I., Kobiyashi, T., and Ståhl, G., 1986, Lack of genetic stock discretion in Pacific cod (*Gadus macrocephalus*), *Can. J. Fish. Aquat. Sci.* **44**:490–498.

Grant, W. S., Becker, I. I., and Leslie, R. W., 1987, Genetic stock structure of the southern African hakes *Merluccius capensis* and *M. paradoxus, Mar. Ecol. Prog. Ser.* **41**:9–20.

Grant, W. S., Becker, I. I., and Leslie, R. W., 1988, Evolutionary divergence between sympatric species of southern African hakes, *Merluccius capensis* and *M. paradoxus*. I. Electrophoretic analysis of proteins, *Heredity* **61**:13–20.

Green, D. M., 1983, Allozymic variation through a clinal hybrid zone between the toads *Bufo americanus* and *Bufo hemiophrys* in southeastern Manitoba, (Canada), *Herpetologica* **39**:28–40.

Greenbaum, I. F., and Baker, R. J., 1976, Evolutionary relationships in *Macrotus* (Chiroptera): Biochemical variation and karyology, *Syst. Zool.* **25**:15–25.

Grudzien, T. A., and Turner, B. J., 1983, Biochemical systematics of *Allodichthys:* A genus of goodeid fishes, *Biochem. Syst. Ecol.* **11**:383–388.

Grudzien, T. A., and Turner, B. J., 1984, Genetic identity and geographic differentiation of trophically dichotomous *Ilyodon* (Teleostei: Goodeidae), *Copeia* **1984**:102–107.

Gutierrez, R. S., Zink, R. M., and Yang, S. Y., 1983, Genic variation, systematic and biogeographic relationships of some galliform birds, *Auk* **100**:33–47.

Guttman, S. I., and Karlin, K. A., 1986, Hybridization of cryptic species of two-lined salamanders (*Eurycea bislineata*) complex. *Copeia* **1986**:96–108.

Guttman, S. I., Grau, G. H., and Karlin, K. A., 1980, Genetic variation in Lake Erie great blue herons (*Ardea herodias*), *Comp. Biochem. Physiol.* **66B**:167–169.

Guttman, S. I., Wood, T. K., and Karlin, K. A., 1981, Genetic differentiation along host plant lines in the sympatric *Enchenopa binotata* Say complex (Homoptera: Membracidae), *Evolution* **35**:205–217.

Hafner, M. S., Hafner, J. C., Patton, J. L., and Smith, M. F., 1987, Macrogeographic patterns of genetic differentiation in the pocket gopher *Thomomys umbrinus, Syst. Zool.* **36**:18–34.

Hafner, D. J., Petersen, K. E., and Yates, T. L., 1981, Evolutionary relationships of jumping mice (genus *Zapus*) of the southwestern U.S.A., *J. Mammal.* **62**:501–512.

Hafner, D. S., and Yates, T. L., 1983, Systematic status of the mojave ground squirrel, *Spermophilus mohavensis* (subgenus Xenospermophilus), *J. Mammal.* **64**:397–404.

Halliday, R. B., Bouton, N. H., and Hewitt, G. M., 1983, Electrophoretic analysis of a chromosomal hybrid zone in the grasshopper *Podisina pedestris, Biol. J. Linn. Soc.* **19**:51–62.

Hamilton, M. J., and Kennedy, M. L., 1986, Genic variation in the coyote, *Canis latrans,* in Tennessee, U.S.A., *Genetica* **71**:167–173.

Hanken, J., 1983, Genetic variation in a dwarfed lineage, the Mexican salamander lineage, genus *Thorius* (Amphibia: Plethodontidae), taxonomic, ecological and evolutionary implications, *Copeia* **1983**:1051–1073.

Hanken, J., and Wake, D. B., 1982, Genetic differentiation among plethodontid salamanders, (genus *Bolitoglossa*) in central and south America: Implications for the South American invasion, *Herpetologica* **38**:272–287.

Hanlin, H. G., 1980, Geographic variation in Dunn's salamander, *Plethodon dunni* Bishop (Amphibia: Caudata: Plethodontidae), Ph.D. thesis, Oregon State University.

Harrison, R. G., 1979, Speciation in North American field crickets: Evidence from electrophoretic comparisons, *Evolution* **33:**1009–1023.

Hartl, G. B., 1986, Steinbock und Gemse in Alpenraum—genetische Variabilitat und biochemische Differenzierung zweischen den Arten, *Z. Zool. Syst. Evolutionsforsch.* **24:**315–320.

Hartl, G. B., and Reimoser, F., 1988, Biochemical variation in roe deer (*Capreolus capreolus* L.): Are *r*-strategists among deer genetically less variable than *K*-strategists?, *Heredity* **60:**221–227.

Hartl, G. B., Willing, R., Grillitsch, M., and Klansek, E., 1988, Biochemical variation in Mustelidae: Are carnivores genetically less variable than other mammals?, *Zool. Anz.* **221:**81–90.

Hatfield, J. S., Wissing, T. E., Guttman, S. I., and Farrell, M. P., 1982, Electrophoretic analysis of gizzard shad (*Dorosoma cepedianum*) from the lower Mississippi river and Ohio, (U.S.A.), *Trans. Am. Fish. Soc.* **111:**742–748.

Hedgecock, D., 1979, Biochemical genetic variation and evidence of speciation in *Chthalamus* barnacles of the tropical eastern Pacific Ocean, *Mar. Biol.* **54:**207–214.

Hedgecock, D., and Ayala, F. J., 1974, Evolutionary divergence in the genus *Taricha.* (Salamandridae), *Copeia* **1974:**738–746.

Hedgecock, D., Nelson, K., Simons, J., and Shleser, R., 1977, Genic similarity of American and European species of the lobster *Homarus, Biol. Bull.* **152:**41–50.

Herd, R. M., and Brockfenton, M., 1984, An electrophoretic, morphological and ecological investigation of a putative hybrid zone between *Myotis lucifugus* and *Myotis yumanensis* (Chiroptera: Vespertillionidae), *Can. J. Zool.* **61:**2029–2050.

Hertz, P. E., and Zouros, E., 1982, Genetic variability in two West Indian anoles (Reptilia, Iguanidae): Relation to field thermal biology, *J. Zool.* **196:**499–518.

Higby, P. K., and Stock, M. W., 1982, Genetic relationships between two sibling species of bark beetle (Coleoptera: Scolytidae), Jeffrey pine beetle (*Dendroctonus jeffreyi*) and mountain pine beetle (*D. ponderosae*), in northern California, U.S.A., *Ann. Entomol. Soc. Am.* **75:**668–674.

Highton, R., 1984, A new species of woodland salamander of the *Plethodon glutinosus* group from the southern Appalachian mountains, *Brimleyana* **9:**1–20.

Highton, R., and MacGregor, J. R., 1983, *Plethodon kentucki:* A valid species of Cumberland (Kentucky, USA) plateau woodland salamander, *Herpetologica* **39:**189–200.

Highton, R., and Webster, T. P., 1976, Geographic protein variation and divergence in populations of the salamander *Plethodon cinereus, Evolution* **30:**33–45.

Hilburn, L. R., 1980, Population genetics of *Chironomus stigmaterus* Say (Diptera: Chironomidae): II. Protein variability in populations of the south-western United States, *Evolution* **34:**696–704.

Hilburn, L. R., and Sattler, P. W., 1986, Electrophoretically detectable protein variation in natural populations of the lone star tick, *Amblyomma americanum* (Acari: Ixodidae), *Heredity* **56:**67–74.

Hill, A., Johnson, M. S., and Merifield, H., 1983, An electrophoretic and morphological examination of *Bothriembryon kendricki,* new species (Pulmonata: Bulimulidae), a new species previously considered conspecific with *B. bulla* (Menke), *Aust. J. Zool.* **31:**227–242.

Hoagland, K. E., 1984, Use of molecular genetics to distinguish species of the gastropod genus *Crepidula* (Prosobranchia: Calptraeidae), *Malacologia* **25:**607–628.

Hoagland, K. E., 1986, Genetic variation in seven wood-boring teredinid and pholadid bivalves with different patterns of life history and dispersal, *Malacologia* **27:**323–339.

Hornbach, D. J., McLeod, M. J., and Guttman, S. I., 1980, On the validity of the genus *Musculium* (Bivalvia: Spaeriidae); electrophoretic evidence, *Can. J. Zool.* **58:**1703–1707.

Hornbach, D. J., McLeod, M. J., Guttman, S. I., and Seilkop, S. K., 1980, Genetic and morphological variation in the freshwater clam *Sphaerium* (Bilvalvia: Sphaeriidae), *J. Molluscan. Stud.* **46:**158–170.

Howard, D. J., 1982, Speciation and coexistence in a group of closely related ground crickets, Ph.D. dissertation, Yale University, New Haven, Connecticut.

Howard, D. J., 1983, Electrophoretic survey of eastern North American *Allonemobius* (Orthoptera: Gryllidae): Evolutionary relationships and the discovery of three new species, *Ann. Entomol. Soc. Am.* **76:**1014–1021.

Howard, J. H., and Wallace, R. L., 1981, Microgeographical variation of electrophoretic loci in populations of *Ambystoma macrodactylum columbionum, Copeia* **1981:**466–471.

Howard, J. H., Wallace, R. L., and Larsen, J. M., 1983, Genetic variation and population divergence in the louch mountain salamander, (*Plethodon lovselli*), *Herpetologica* **29:**41–47.

Jacobs, J. F., 1987, A preliminary investigation of geographic genetic variation and systematics of the two-lined salamander, *Eurycea bislineata* (green), *Herpetologica* **43:**423–446.

Johnson, G. L., and Packard, R. L., 1974, Electrophoretic analysis of *Peromyscus comanche,* Blair, with comments on its systematic status, *Occ. Pap. Mus. Tex. Tech. Univ.* **24:**1–16.

Johnson, M. S., 1975, Biochemical systematics of the atherinid genus *Menidia, Copeia* **1975:**662–691.

Johnson, M. S., Murray, J., and Clarke, B. C., 1986, An electrophoretic analysis of phylogeny and evolutionary rates in the genus *Partula* from the Society Islands (French Polynesia), *Proc. R. Soc. Lond.* **227:**161–177.

Johnson, M. S., Murray, J., and Clarke, B. C., 1986, Allozyme similarities among species of *Partula* on Moorea (French Polynesia), *Heredity* **56:**319–328.

Johnson, N. K., and Zink, R. M., 1983, Speciation in sapsuckers (*Sphyrapicus*). I: Genetic differentiation, *Auk* **100:**871–884.

Johnson, W. E., and Selander, R. K., 1971, Protein variation and systematics in kangaroo rats (genus *Dipodomys*), *Syst. Zool.* **20:**377–405.

Johnson, W. E., Selander, R. K., Smith, M. H., and Kim, Y. J., 1972, Biochemical genetics of sibling species of the cotton rat (*Sigmodon*), *Stud. Genet. VII Univ. Tex. Pub.* **7213:**297–305.

Johnston, P. G., Potter, I. C., and Robinson, E. S., 1987, Electrophoretic analysis of populations of the southern hemisphere lampreys *Geotria australis* and *Mordacia mordax, Genetica* **74:**113–117.

Kalezic, M. L., and Tucic, N., 1984, Genetic diversity and population genetic structure of *Triturus vulgaris* (Urodela, Salamandridae), *Evolution* **38:**389–401.

Karlin, A. A., and Guttman, S. I., 1986, Systematics and geographic isozyme variation in the plethodontid salamander *Desmognathus fuscus* (Rafinesque), *Herpetologica* **42:**282–301.

Kartavtsev, Y. P., and Mamontov, A. M., 1983, Electrophoretic study of protein variability and similarity of *Coregonus autumnalis,* two forms of whitefish (Coregonidae) and greyling (Thymallidae) from the Baikal Lake, *Genetika* **19:**1895–1902.

Kartavtsev, Y. F., Gulbukovskii, M. K., and Chereshnev, I. A., 1983, Genetic differentiation and the level of variability of two sympatric charr species (*Salvelinus:* Salmonidae) from Chukat National Okrug (USSR), *Genetika* **19:**584–593.

Kawamoto, Y., and Ischak, T. M., 1981, Genetic differentiation of the Indonesian crab-eating macaque (*Macaca fascicularis*): I. Preliminary report on blood protein, *Primates* **22:**237–252.

Kawamoto, Y., Shotake, T., and Nozawa, K., 1982, Genetic differentiation among three genera of the family Cercopithecidae, *Primates* **23**:272–286.

Kijima, A., Moni, K., and Fujio, Y., 1984, Population differences in gene frequency of the Japanese scallop *Patinopecten yessoensis* on the Okhotsk Sea coast of Hokkaido (Japan), *Bull. Jpn. Soc. Sci. Fish.* **50**:241–248.

Kijima, A., Taniguchi, N., Makino, H., and Ochai, A., 1985, Degree of genetic divergence and breeding structure in jack mackerel, *Trachurus japonicus, Rep. Usa Mar. Biol. Inst. Kochi Univ.* **1985**:49–60.

Kijima, A., Taniguchi, N., and Ochiai, A., 1986, Genetic divergence and morphological difference between the spotted mackerel and the common mackerel, *Jpn. J. Ichythol.* **33**:151–161.

Kilpatrick, C. W., 1984, Molecular evolution of the Texas mouse, *Peromyscus attwateri,* in: *Festschrift for Walter W. Dalquest in Honour of his 65th Birthday* (N. V. Horner, ed.), Department of Biology, Midwestern State University, Witchita Falls, Texas, pp. 87–96.

Kilpatrick, C. W., and Crowell, K. L., 1985, Genic variation of the rock vole, *Microtus chrotorrhinus, J. Mammal.* **66**:94–101.

Kilpatrick, C. W., and Zimmerman, E. G., 1975, Genetic variation and systematics of four species of mice of the *Peromyscus boylii* species group, *Syst. Zool.* **24**:143–162.

Kilpatrick, C. W., and Zimmerman, E. G., 1976, Biochemical variation and systematics of *Peromyscus pectoralis, J. Mammal.* **57**:506–522.

Kim, Y. J., Gorman, G. C., Papenfuss, T., and Roychoudhury, A. K., 1976, Genetic relationships and genetic variation in the amphisbaenian genus *Bipes, Copeia* **1976**:120–124.

Kimura, M., and Yamamoto, S., 1982, Protein polymorphism in a feral population of the pigeon *Columba livia domestica, Anim. Blood Groups Biochem. Genet.* **13**:299–304.

Kimura, M., Oniwa, K., and Shinichi, I., 1984, Protein polymorphism in two populations of the wild quail *Coturnix coturnix japonica, Anim. Blood Groups Biochem. Genet.* **15**:13–22.

King, M.-C., and Wilson, A. C., 1975, Evolution at two levels in humans and chimpanzees, *Science* **188**:107–116.

Kirkpatrick, M., and Selander, R. K., 1979, Genetics of speciation in lake white fishes in the Allegash Basin, *Evolution* **33**:478–485.

Knight, A. J., Hughes, R. N., and Ward, R. D., 1987, A striking example of the founder effect in the mollusc *Littorina saxatilis, Biol. J. Linn. Soc.* **32**:417–426.

Kornfield, I. L., Ritte, V., Richler, C., and Wahrman, S., 1979, Biochemical and cytological differentiation among cichlid fishes of the Sea of Galilee, *Evolution* **33**:1–14.

Kornfield, I. L., Beland, K. F., Moring, J. R., and Kircheis, F. W., 1981, Genetic similarity among endemic Arctic char (*Salvelinus alpinus*), and implications for their management, *Can. J. Fish. Aquat. Sci.* **38**:32–39.

Kovacic, D. A., and Guttman, S. I., 1979, Electrophoretic comparison of genetic variability between eastern and western populations of the opossum (*Didelphis virginiana*), *Am. Midl. Nat.* **101**:269–277.

Krepp, S. R., and Smith, M. H., 1974, Genic heterozygosity in the 13-year cicada, *Magicicada, Evolution* **28**:396–401.

Kuhl, S., and Schneppenheim, R., 1986, Electrophoretic investigation of genetic variation in two krill species *Euphausia superba* and *Euphausia crystallorophias* (Euphaisiidae), *Polar Biol.* **6**:17–23.

Lambert, D. M., 1983, A population genetic study of the African mosquito *Anopheles marshallii* (Theobold), *Evolution* **37**:484–495.

Larson, A., 1980, Paedomorphosis in relation to rates of morphological and molecular evolution in the salamander *Aneides flavipunctatus* (Amphibia, Plethodontidae), *Evolution* **34**:1–17.

Larson, A., 1983, A molecular phylogenetic perspective on the origins of a lowland tropical salamander fauna: I. Phylogenetic influences from protein comparisons, *Herpetologica* **39**:85–99.

Larson, A., and Highton, R., 1978, Geographic protein variation and divergence in the salamanders of the *Plethodon welleri* group (Amphibia, Plethodontidae), *Syst. Zool.* **27**:431–448.

Lawson, R., and Dessauer, H. C., 1979, Biochemical genetics and systematics: Garter snakes of the *Thamnophis elegans, Thamnophis couchii* and *Thamnophis ordinoides* complex, *Occ. Pap. Mus. Zool. Louisiana* **56**:1–24.

Leary, R., and Booke, H. E., 1982, Genetic stock analysis of yellow perch from Green Bay and Lake Michigan, U.S.A., *Trans. Am. Fish. Soc.* **111**:52–57.

Lees, J. H., and Ward, R. D., 1987, Genetic variation and biochemical systematics of British Nemouridae, *Biochem. Syst. Ecol.* **15**:117–125.

Lemus, L. G. L., 1988, Biochemical systematic analysis of evolutionary relationships of groupers from the Gulf of California, *Biochem. Syst. Ecol.* **16**:79–87.

Levenson, H., Hoffmann, R. S., Nadler, C. S., Deutsch, L., and Freeman, S. D., 1985, Systematics of the holarctic chipmunks (*Tamias*), *J. Mammal.* **66**:219–242.

Liebherr, J. K., 1986, Comparison of genetic variation in two carabid beetles (Coleoptera) of differing vagility, *Ann. Entomol. Soc.* **79**:424–433.

Loudenslager, E. J., and Gall, G. A. E., 1980, Geographic patterns of protein variation and subspeciation in cutthroat trout, *Salmo clarki, Syst. Zool.* **29**:27–42.

Loudenslager, E. J., Rinne, J. N., Gall, G. A. E., and David, R. E., 1986, Biochemical genetic studies of native Arizona and New Mexico trout, *Southwest. Nat.* **31**:221–234.

Lynch, J. F., Yang, S. Y., and Papenfuss, T. J., 1977, Studies of neotropical salamanders of the genus *Pseudoeurycea.* 1: Systematic status of *Pseudoeurycea unguidentis, Herpetologica* **33**:46–52.

Lynch, J. F., Wake, D. B., and Yang, S. H., 1983, Genic and morphological differentiation in Mexican *Pseudoeurycea* (Caudata: Plethodontidae) with a description of a new species, *Copeia* **1983**:884–894.

Magnusson, K. P., and Ferguson, M. M., 1987, Genetic analysis of four sympatric morphs of Arctic charr, *Salvelinus alpinus,* from Thingvallavatn, Iceland, *Environ. Biol. Fishes* **20**:67–73.

Manlove, M. N., Baccus, R., Pelton, M. R., Smith, M. H., and Grabster, D., 1978, Biochemical variation in the black bear, in: *Bears—Their Biology and Management* (C. J. Montuika and K. L. McArthur, eds.), pp. 37–41, U. S. Government Printing Office, Washington, D.C.

Marten, J. A., and Johnson, N. K., 1986, Genetic relationships of North American cardueline finches, *Condor* **88**:409–420.

Mascarello, J. T., 1978, Chromosomal, biochemical, mensural, penile and cranial variation in desert woodrats (*Neotoma lepida*), *J. Mammal.* **59**:477–495.

Matthews, T. C., 1975, Biochemical polymorphism in populations of the Argentine toad, *Bufo arenarum, Copeia* **1975**:454–465.

Matthews, T. C., and Craig, G. B., 1980, Genetic heterozygosity in natural populations of the tree-hole mosquito *Aedes triseriatus, Ann. Entomol. Soc. Am.* **73**:739–743.

Matthews, T. C., and Munstermann, L. E., 1983, Genetic diversity and differentiation in northern populations of the tree-hole mosquito *Aedes hendersoni* (Diptera: Culicidae), *Ann. Entomol. Soc. Am.* **76**:1005–1010.

Mayer, Von, W., 1981, Electrophoretic investigations on European species of the genus *Lacerta* and *Podarcis.* III. *Podarcis tiliguerta*—Species or subspecies?, *Zool. Anz.* **207**:151–157.

Mayer, W., and Tiedemann, F., 1981, Electrophoretic investigations on European species of the

genus *Lacerta* and *Podarcis*. II. To the systematic status of the lizards of the island Piperi (N. Sporades, Greece), *Zool. Anz.* **207**:143–150.
McCleod, M. J., Wynes, D. L., and Guttman, S. I., 1980, Lack of biochemical evidence for hybridization between two species of darters, *Comp. Biochem. Physiol.* **67B**:323–325.
McKaye, K. R., Kocher, T., Reinthal, P., and Kornfield, I., 1982, A sympatric sibling species complex of *Petrotilapia trewavas* from Lake Malawi analysed by enzyme electrophoresis (Pisces: Cichlidae), *Zool. J. Linn. Soc.* **76**:91–96.
McKinney, C. O., Selander, R. K., Johnson, W. E., and Yang, A. S., 1972, Genetic variation in the side-blotched lizard, *Uta stansburiana, Stud. Genet. VII Univ. Tex. Pub.* **7213**:307–318.
Medwaldt, W. T., and Jenkins, S. H., 1986, Genetic variation of woodrats (*Neotoma cinerea*) and deer mice (*Peromyscus maniculatus*) on montane habitat islands in the Great Basin, *Great Basin Nat.* **46**:577–580.
Melnick, D. J., Jolly, C. J., and Kidd, K. K., 1986, Genetics of a wild population of rhesus monkeys (*Macaca mulatta*): II. The Dunga Gali population in a species-wide perspective, *Am. J. Phys. Anthropol.* **71**:129–140.
Meredith, M., Meredith, A., and Sin, F. Y. T., 1988, Genetic variation of four populations of the Little Blue Penguin, *Eudyptula minor, Heredity* **60**:69–76.
Merrit, R. B., Rogers, J. F., and Kurz, B. J., 1978, Genic variability in the longnose dace, *Rhinichthys cataractae, Evolution* **32**:116–124.
Millet, M. C., Britton-Davidison, J., and Orsini, P., 1982, Comparative biochemical genetics of *Macrotus cabrerae* and three other species of Mediterranean Arvicolidae, *Mammalia* **46**:381–388.
Miyamoto, M. M., Hayes, M. P., and Tennant, M. R., 1986, Biochemical and morphological variation in Floridian populations of the bark anole (*Anolis distichus*), *Copeia* **1986**:76–86.
Morafka, D. J., and Adest, G. A., 1984, Biochemical, behavioral and body size differences between *Rana aurora aurora* and *R. a. draytoni, Copeia* **1984**:1018–1022.
Moran, C., Wilkinson, P., and Shaw, D. D., 1980, Allozyme variation across a narrow hybrid zone in the grasshopper, *Caledia captiva, Heredity* **44**:69–81.
Morgan, R. P., Block, S. B., Vlanowicz, N. I., and Buys, C., 1978, Genetic variation in the soft-shelled clam *Mya arenaria, Estuaries* **1**:255–258.
Mork, J., and Fris-Sorensen, E., 1983, Genetic varaition in capelin *Mallotus villosus* from Norwegian waters, *Mar. Ecol. Prog. Ser.* **12**:199–206.
Mulley, J. C., and Latter, B. D. H., 1980, Genetic variation and evolutionary relationships within a group of thirteen species of Penaeid prawns, *Evolution* **34**:904–916.
Murphy, R. W., Cooper, W. E., Jr., and Richardson, W. S., 1983, Phylogenetic relationships of the North America five lined skinks, genus *Eumeces* (Sauria:Iguanidae), *Herpetologica* **39**:200–211.
Murphy, R. W., McCollum, F. C., Gorman, G. C., and Thomas, R., 1984, Genetics of hybridizing populations of Puerto Rican *Sphaenodactylis, J. Herpetol.* **18**:93–105.
Nascetti, G., Paggi, L., Orecchia, P., Smith, J. W., Mattiucci, S., and Bullini, L., 1986, Electrophoretic studies on the *Anisakis simplex* complex (Ascaridida: Anisakidae) from the Mediterranean and North Atlantic, *Int. J. Parasitol.* **16**:633–640.
Nemeth, S. T., and Tracey, L. T., 1979, Allozyme variability and relatedness in six crayfish species, *J. Hered.* **70**:37–43.
Nevo, E., 1981, Genetic variation and climatic selection in the lizard *Agama stellio* in Israel and Sinai, *Theor. Appl. Genet.* **60**:369–380.
Nevo, E., and Shaw, C. R., 1972, Genetic variation in a subterranean mammal, *Spalax ehrenbergi, Biochem. Genet.* **7**:235–241.
Nevo, E., and Yang, S. Y., 1979, Genetic diversity and climatic determinants of tree frogs in Israel, *Oecologia* **41**:47–63.

Nevo, E., and Yang, S. Y., 1982, Genetic diversity and ecological relationships of marsh frog populations in Israel, *Theor. Appl. Genet.* **63**:317–330.

Nevo, E., Kim, Y. S., Shaw, C. R., and Thaeler, Jr., C. S., 1974, Genetic variation, selection and speciation in *Thomomys talpoides* pocket gophers, *Evolution* **28**:1–23.

Nevo, E., Dessauer, H. C., and Chuang, K. C., 1975, Genetic variation as a test of natural selection, *Proc. Natl. Acad. Sci. USA* **72**:2145–2149.

Nevo, E., Bar-El, C., Bar, Z., and Beiles, A., 1981, Genetic structure and climatic correlates of desert landsnails, *Oecologia* **48**:199–208.

Nicklas, N. L., and Hoffmann, R. J., 1979, Genetic similarity between two morphologically similar species of Polychaetes, *Mar. Biol.* **52**:53–59.

Nottebohm, F., and Selander, R. K., 1972, Vocal dialects and gene frequencies in the chingolo sparrow (*Zonotrichia capensis*), *Condor* **74**:137–143.

Nozawa, K., Shotake, T., Ohkura, Y., and Tanabe, Y., 1977, Genetic variation within and between species of Asian macaques, *Jpn. J. Genet.* **52**:15–30.

O'Brien, S. J., Martenson, J. S., Packer, C., Herbst, L., de Vos, V., Joslin, P., Ott-Joslin, J., Wildt, D. E., and Bush, M., 1987, Biochemical genetic variation in geographical isolates of African and Asiatic lions, *Natl. Geogr. Res.* **3**:114–124.

Okazaki, T., 1982, Genetic study on population structure in chum salmon (*Oncorhynchus keta*), *Bull. Far Seas Fish. Res.* **19**:25–116.

Oniwa, K., and Kimura, M., 1986, Genetic variability and relationships in six snail species of the genus *Semisulcospira, Jpn. J. Genet.* **61**:503–514.

Pashley, D. P., 1983, Biosystematic study in Tontricidae (Lepidoptera) with a note on evolutionary rates of allozymes, *Ann. Entomol. Soc. Am.* **76**:139–148.

Pashley, D. P., and Johnson, S. J., 1986, Genetic population structure of migratory moths: The velvetbean caterpillar (Lepidoptera: Noctuidae), *Ann. Entomol. Soc. Am.* **79**:26–30.

Pashley, D. P., Johnson, S. J., and Sparks, A. N., 1985, Genetic population structure of migrating moths: The fall armyworm (Lepidoptera: Noctuidae), *Ann. Entomol. Soc. Am.* **78**:756–762.

Patton, J. C., and Avise, J. C., 1985, Evolutionary genetics of birds: IV. Rates of protein divergence in waterfowl (Anatidae), *Genetica* **68**:129–144.

Patton, J. L., Selander, R. K., and Smith, M. H., 1972, Genic variation in hybridizing populations of gophers (genus *Thomomys*), *Syst. Zool.* **21**:263–270.

Patton, J. L., Yang, S. Y., and Myers, P., 1975, Genetic and morphological divergence among introduced rat populations (*Rattus rattus*) of the Galapagos archipelago, Ecuador, *Syst. Zool.* **24**:296–310.

Patton, J. L., Macarthur, M., and Yang, S. Y., 1976, Systematic relationships of the four-toed populations of *Dipodomys heermanni, J. Mammal.* **59**:159–163.

Patton, J. L., Sherwood, S. W., and Yang, S. Y., 1981, Biochemical systematics of chaetodipine pocket mice, genus *Perognathus, J. Mammal.* **62**:477–492.

Pemberton, J. M., and Smith, R. H., 1986, Lack of biochemical polymorphism in British fallow deer, *Heredity* **55**:199–204.

Philipp, D. P., Childers, W. F., and Whitt, G. S., 1982, Biochemical Genetics of Large Mouth Bass, Final Report, Research project 1063-1, Electric Power Research Institute, Palo Alto, California.

Picariello, O., and Scillitani, G., 1988, Genetic distances between the populations of *Cyrtodactylus kotschyi* (Squamata: Gekkonidae) from Apulia and Greece, *Amphibia-Reptilia* **9**:245–250.

Pope, M., and Highton, R., 1980, Geographic genetic variation in the Sacramento mountain salamander (*Aneides hardii*), *J. Herpetol.* **14**:343–346.

Prakash, S., 1977, Genetic divergence in closely related sibling species *Drosophila pseudoobscura, Drosophila persimilis,* and *Drosophila miranda, Evolution* **31:**14–23.

Present, T. M. C., 1987, Genetic differentiation of disjunct Gulf of California and Pacific outer coast populations of *Hypsoblennius jenkinsi, Copeia* **1987:**1010–1024.

Rafinski, J., and Arntzen, J. W., 1987, Biochemical systematics of the Old World newts, genus *Triturus:* Allozyme data, *Herpetologica* **43:**446–457.

Ramsey, P. R., and Wakeman, J. M., 1987, Population structure of *Sciaenops ocellatus* and *Cynoscion nebulosus* (Pisces: Sciaenidae): Biochemical variation, genetic subdivision and dispersal, *Copeia* **1987:**682–695.

Ramsey, P. R., Avise, J. C., Smith, M. H., and Urbston, D. F., 1979, Biochemical variation and genetic heterogeneity in South Carolina deer populations, *J. Wildl. Manage.* **43:**136–142.

Renaud, C. B., Speers, L. J., Qadri, S. V., and McAllister, D. Z., 1986, Biochemical evidence of speciation in the cod genus *Gadus, Can. J. Zool.* **64:**1563–1566.

Rennert, P. D., and Kilpatrick, C. W., 1987, Biochemical systematics of *Peromyscus boylii.* II. Chromosomally variable populations from eastern and southern Mexico, *J. Mammal.* **68:**799–811.

Rice, M. C., Gardner, M. B., and O'Brien, S. J., 1980, Genetic diversity in leukemia-prone feral house mice infected with murine leukemia virus, *Biochem. Genet.* **18:**915–928.

Richardson, B. J., Rogers, P. M., and Hewitt, G. M., 1980, Ecological genetics of the wild rabbit, north Australia. II. Protein variation in British, French and Australian rabbits and the geographical distribution of the variation in Australia, *Aust. J. Biol. Sci.* **33:**371–383.

Ritte, V., and Pashtan, A., 1982, Extreme levels of genetic variability in two Red Sea *Cerithium* species (Gastropoda: Cerithidae), *Evolution* **36:**403–407.

Roed, K. H., 1986, Genetic variability in Norwegian wild reindeer (*Rangifer tarandus*), *Hereditas* **104:**293–298.

Ryman, N., Reuterwall, C., Nygren, K., and Nygren, T., 1980, Genetic variation and differentiation in Scandinavian moose (*Alces alces*): Are large mammals monomorphic?, *Evolution* **34:**1039–1049.

Saavedra, C., Zapata, C., Guerra, A., and Alvarez, G., 1987, Genetic structure of populations of flat oyster (*Ostrea edulis* [Linneo, 1758]) from the NW of the Iberian penninsula, *Invest. Pesq.* **51:**225–241.

Sage, R. D., and Selander, R. K., 1975, Trophic radiation through polymorphism in cichlid fishes, *Proc. Natl. Acad. Sci. USA* **72:**4669–4673.

Sakaizumi, M., 1986, Genetic divergence in wild populations of medaka, *Oryzias latipes* (Pisces: Oryziatidae) from Japan and China, *Genetica* **69:**119–125.

Salini, J., and Shaklee, J. B., 1988, Genetic structure of barramundi (*Lates calcarifer*) stocks from Northern Australia, *Aust. J. Mar. Freshwater Res.* **39:**317–329.

Saura, A., Halkka, O., and Lokki, J., 1973, Enzyme gene heterozygosity in small island populations of *Philaenus spumarius* (L.) (Homoptera), *Genetica* **44:**459–473.

Sbordoni, J., Allegrucci, G., Caccone, A., Cesaroni, D., Sbordoni, M. C., and De Matthaeis, E., 1981, Genetic variability and divergence in cave populations of *Troglophilus cavicola* and *Troglophilus andreinii* (Orthoptera, Rhaphidophoridae), *Evolution* **35:**226–233.

Sbordoni, S., Allegrucci, G., Cesaroni, D., Sbordoni, M. C., and De Matthaeis, E., 1985, Genetic structure of populations and species of *Dolichopoda* cave crickets: Evidence of peripatric divergence, *Boll. Zool.* **52:**139–156.

Schmitt, L. H., 1975, Genetic variation in isolated populations of the Australian bush rat, *Rattus fuscipes, Evolution* **32:**1–14.

Schneppenheim, R., and MacDonald, C. M., 1984, Genetic variation and population structure of krill (*Euphasia superba*) in the Atlantic sector of Antarctic waters off the Antarctic peninsula, *Polar Biol.* **3**:19–28.

Scribner, K. T., and Warren, R. J., 1986, Electrophoretic and morphological comparisons of *Sylvilagus floridanus* and *Sylvilagus audubonii* in Texas, *Southwest. Natl.* **31**:65–71.

Seeb, L. W., and Gunderson, D. R., 1988, Genetic variation and population structure of Pacific ocean perch (*Sebastes alutus*), *Can. J. Fish. Aquat. Sci.* **45**:78–88.

Seeb, J. E., Seeb, L. W., Oates, D. W., and Utter, F. M., 1987, Genetic variation and postglacial dispersal of populations of northern pike (*Esox lucius*) in North America, *Can. J. Fish. Aquat. Sci.* **44**:556–561.

Seidel, M. E., Reynolds, S. L., and Lucchino, R. V., 1981, Phylogenetic relationships among musk turtles (genus *Sternotherus*), and genic variation in *Sternotherus odoratus, Herpetologica* **37**:161–165.

Selander, R. K., Hunt, W. G., and Yang, S. Y., 1969, Protein polymorphism and genic heterozygosity in two European subspecies of the house mouse, *Evolution* **23**:379–390.

Selander, R. K., Yang, S. Y., Lewontin, R. C., and Johnson, W. E., 1970, Genetic variation in the horseshoe crab (*Limulus polyphemus*); a phylogenetic "relic", *Evolution* **24**:402–414.

Selander, R. K., Smith, M. H., Yang, S. Y., Johnson, W. E., and Gentry, J. B., 1971, Biochemical polymorphism and systematics in the genus *Peromyscus*. 1. Variation in the old-field mouse (*Peromyscus polionotus*), *Stud. Genet. VI Univ. Tex. Tech. Pub.* **7103**:49–90.

Sene, F. M., and Carson, H. L., 1977, Genetic variation in Hawaiian *Drosophila*. IV. Allozymic similarity between *Drosophila silvestris* and *Drosophila heteroneura* from the islands of Hawaii, *Genetics* **86**:187–198.

Shaffer, H. B., 1983, Biosystematics of *Ambystoma rosaceum* and *Ambystoma tigrinum* in north-west Mexico, *Copeia* **1983**:67–78.

Shaklee, J. B., and Tamuru, C. S., 1981, Biochemical and morphological evolution of Hawaiian bonefishes (*Albula*), *Syst. Zool.* **30**:125–146.

Shotake, T., 1981, Population genetic study of natural hybridization between *Papio anubis* and *Papio hamadryas, Primates* **22**:285–308.

Siegismund, H. P., Simonsen, V., and Kolding, S., 1985, Genetic studies of *Gammarus*. I. Genetic differentiation of local populations, *Hereditas* **102**:1–13.

Siebenaller, J. F., 1978, Genetic variability in deep-sea fishes of the genus *Sebastolobus* (Scorpaenidae), in: *Marine Organisms: Genetics, Ecology and Evolution* (B. Battaglia and J. A. Beardmore, eds.), pp. 95–122, Plenum Press, New York.

Simonsen, V., 1982, Electrophoretic variation in large mammals. II. The red fox, *Vulpes vulpes,* the stoat, *Mustela erminea,* the weasel, *Mustela nivalis,* the pole cat, *Mustela putorius,* the pine marten, *Martes martes,* the beech marten, *Martes foina,* and the badger, *Meles meles, Hereditas* **96**:299–305.

Sin, F. Y. T., and Jones, M. B., 1983, Enzyme variation in marine and estuarine populations of a mud crab *Macrophthalamus hurtipes* (Geypodidae), *N. Z. J. Mar. Freshwater Res.* **17**:367–372.

Singh, R. S., Choudhary, M., and David, J. R., 1987, Contrasting patterns of geographic variation in the cosmopolitan sibling species *Drosophila melanogaster* and *Drosophila simulans, Biochem. Genet.* **25**:27–40.

Sites, J. W., Jr., and Greenbaum, I. F., 1983, Chromosome evolution in the iguanid lizard *Sceloporus grammicus:* II. Allozyme variation, *Evolution* **37**:54–65.

Skadsheim, A., and Siegismund, H. R., 1986, Genetic relationships among North-Western European Gammaridiae (Amphipoda), *Crustaceana* **51**:163–175.

Sluss, T. P., Sluss, E. S., Graham, H. M., and Dubois, M., 1978, Allozyme differences between *Heliothis virescens* and *Heliothis zea, Ann. Entomol. Soc. Am.* **32**:191–195.

Sluss, T. P., Graham, H. M., and Sluss, E. S., 1982, Morphometric, allozymic and hybridization comparisons of four *Lygus* species. (Hemiptera: Heteroptera: Miridae), *Ann. Entomol. Soc. Am.* **75**:448–456.

Smith, A. M. A., 1983, The subspecific biochemical taxonomy of *Autechinus minimus, Autechinus swainsonii* and *Sminthopsis leucopus* (Marsupials: Dosyuvidae), *Aust. J. Zool.* **32**:753–762.

Smith, J. K., and Zimmerman, E. G., 1976, Biochemical genetics and evolution of North American blackbirds, (family Icteridae), *Comp. Biochem. Physiol.* **53B**:319–324.

Smith, M. F., 1979, Geographic variation in genic and morphological characters in *Peromyscus californicus, J. Mammal.* **60**:705–722.

Smith, M. F., and Patton, J. L., 1980, Relationships of pocket gophers (*Thomomys bottae*) populations of the lower Colorado river, *J. Mammal.* **61**:681–696.

Smith, M. H., Selander, R. K., and Johnson, W. E., 1973, Biochemical polymorphism and systematics in the genus *Peromyscus.* III Variation in the Florida deer mouse (*Peromyscus floridanus*), a Pleistocene relict, *J. Mammal.* **54**:1–13.

Smith, M. H., Britton, J., Burke, P., Chesser, R. K., Smith, M. W., and Hagen, J., 1979, Genetic variation in *Corbicula,* an invading species, in: *Proceedings First International Corbicula Symposium. Texas Christian University,* pp. 243–248.

Smith, P. J., 1986, Genetic similarity between samples of the orange roughy *Hoplostethus atlanticus* from the Tasman Sea, southwest Pacific Ocean and northeast Pacific Ocean, *Mar. Biol.* **91**:173–180.

Smith, P. J., Francis, R. I. C. C., and Paul, L. J., 1978, Genetic variation and population structure in the New Zealand snapper, *N. Z. J. Mar. Freshwater Res.* **12**:343–350.

Smith, P. J., McKoy, J. L., and Machin, P. J., 1980, Genetic variation in the rock lobsters *Jasus edwardsii* and *Jasus novaehollandiae, N. Z. J. Mar. Freshwater Res.* **14**:55–63.

Smith, P. J., Wood, B. A., and Benson, P. G., 1980, Electrophoretic and meristic separation of blue maomao (*Scorpis violaceus*) and sweep (*Scorpis aequipinnis*), *N. Z. J. Mar. Freshwater Res.* **13**:549–551.

Steiner, W. W., Kitzmiller, J. B., and Osterbur, D. L., 1980, Gene differentiation in chromosome races of *Anopheles nuneztovari, Mosq. Syst.* **12**:306–319.

Stock, M. W., and Castroville, P. J., 1981, Genetic relationships among representative populations of five *Choristoneura* species, *C. occidentalis, C. retiniana, C. biennis, C. lambertiana,* and *C. fumiferana* (Lepidoptera: Tortricidae), *Can. Entomol.* **113**:857–865.

Stoneking, M., Wagner, D. J., and Hildebrand, A. C., 1981, Genetic evidence suggesting subspecific differences between northern and southern populations of brook trout (*Salvelinus fontinalis*), *Copeia* **1981**:810–819.

Straney, D. O., O'Farrell, M. J., and Smith, H., 1976, Biochemical genetics of *Myotis californicus* and *Pipistrellus hesperus* from southern Nevada, *Mammalia* **40**:344–347.

Sullivan, R. M., 1985, Phyletic, biogeographic and ecologic relationships among montane populations of least chipmunks (*Eutamias minimus*) in the southwest (U.S.A.), *Syst. Zool.* **34**:419–448.

Sumantadinata, K., and Taniguchi, N., 1982, Biochemical genetic variations in Black Sea bream, *Bull. Jpn. Soc. Sci. Fish.* **48**:143–150.

Sweeney, B. W., Funk, D. H., and Vanotte, R. L., 1986, Population genetic structure of two mayflies (*Ephemerella subvaria, Eurylophella verisimilis*) in the Delaware river drainage basin, *J. N. Am. Benthol. Soc.* **5**:253–262.

Taniguchi, N., Seki, S., and Inada, Y., 1983, Genic variability and differentiation of amphidromous, landlocked and hatchery populations of ayu, *Plecoglossus altivelis, Bull. Jpn. Soc. Sci. Fish.* **49**:1655–1663.

Taylor, C. E., and Gorman, G. C., 1975, Population genetics of a "colonizing" lizard: Natural selection for allozyme morphs in *Anolis grahami, Heredity* **35**:241–247.

Thompson, P., and Sites, J. W., 1986, Comparison of population structure in chromosomally polytypic and monotypic species of *Sceloporus* (Saura; Iguanidae) in relation to chromosomally-mediated speciation, *Evolution* **40**:303–314.

Tilley, S. G., and Schwerdtfeger, P. M., 1981, Electrophoretic variation in Appalachian populations of the *Desmognathus fuscus* complex (Amphibia: Plethodontidae), *Copeia* **1981**:109–119.

Tolliver, D. K., Choate, J. R., Kaufman, D. W., and Kaufman, G. A., 1986, Microgeographic variation of morphometric and electrophoretic characters in *Peromyscus leucopus, Am. Midl. Nat.* **117**:420–427.

Tracey, M. L., Nelson, K., Hedgecock, D., Shlesser, R. A., and Pressick, M. L., 1975, Biochemical genetics of lobsters: Genetic variation and the structure of American lobster (*Homarus americanus*) populations, *J. Fish. Res. Board Can.* **32**:2091–2107.

Turner, K., and Lyerla, T. M., 1980, Electrophoretic variation in sympatric mud crabs from North Inlet, South Carolina, U.S.A., *Biol. Bull.* **159**:418–427.

Van Wagner, C. E., and Baker, A. J., 1986, Genetic differentiation in populations of Canada geese (*Branta canadensis*), *Can. J. Zool.* **64**:940–947.

Verheyen, E., van Rompaey, J., and Selens, M., 1985, Enzyme variation in haplochromine cichlid fishes from Lake Victoria (Central Africa), *Neth. J. Zool.* **35**:469–478.

Vuorinen, J., Himberg, M. K. J., and Lankinen, P., 1981, Genetic differentiation in *Coregonus albula* (L.) (Salmonidae), *Hereditas* **94**:113–122.

Wake, D. B., and Yanev, K. P., 1986, Geographic variation in allozymes in a "ring species", the plethodontid salamander *Ensatina eschscholtzii* of western North America, *Evolution* **40**:702–715.

Wake, D. B., Maxson, L. R., and Wurst, G. F., 1978, Genetic differentiation, albumin evolution, and their biogeographic implications in plethodontid salamanders of California and southern Europe, *Evolution* **32**:529–539.

Ward, R., Rutledge, C. J., and Zimmerman, E. G., 1987, Genetic variation and population subdivision in the cricket frog *Acris crepitans, Biochem. Syst. Ecol.* **15**:377–384.

Ward, R. D., and Beardmore, J. A., 1977, Protein variation in the plaice, *Pleuronectes platessa* L., *Genet. Res.* **30**:45–62.

Ward, R. D., and Galleguillos, R. A., 1978, Protein variation in the plaice, dab and flounder and their genetic relationships, in: *Marine Organisms: Genetics, Ecology and Evolution* (B. Battaglia and J. A. Beardmore, eds.), pp. 71–94, Plenum Press, New York.

Ward, R. D., and Warwick, T., 1980, Genetic differentiation in the molluscan species *Littorina rudis* and *Littorina arcana* (Prosobranchia: Littorinidae), *Biol. J. Linn. Soc.* **14**:417–428.

Webster, T. P., 1975, An electrophoretic characterization of the hispaniolan lizards *Anolis cybotes* and *Anolis marcanoi, Breviora* **431**:1–8.

Webster, T. P., Selander, R. K., and Yang, S. Y., 1972, Genetic variability and similarity in the *Anolis* lizards of Bimini, *Evolution* **26**:523–535.

Wilson, R. R., Jr., and Waples, R. S., 1984, Electrophoresis and biometric variability in the abyssal grenadier *Coryphaenoides armatus* of the western north Atlantic, eastern south Pacific and eastern north Pacific oceans, *Mar. Biol.* **80**:227–238.

Winans, G. A., 1980, Geographic variation in the milkfish, *Chanos chanos:* I. Biochemical evidence, *Evolution* **34**:558–574.

Woodruff, D. S., McMeekin, L. L., Mulvey, M., and Carpenter, M. P., 1986, Population genetics of *Crepidula onyx:* Variation in a Californian slipper snail recently established in China, *Veliger* **29**:53–63.

Wright, J., Tennant, B. C., and May, B., 1987, Genetic variation between woodchuck populations with high and low prevalence rates of woodchuck hepatitus virus infection, *J. Wildl. Dis.* **23:**186–191.

Yanev, K. P., and Wake, D. B., 1981, Genic differentiation in a relict desert salamander, *Batrachoseps campi, Herpetologica* **37:**16–28.

Yang, S. Y., and Patton, J. L., 1981, Genic variability and differentiation in the Galapagos finches, *Auk* **98:**230–242.

Yang, S. Y., Wheeler, L. L., and Bock, I. R., 1972, Isozyme variations and phylogenetic relationships in the *Drosophila bipectinata* species complex, *Stud. Genet. VII Univ. Tex. Pub.* **7213:**213–227.

Yang, S. Y., Soulé, M., and Gorman, C., 1974, *Anolis* lizards of the eastern Carribean: A case study in evolution. I. Genetic relationships, phylogeny, and colonization sequence of the *roquet* group, *Syst. Zool.* **23:**387–399.

Yates, T. L., and Greenbaum, I. F., 1982, Biochemical systematics of North American moles (Insectivora: Talipidae), *J. Mammal.* **63:**368–374.

Zera, A. J., 1981, Genetic structure of two species of water-striders (Gerridae; Hemiptera) with differing degrees of winglessness, *Evolution* **35:**218–225.

Zimmerman, E. G., and Nejtek, M. E., 1977, Genetics and speciation of three semispecies of *Neotoma, J. Mammal.* **58:**391–402.

Zimmerman, E. G., and Wooten, M. C., 1981, Allozymic variation and natural hybridization in sculpins, *Cottus confusus* and *Cottus cognatus, Biochem. Syst. Ecol.* **9:**341–346.

Zink, R. M., 1982, Patterns of genic and morphologic variation among sparrows in the genera *Zonotrichia, Nelospiza, Juneo* and *Passerella, Auk* **99:**632–649.

Zink, R. M., and Johnson, N., 1984, Evolutionary genetics of flycatchers: I. Sibling species in the genera *Empidonax* and *Contopus, Syst. Zool.* **33:**205–316.

Zink, R. M., and Winkler, D. W., 1983, Genic and morphological similarity of two Californian gull (*Larus californicus*) populations with different life-history traits, *Biochem. Syst. Ecol.* **11:**397–404.

Zink, R. M., Lott, D. F., and Anderson, D. W., 1987, Genetic variation, population structure, and evolution of California quail, *Condor* **89:**395–405.

Zouros, E., 1973, Genic differentiation associated with early stages of speciation in the *mulleri* subgroup of *Drosophila, Evolution* **27:**601–621.

4

Accumulation of Lethals and Suppression of Genetic Load in Irradiated Populations of *Drosophila melanogaster*

HORST NÖTHEL

INTRODUCTION

Populations carry their load of mutations. It is considerably increased by increased exposures to environmental mutagens. This is considered the most serious consequence of such mutagens in the long run (Morton *et al.*, 1956). Genetic load means that the optimal fitness for any genetic locus is reduced by mutations that are deleterious if heterozygous and/or homozygous. The concept of genetic load has been developed by Muller (1950). He has employed it to underscore his urgent warnings against the enormous genetic hazards by the invisible ionizing radiations. It took two decades from Muller's first proof of the mutagenic action of X-rays (Vth International Congress of Genetics, Berlin, 1927; see Muller, 1927, 1928*a,b*) to his winning the Nobel prize (after the explosion of atomic bombs raised public interest). It took another two decades until public interest and legislation began to center on artificial radiation sources and their potential dangers and still another decade to initiate the almost total ban on radiations that governs present-day politics. A focus of the argumentation was the dangerous in-

---

HORST NÖTHEL • Institut für Genetik, Freie Universität Berlin, D 1000 Berlin-33 (Dahlem), Germany.

Evolutionary Biology, Volume 26, edited by Max K. Hecht *et al.* Plenum Press, New York, 1992.

crease in genetic load. The concept of genetic load, its necessity, and its limits have been outlined by Wallace (1970) in an excellent monograph. Experimental evidence on load and on viabilities came mainly from work on *Drosophila* (Bonnier *et al.,* 1958, 1959; Carson, 1964; Cordeiro *et al.,* 1973; Eiche, 1972; Falk, 1967; Kratz, 1975; Marques, 1973; Sankaranarayanan, 1964, 1965; Salceda, 1967; Wallace, 1956, 1958, 1963) with some additional data from mammals [see Green (1968) for a review]. These data have not been substantially enlarged since then, but have been revisited with interesting new views (Wallace, 1989; Wallace and Blohowiak, 1985*a,b*).

This chapter describes the genetic load in experimental populations of *Drosophila melanogaster* that have been exposed to high X-ray levels for 800 generations. Its aim is to present a very generous amount of material in a form as compressed as possible and to restrict discussions to specific and new aspects. The data will corroborate a tremendous gain in lethal frequencies, but they will indicate that only part of these lethals are expressed in genomes adapted to increased mutational yields.

MATERIALS AND METHODS

The X-irradiated populations tested are the *RÖ* populations. The phylogeny of these populations started with the basic population *Berlin wild.* The pedigree is illustrated in Fig. 1. Test generations are characterized by the number of generations between nodes in the tree. For example, generation 255+50+2 of $RÖI_4$ denotes 255 generations of *RÖI* history plus 50 generations of $RÖI_4$ history and 2 generations without exposures prior to test. For brevity, the corresponding *RÖI* count may be given: number of generations since *RÖI* was started, i.e., 307 in the above example.

The following balancer stocks were used to isolate chromosomes X, 2, and 3, respectively, from the populations [see Lindsley and Grell (1968) for description]:

- X $In(1)sc^{S1L}sc^{8R} + S, sc^{S1}\ sc^{8}\ w^{a}\ B$ (=*M-5*, males with a free Y)
- X $C(1)RM,\ y\ v\ shi^{ts1}/O/Y^{S}.X.Y^{L},\ In(1)EN + dl\text{-}49,\ y\ v\ f\ car$ (no free Y)
- 2 $In(2LR)SM5,\ al^{2}\ Cy\ lt^{v}\ cn^{2}\ sp^{2}/Bl\ L^{2}$
- 3 $In(3LR)TM3,\ y^{+}\ ri\ p^{p}\ sep\ Sb\ bx^{34e}\ e^{s}\ Ser/D^{2}$
- 3 $bw;\ In(3LR)TM3,\ y^{+}\ ri\ p^{p}\ sep\ Sb\ bx^{34e}\ e^{s}\ Ser/D^{2}$

RÖ chromosomes were isolated from females (X) or males (autosomes) using conventional mating schemes. Only one chromosome was isolated per

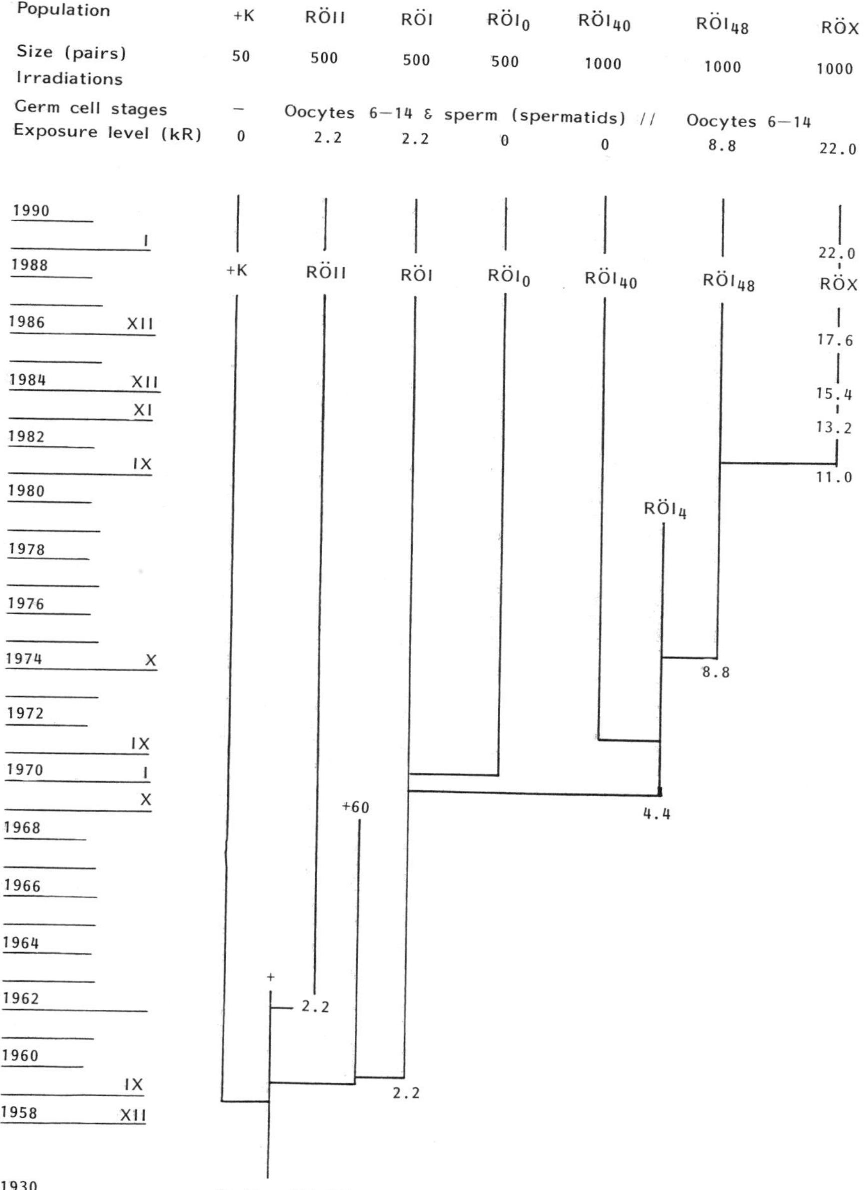

FIG. 1. Pedigree of *Berlin-wild* sublines and of the X-irradiated populations *RÖII* and *RÖI* with its various subpopulations. The two left columns give year (Arabic numbers) and month (Roman numbers) when populations/subpopulations were started or terminated. Twenty six generations were reared per year. Any *RÖ* population was kept in twenty 250-ml bottles with progeny thoroughly mixed in every generation. Population sizes and acute X-ray exposures are listed in the top rows; oocyte stages exposed are given in the terminology of King (1970).

parental fly. All presumed lethals were retested to meet the specifications for lethals *sensu stricto*.

X chromosomes were screened for lethality in males without a free Y chromosome. All lethals were then tested with *M-5* males for Y suppression.

In one experiment, effects of *RÖ* second chromosomes on the lethality of *RÖ* third chromosomes were tested using *bw*; *TM3/D* females in the parental and in the first generation (backcross). Second generation crosses were with any *RÖ* third chromosome (#$RÖ_1$, etc) (1) *bw/bw*; $TM3/RÖ_1$ × *bw/bw*; $TM3/RÖ_1$ and (2) *bw/RÖ*; $TM3/RÖ_1$ × *bw/RÖ*; $TM3/RÖ_1$. Progeny were scored for third-chromosome lethality in flies with or without second chromosomes of *RÖ* origin.

Part of the autosomal lethals were tested for complementation in crosses of the type $TM3/l_1 \times TM3/l_2$, where l_1 and l_2 represent lethal chromosomes from different parents. All tests were performed in reciprocal crosses. They were repeated in cases of contradictions between reciprocals so that results finally included in the data are valid.

Third-chromosome lethals were mapped using *ru cu ca*, a stock carrying the recessive markers (symbols and map positions in parentheses) *roughoid* (*ru* 3-0.0) *hairy* (*h* 3-26.5) *thread* (*th* 3-43.2) *scarlet* (*st* 3-44.0) *curled* (*cu* 3-50.0) *stripe* (*sr* 3-62.0) *ebony* (e^s 3-70.7) *claret* (*ca* 3-100.7). Two-day-old females heterozygous for a lethal l_1 and *ru cu ca* were "back"crossed to *ru cu Pr ca* males. These carry in addition to the above factors the dominant *Prickly* (*Pr* 3-90.0) and the chromosome is balanced over *In(3LR)TM6, Ubx e*. The *Pr* progeny were scored for crossover rates. Male progeny were then mated individually to $TM3/l_1$ to test for lethality of the various recombinant chromosomes.

Culture conditions were as described elsewhere (Nöthel, 1967). They included a corn–agar–molasses medium with some live yeast added. Sampling of virgins or large numbers of males was done from 250-ml bottles reared with 25 pairs each. In experiments, single-pair matings were done in vials. Oviposition rates and egg-to-adult survival were measured in egg-laying units with four pairs each: Eggs were counted on days 3 and 4 after mating of 1- to 2-day-old virgins; egg samples were transferred to bottles to test development.

Temperature was always 25°C except for the *C(1)RM* stock, which was maintained at 18°C. After shifting to 25°C it produced "automatic males" due to the effects of the temperature-sensitive lethal shi^{ts1} in females (Poodry *et al.*, 1973).

Statistics were performed with the programs of $SPSS/PC^+$ (SPSS Inc., Chicago, Illinois). Specific calculations are outlined when used.

RESULTS

Mutation Rates and Lethal Frequencies

Mutagenic effects of X-irradiations administered to the various *RÖ* populations are depicted by exposing the unirradiated control line *Berlin wild* to similar irradiation treatments. Results are summarized in Table I. The data on X-chromosome lethals are based on large numbers tested. They are in line with the literature (Timoféeff-Ressovsky and Zimmer, 1947; Sankaranarayanan and Sobels, 1976). They give dose–response regressions that are linear and equal in both sexes with rates of 0.24% in unirradiated controls and an average induction (Poisson-corrected) of 2.5%/kR. Data on major autosomes are based on limited material tested with respect to exposure, chromosome, and sex. Pooled data on sexes and chromosomes (Table I) yield excellent linear dose–response relationships. Frequency of recessive lethals is 0.9% per chromosome in unirradiated controls. Induction (Poisson-corrected) is 6.3%/kR. These results, again, are in line with the literature, as is the ratio of 2.5 between autosomal and X-chromosomal lethals.

Genetic load was studied in the course of irradiation histories of the *RÖ* populations and in *Berlin wild.* Test generations, number of chromosomes

TABLE I. Recessive Lethal Frequencies in the Major Chromosomes of *Berlin Wild K* Males and Females, Respectively, after X-Ray Exposures under *RÖI*/*RÖI*$_4$ Conditions with Respect to Germ-Cell Stages Exposed and Doses Administered[a]

Exposure level (kR)	X chromosome						Chromosomes 2 and 3 totals		
	Females		Males		Totals				
	N	%	*N*	%	%	Corr.%	*N*	%	Corr.%
0	4061	0.30	4800	0.19	0.24	—	651	0.92	—
2	1499	5.0	1943	4.8	4.92	4.79	535	12.34	12.13
4	1424	8.8	694	12.0	9.82	10.07	441	23.72	25.88

[a] Either males or females were exposed. +*K* was tested for X-chromosomal lethals. Newly synthesized "lethal-free" lines of +*K* origin were tested for autosomal lethals. Frequency of lethal chromosomes (*sensu stricto*) is given in percentage (%) of chromosomes tested (N). Frequency of lethal factors results from corrections (Corr.%) for unirradiated controls (0 R) and for a Poisson distribution of lethals per chromosome.

tested, and frequencies of lethal chromosomes obtained are summarized in Table II. Third-chromosome lethals were obtained in test intervals of about 100–200 generations; only test 3 was intended to follow the frequency in one population in shorter intervals. Second chromosomes were mainly tested in early population history. In general, data obtained for one population are equivalent with respect to both major autosomes and to various test generations. There are some fluctuations that have to be expected in view of the limited material studied and, in the case of *Berlin wild,* because of the small population size. But there is only one significant deviation and this is between frequencies of second- and third-chromosome lethals of *RÖI*$_{40}$ in test 3 (see below). With the exception of the latter case, pooled data for both major chromosomes and all tests were calculated. These *observed* lethal frequencies are compared, in Table III with frequencies *expected* under the

TABLE II. Lethal Major Autosomes in Percentage (%) of Chromosomes Tested (*N*) from *Berlin Wild* Lines and from Various *RÖ* Populations in Course of Irradiation History

		Generations with exposures of				Chromosome 2		Chromosome 3	
Test	Population	2.2 +	4.4 +	8.8 +	0 kR	*N*	%	*N*	%
1	*RÖI*	103–128			2	98	92.9	44	90.9
	RÖII	42–67			2	106	83.9	50	74.0
	+					166	1.2	130	0.0
	+60					178	5.6	74	13.5
	+K					92	23.9	94	2.1
2	*RÖI*$_{4}$	255	50		2	392	97.5	96	94.8
	RÖI$_{4}$	255	65		2	113	98.2	—	—
	+K					72	11.1	81	17.3
3	*RÖI*$_{40}$	255	48		150	170	14.7	100	83.0
	RÖI$_{40}$	255	48		175	177	7.9	188	71.8
	RÖI$_{40}$	255	48		187	—	—	315	72.4
4	*RÖI*$_{40}$	255	48		217	—	—	56	76.8
	RÖI$_{48}$	255	120	146	2	—	—	97	72.2
	RÖI	529			2	—	—	107	93.5
	RÖI$_{0}$	260			271	—	—	164	57.3
	RÖII	470			2	—	—	105	99.0
	+K					175	5.7	170	5.3
5	*RÖI*	625			2	—	—	155	84.5
	RÖI$_{0}$	260			370	—	—	169	16.6
	RÖI$_{40}$	255	48		321	—	—	150	16.0
	RÖI$_{48}$	255	120	243	2	—	—	104	76.9
	RÖII	570			2	—	—	140	94.3

TABLE III. Frequencies of Recessive, Autosomal Lethals in *Berlin Wild K* and Various *RÖ* Populations in Mutation-Elimination Equilibria: Observed Data Compared with Expectations under the Sole Actions of Mutation and Random Drift

	Conditions in any generation					Data for mutation-elimination equilibria				
								Lethal chromosomes		
Population	Dose (kR)	Sex exposed	DRF in females	Mutations expected (%)[a]	Population size (N)	Time t^b	Lethals expected (%)[c]	Expected (%)[d]	Observed (%)[e]	Difference a^f
$+K$	—	—	1.0	0.90	100	10.6	9.54	9.1	7.1 (1232)	**
RÖII	2.2	Both	1.53	12.46	1000	15.2	187.98	84.7	90.3 (401)	**
RÖI	2.2	Both	1.72	11.86	1000	15.2	180.26	83.5	89.6 (404)	**
$RÖI_4$	4.4	Both	3.24	19.04	2000	16.6	315.84	95.8	97.2 (601)	*
$RÖI_{48}$	8.8	Female	3.24	9.46	2000	16.6	156.87	79.2	74.6 (201)	*
			7.5	4.60	2000	16.6	76.25	53.4		***
$RÖI_0{}^g$	0	(3 & <300)		0.90	1000	15.2	13.68	12.8	57.3 (164)	***
	0	(3 & >300)						12.8	16.6 (169)	*
$RÖI_{40}{}^g$	0	(3 & <300)		0.90	2000	16.6	14.9	13.8	75.9 (659)	***
	0	(2 & <300)						13.8	11.2 (347)	*
	0	(3 & >300)		0.90	2000	16.6	14.9	13.8	16.0 (150)	*

[a] Lethal mutations per major autosome (see Table I): 0.9% (0 R), 6.3%/kR; new lethals per generation and major autosome: (male-kR × 6.3% + female-kR × 6.3%/DRF)/2 + 0.9%.

[b] Time t is the number generations until the random loss of a newly induced lethal; $t = 2\,(N_e/N_a)\ln(2N_a)$, where N_e is the effective population size and N_a is the actual population number (Hartl, 1980). To avoid underestimates of t and in line with observations, $N_e = N_a = N$ and, hence, $t = 2\ln(2N)$.

[c] Total recessive lethals expected in equilibrium is $t \times \%$, the product between frequency (%) of new lethals/generation (see note *a*) and mean number of generations (t) until random loss (see note *b*).

[d] Expected frequency of lethal chromosomes is the Poisson-corrected frequency of recessive lethal factors (see note *c*).

[e] Observed frequencies of lethal major autosomes are pooled from Table II for all generations studied. Total number of chromosomes tested is given in parentheses. Exceptions are $RÖI_0$ and $RÖI_{40}$ (see note *g*).

[f] Significance between observed and expected frequencies according to χ^2 test: *, $a > 0.05$; **, $0.05 > a > 0.001$; ***, $a < 0.001$.

[g] In the formerly irradiated populations $RÖI_0$ and $RÖI_{40}$, chromosomes 2 and 3 and <300 versus >300 generations without exposures are given separately: (3 & <300) means chromosome 3 and <300 generations without exposures, etc.

following assumptions: (1) Mutation rate is 0.9% in unirradiated controls and 6.3%/kR (see Table I); actual values are calculated using the per-generation exposure of a population and the gain in relative radioresistance as characterized by the respective dose–reduction factor (DRF). (2) Newly induced mutations are lost from a population by chance but not by selection; the mean number of generations between induction and loss is $t = 2 \ln(2N)$, where N is the actual population number N_a, which is assumed to equal the effective population size N_e (Hartl, 1980). (3) In a mutation-elimination equilibrium, the expected genetic load is $t \times$ mutation rate (for the calculation of the mutation rate see assumption 1). (4) This load is based on lethal factors, whereas experimental data are on lethal chromosomes; both values are simply connected by the Poisson distribution of lethals over chromosomes.

Lethal frequencies expected under these assumptions in the *continuously irradiated populations* are slightly lower than those actually observed, even if this discrepancy is most often insignificant (Table III). In case of $RÖI_{48}$, the observed frequency is between those expected with DRFs of 3.24 (in $RÖI_4$) and 7.5 (final value in $RÖI_{48}$), respectively, and this corresponds to a new part of resistance in *status nascendi* (i.e., selection of the resistance factor *rar-4*). Lethal frequencies expected under the above assumptions are far lower than those observed in the *previously irradiated populations* $RÖI_0$ and $RÖI_{40}$. This holds especially for chromosome 3 up to test 4, but not for test 5 and not for chromosome 2 in an earlier test.

The main results of the above tests are the higher than expected lethal frequencies. They are quite evident in subpopulations without further X-ray exposures. This is so even if expectations are solely based on chance effects and not on any purifying selection. A first attempt to analyze these findings is the evaluation of the frequency of specific lethals in complementation tests. These tests may uncover heterozygous advantages of some lethals. The procedure is illustrated in Fig. 2.

Transient Lethals in Chromosome 2

In Fig. 2, analysis of second-chromosome lethals from $RÖI_4$ (generation 255+50+2) is outlined. Twenty nine chromosomes were tested. They are listed with their respective numbers (#) in the top row and in the right-hand ordinate. The diagonal represents the homozygotes for each of the 29 lethals, and their lethality may serve as a control. The other lethal combinations given above the diagonal indicate allelism between lethal chromosomes derived from different parents. The term allelism is used in an operational way and covers true alleles (recurrence of the same variant), pseudoalleles (inde-

Current numbers of lethal chromosomes tested for allelism of lethal factors (1)

	01	02	13	18	19	21	36	42	43	44	45	47	51	55	60	72	73	76	78	81	86	87	88	89	92	93	97	98	99
01	1																												
02		1						1								1	1		1			1	1						
13			1	1																						1			
18				1						1	1															1			
19					1				1																	1			1
21						1																							
36							1													1									
42								1								1	1		1			1	1						
43									1					1				1											
44										1	1																		
45											1																		
47												1		1				1											
51													1									1	1						
55														1				1							1				
60															1									1				1	
72																1	1		1			1	1						
73																	1		1			1	1						
76																		1											
78																			1			1	1						
81																				1							1		
86																					1								
87																						1	1						
88																							1						
89																								1				1	
92																									1				
93																										1			
97																											1		
98																												1	
99																													1

chromosomes detected			Lethal factors detected															
type	#l/ chr.	#of cases	01	02	03	04	05	06	07	08	08	10	11	12	13	14	15	16
A	1	1	01															
B	1	5		02 42 72 73 78														
C	2	2		87 88	87 88													
D	1	1			51													
E	1	1				21												
F	1	1					36											
G	2	1					81	81										
H	1	1						97										
J	1	1							86									
K	1	3								60 89 98								
L	1	1									92							
M	3	1									55	55	55					
N	1	1										47						
O	2	1										76	76					
P	2	1											43	43				
Q	3	1												19	19	19		
R	1	1													99			
S	2	1														93	93	
T	1	1															13	
U	2	1															18	18
V	1	2																44 45
Z	0	.74																
22		29.74																
Frequency of these factors			.03	.24	.10	.03	.07	.07	.03	.10	.07	.10	.10	.07	.07	.07	.10	.10

FIG. 2. Complementation tests with 29 lethal second chromosomes from $RÖl_{40}$ (generation 255 + 50 + 2). Above diagonal: Current number of lethal chromosomes is given on the coordinates. Lethal combinations are denoted by *l.* The diagonal represents homozygous, lethal combinations. Besides the latter, 43 combinations are lethal, i.e., rate of "allelism" equals $43/0.5(29^2 - 29) = 0.106$. Below diagonal: 16 different (i.e., nonallelic) lethals were obtained in the above tests. They are listed on the abscissa in numerical order. Their respective frequencies are given in the bottom row. Lethal chromosomes are characterized by the lethal factors they carry. The different types of chromosomes thus detected are denoted on the left-hand ordinate by letters A–V. Number of lethals per any chromosome (#l/chr.) and number of cases occurring with any type (# of cases) are added. An additional type are lethal-free chromosomes (type Z), whose number of cases is estimated from their rate in the original test (2.5%). The total number of lethals is equal to the sum of (#l/chr.) × (# of cases) = 40. The mean number of lethals per lethal chromosome is equal to 40/29 = 1.38, and the mean number of lethals per chromosome is equal to 40/29.74 = 1.34.

pendent and slightly different mutations of the same vital locus), and also point mutations in one and deletions in the other chromosome. The rate of this allelism is calculated as $43/0.5(29^2 - 29) = 0.106$. Several different chromosomes may be allelic to each other but to no additional chromosomes. This is the case, for example, with #60, 89, and 98. Any of these three chromosomes carries just the one lethal factor that probably originated from a single mutational event. Other chromosomes may be allelic to one another and, in addition, to other chromosomes that are not allelic to each other. For example, #81 is allelic to 36 and 97, but 36 and 97 are not allelic. In this case, 81 carries two lethals, one of which is considered allelic to 36, the other to 97. The number of different lethal factors per chromosome thus revealed is given in the left-hand ordinate for the diagram below the diagonal. The different lethal factors thus detected are listed on the abscissa in numerical order. Their respective frequencies are given in the bottom row. In no case were more than three distinct lethals detected per chromosome. Total number of lethals equals 40, with mean number of lethals per lethal chromosome being $40/29 = 1.38$. This corresponds to $40/29.74 = 1.34$ lethals per chromosome tested (including 2.5% nonlethal chromosomes). The number of different lethal loci detected is 16. The frequencies of these 16 lethals are of considerable variability, with a high of 0.24 (lethal #02 in 7 of 29.74 chromosomes) and a mean of $1.34/16 = 0.084$. This is outlined in Table IV. In addition, observed numbers are compared with Poisson expectations of number of lethals per chromosome and of recurrence of identical lethals (allels) among different chromosomes, respectively.

The data of Table IV show, with respect to the 29 second-chromosome lethals of generation 255+50+2 of $RÖI_4$, a slight deviation between the Poisson distribution and the observed distribution of lethals per chromosome but no deviation with respect to the recurrence of lethals. In generation 255+65+2 of $RÖI_4$, 78 lethal chromosomes were tested. With more lethals tested, some values are elevated as expected: number of different lethals detected (76), mean number of lethals per chromosome (2.93), and mean absolute number of identical lethals (3.05). Number of lethals per chromosome tested does not deviate significantly from a Poisson distribution. Recurrence of allelic lethals is slightly at variance with Poisson expectations, but this is due to shifts between classes 0 (too few cases) and 1 (too many) and not to high frequencies of a few specific lethals (Table IV). Therefore, there is no evidence for heterotic lethals. Likewise, the high frequency of lethal #02 in generation 255+50+2 (24% of the chromosomes tested) decreased to 6.9% (102 chromosomes tested) in generation 255+65+2. In a subpopulation of $RÖI_4$ run without further irradiation during the 15 generations between the two tests, a corresponding decrease was documented in steps of 5 generations: 21% (107 chromosomes tested), 13% (119), 6.4% (110). This decline in

TABLE IV. Number of Lethals per Chromosome and Frequency of Specific Lethals According to Complementation Tests Performed with Second-Chromosome Lethals from *RÖI*$_4$[a]

Generations tested	255 + 50 + 2	255 + 65 +2
Lethal chromosomes tested for complementation	29	78
Total chromosomes "tested"[b]	29.74	79.27
Total number of lethals detected	40	232
Different lethal factors detected	16	76
Mean number of different lethals per chromosome	40/29.74 = 1.34	232/79.27 = 2.93
Mean occurrence of identical lethals in different chromosomes	40/16 = 2.50	232/76 = 3.05
N_1		
0	0.74 (7.7)[c]	1.3 (4.2)
1	20.0 (10.4)	15.0 (12.4)
2	7.0 (7.0)	25.0 (18.2)
3	2.0 (3.1)	12.0 (17.8)
4	0.0 (1.1)	14.0 (13.0)
5	0.0 (0.3)	3.0 (7.6)
6	0.0 (0.1)	4.0 (3.7)
7	0.0 (0.0)	4.0 (1.6)
8	0.0 (0.0)	1.0 (0.6)
9	0.0 (0.0)	0.0 (0.2)
Sum	29.7 (29.7)	79.3 (70.3)
Difference of observed versus expected,[c] *a*	0.001	0.09
N_2		
0	3 (3.6)	1 (9.8)
1	6 (5.4)	30 (20.1)
2	6 (4.0)	23 (20.6)
3	0 (2.0)	17 (14.0)
4	0 (0.8)	1 (7.2)
5	0 (0.2)	1 (3.0)
6	1 (0.0)	1 (1.0)
7	0 (0.0)	2 (0.3)
Sum	16 (16.0)	76 (76.0)
Difference of observed versus expected,[c] *a*	0.4	0.002

[a] See Table II, test 2, for frequencies of lethals obtained, and Fig. 2 for complementation test and treatment of data. N_1 is the number of different lethals per chromosome; 0 = chromosomes without lethals = total chromosomes − lethal chromosomes tested. N_2 is the recurrence of identical or allelic lethals in different chromosomes; 0 = no allelism, i.e., one lethal; 1 = one recurrence, i.e., occurrence in two chromosomes; etc.

[b] Total chromosomes = (lethal chromosomes tested for complementation)/(frequency of lethals in test).

[c] Frequencies expected according to Poisson distributions are given in brackets.

frequency is what is to be expected with selection acting against a completely recessive lethal: The frequency q of the lethal decreases from generation 0 (q_0) to 15 (q_{15}) according to the formula $q_{15} = q_0/(1 + 15q_0)$. This yields 5.2%. It is in line with the observed mean (14 lethals/212 chromosomes = 6.6%) from both populations according to the χ^2 test ($a > 0.3$). It corresponds to the decrease in frequency of lethal #02 that the rate of allelism in the second test is only 5.2% compared to 10.6% in the first test and that this latter value is reduced to 6.1% when omitting lethal #02.

Balanced Lethals in Chromosome 3

Analysis of third-chromosome lethals yields a completely different picture. In Fig. 3, results of complementation tests with lethals from generation 255+48+150 of $RÖI_{40}$ are presented. This subpopulation of $RÖI_4$ had then been unirradiated for 150 generations and had frequencies of second-chromosome lethals reduced as expected according to effects of chance and purifying selection (7.9%; see Table II). It had, however, frequencies of third-chromosome lethals (83%) that were still almost as high as with continuous irradiation. In subsequent tests, attention was therefore focused on chromosome 3. Complementation tests were performed with all of the isolated lethal chromosomes from the above experiment, i.e., with 83 from 100 chromosomes. Data on noncomplementation between lethal chromosomes were extracted from a 83 × 83 combination square in a way corresponding to that outlined in Fig. 2. These results are summarized in Fig. 3. They show that the 83 chromosomes can be grouped into 22 types according to the lethals detected (an additional type are the 17 nonlethal chromosomes). There are only 15 different lethals and the *lethals* that constitute a type of chromosome are denoted by current numbers in the right part of Fig. 3. (Note the difference to the otherwise similar Fig. 2, where numbers denote *chromosomes.*) The rate of allelism is calculated as above and yields $(1649 - 83)/(83^2 - 83) = 0.230$. The number of lethals per chromosome (#l/chromosome) is moderate and is in excellent agreement with a Poisson distribution ($a = 0.5$, including the 17 nonlethal chromosomes). The different chromosome types occur on average 83/22 = 3.8 times. Their distribution has pronounced extremes on both sides and deviates highly significantly from a Poisson distribution ($a < 10^{-10}$). Apparently, the most frequent types—A, G, M, and S, with lethals #01 and 02 in A, 04 in G, 05 in M, and 09, 10, 11, and 12 in S—are representatives of four different, basic groups of lethals. None of these four groups occur together in one chromosome. In any of these four groups an original chromosome type may eventually break up: For example, type A lost lethal #01 (resulting in type C) or lethal #02 (type B), respectively, and the four lethals combined in type S are split in various ways into other types

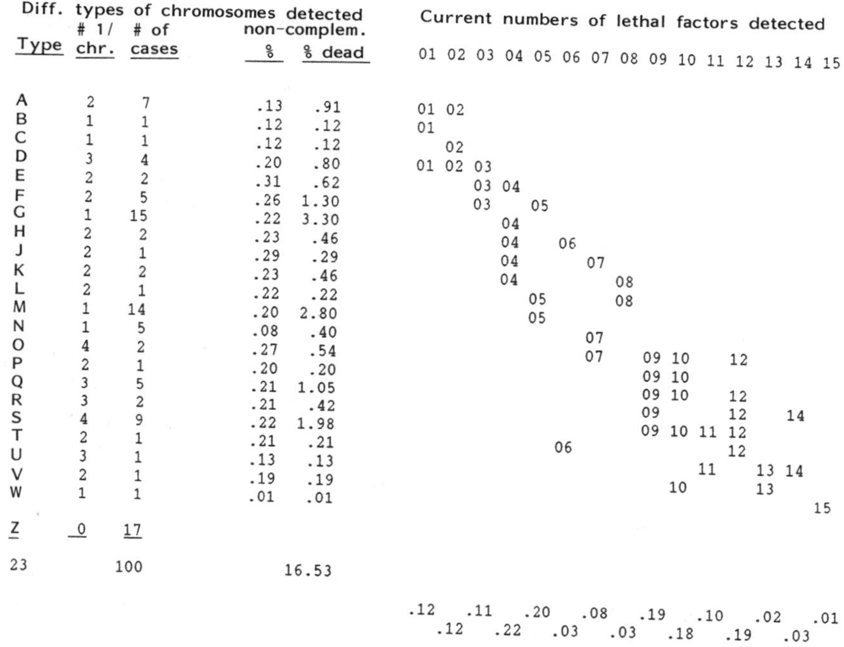

Diff. types of chromosomes detected			non-complem.		Current numbers of lethal factors detected														
Type	# l/ chr.	# of cases	%	% dead	01	02	03	04	05	06	07	08	09	10	11	12	13	14	15
A	2	7	.13	.91	01	02													
B	1	1	.12	.12	01														
C	1	1	.12	.12		02													
D	3	4	.20	.80	01	02	03												
E	2	2	.31	.62			03	04											
F	2	5	.26	1.30			03		05										
G	1	15	.22	3.30				04											
H	2	2	.23	.46				04		06									
J	2	1	.29	.29				04			07								
K	2	2	.23	.46				04				08							
L	2	1	.22	.22					05			08							
M	1	14	.20	2.80					05										
N	1	5	.08	.40							07								
O	4	2	.27	.54							07		09	10		12			
P	2	1	.20	.20									09	10					
Q	3	5	.21	1.05									09	10		12			
R	3	2	.21	.42									09			12		14	
S	4	9	.22	1.98									09	10	11	12			
T	2	1	.21	.21						06						12			
U	3	1	.13	.13											11		13	14	
V	2	1	.19	.19										10			13		
W	1	1	.01	.01															15
Z	0	17																	
23		100		16.53															
Frequencies of lethal factors detected					.12	.12	.11	.22	.20	.03	.08	.03	.19	.18	.10	.19	.02	.03	.01

FIG. 3. Different lethal factors and types of chromosomes revealed from complementation tests with 83 lethal third chromosomes from *RÖl*$_{40}$ (generation 255 + 48 + 150). The 15 different lethal factors obtained from the 83 chromosomes are listed in the right part. They are linked in 22 different types of chromosome (A–W). They are listed on the left-hand ordinate, and the lethal-free chromosomes are added as type Z. Number of lethals per any chromosome (#l/chr.) and number of cases occurring with any type (# of cases) are listed in the second and third columns, respectively. The next two columns consider noncomplementation between any given chromosome and all others, including type Z: For example, any type A chromosome does not complement with types A, B, C, and D and this means with 13 out of 100^2 chromosome combinations (= 13/10,000 = 0.13%); since there are 7 cases of type A, 7 × 0.13% or 0.91% of the 10,000 combinations will die because of this lethal chromosome. Summation over all chromosome types yields 16.53% lethality expected to be expressed. The total number of lethals is equal to the sum over the various types of chromosomes for (#l/chr.) × (# of cases) = 163, and the mean number of lethals per chromosome is equal to 163/100 = 1.63.

ranging from O to V. In addition to the eight lethals of these four main groups, there are two more lethals with relatively high frequencies, namely #03 and 07. They may be associated with specific groups or not (type N). Finally, there are five rare lethals that are either randomly associated with one of the four groups or solitary (#15).

Additional genetic analysis of six frequent third-chromosome lethals (#01, 03, 04, 05, 07, and 09) focused on (1) genetic and subsequent cytological location, (2) frequencies in the course of the further history of $R\ddot{O}_{40}$, and (3) occurrence of these lethals in *Berlin wild* and in the other *RÖ* populations. Results on points 1 and 2 are summarized in Table V, and those on point 3 in Table VI.

1. Lethals #01, 03, and 07 are point mutations without effects on crossover rates. Another point mutation (lethal #04) has slightly reduced crossover frequencies in its vicinity, and this is apparently due to a closely neighboring inversion. Two lethals are located within inversions, lethal #05 in the left and 09 in the right arm. The association of these lethals with inversions was clearly deducible from their thorough effects on crossover rates and was verified cytologically by Wienberg (1983). Locations of lethals and their association with inversions explain the limitations in linkage and the "breaking up" of some closely linked groups of lethals as described above. They point to a balanced polymorphism between inversions. The balanced system may occasionally include point mutations at corresponding sites such as lethal #01 or 03.

2. In line with this interpretation, the high frequencies (about 20% each) of lethals associated with inversions (represented by lethals #04, 05, and 09) did not decrease between 150 and 217 generations of $R\ddot{O}I_{40}$. Moreover, frequencies of the three point mutations considered are higher than 0.01, a value expected with purifying selection according to $q_{217} = q_{150}/(1 + 67q_{150}) = 0.01$.

3. It is an interesting question, whether the system of balanced inversions and lethals evolved as a consequence of and as an adaptation to the irradiation conditions of an *RÖ* population or whether it was already present in the basic population *Berlin wild* and provided all *RÖ* populations with a kind of preadaptation. This was looked at in lethals isolated in tests 4 and 5 (according to Table II), respectively, from various populations (except $R\ddot{O}I_4$, which had then been discarded). They were studied for "allelism" with the six lethals dealt with. The results are summarized in Table VI. (With respect to $R\ddot{O}I_{40}$, the data are the same as in Table V to facilitate comparisons, and only test 5 is added.) It is not surprising that all lethals are also present in $R\ddot{O}I_{48}$, since this population was separated from $R\ddot{O}I_4$ shortly after $R\ddot{O}I_{40}$ and might still carry the same lethals. It is in line with the apparent heterozygous advantage of the lethals associated with inversions that these persisted for more than 180 generations in two sublines of $R\ddot{O}I_{40}$ with population size so drastically reduced (<50 pairs) that even a selectively neutral factor with an initial frequency of 0.2 should have disappeared due to random drift. It is indicative for the existence of the balanced system already in the basic population that it is present not only in *RÖI* and its since long unirradiated

TABLE V. Third-Chromosome Lethals of $RÖI_{40}$: Genetics, Cytogenetics, and Persistence[a]

Chromosome & lethal	Genetic locus of lethal	Crossover[b]			Cytological lethal site	Lethal frequency			
						Generations at 0 R			
		Locus	Left arm	Right arm		150	175	187	217
A & 01	3-43.5	=	=	(+)	Point mutation	0.12	0.06	0.06	0.05
G & 04	3-41.3	(−)	=	(+)	Left of In(3L)74A;76C	0.22	0.10	0.13	0.20
M & 05	3-In(3L)	−	−	(+)	In(3L)67B;79F	0.20	0.19	0.26	0.14
S & 09	3-In(3R)	−	+	−	In(3R)82E;82F & In(3R)87F;94A	0.19	0.27	0.13	0.21
F & 03	3-68.5	=	=	=	Point mutation	0.11	0.07	0.07	0.04
N & 07	3-10.6	=	=	=	Point mutation	0.08	0.10	0.02	0.02

[a] Lethals were isolated from $RÖI_{40}$ after 255 generations at 2.2 kR + 48 generations at 4.4 kR + 150 generations at 0 kR (test 3 in Table II) and tested for further persistence after 175, 187, and 217 generations at 0 R (tests 3 and 4). Chromosomes (letters) and lethals (numbers) are given according to complementation tests (Fig. 3). Genetic location was with markers *ru h th st cu sr e ca* and any locus given is based on >300 flies classified. Cytological data are from Wienberg (1983).

[b] Crossover rates are given for: vicinity of the lethal locus, i.e., between markers neighboring it; the left arm; and the right arm of the chromosome; rates above map values are given by +, below map rates by −, and equal rates by =; parentheses stand for less severe effects (20–40% deviation from map values).

TABLE VI. Noncomplementation between Frequent Third Chromosome Lethals from $RÖI_{40}$[a] and Third Chromosomes from Various $RÖ$ Populations at Later Test Intervals

Population	Generations with exposures of 2.2 +	4.4 +	8.8 +	0.0	Test in Table II	Lethals tested for complementation (*N*) and frequencies of specific lethals[b] *N*	Current no. of 6 specific lethals 01	03	04^{I}	05^{I}	07	09^{I}	Other
$RÖI_{40}$	255	48	0	150	3	83	0.12	0.11	0.22	0.20	0.08	0.19	0.12
	255	48	0	175	3	80	0.06	0.07	0.10	0.19	0.10	0.27	0.06
	255	48	0	187	3	100	0.06	0.07	0.13	0.26	0.02	0.13	0.08
	255	48	0	217	4	43	0.05	0.04	0.20	0.14	0.02	0.21	0.11
	255	48	0	321	5	24	0.01	0.00	0.09	0.03	0.00	0.01	0.02
$RÖI_{40}$[c]	255	48	0	237	—	35	0.00	0.00	0.10	0.11	0.00	0.10	0.04
$RÖI_{40}$[c]	255	48	0	237	—	38	0.00	0.00	0.00	0.35	0.00	0.00	0.03
$RÖI_{0}$	260			271	4	55	0.00	0.00	0.06	0.13	0.00	0.05	0.33
	260			370	5	28	0.00	0.00	0.02	0.00	0.00	0.00	0.15
$RÖI_{48}$	255	120	146	2	4	102	0.18	0.12	0.03	0.00	0.01	0.24	0.43
	255	120	243	2	5	80	0.00	0.00	0.08	0.02	0.00	0.07	0.60
RÖI	529			2	4	100	0.00	0.00	0.03	0.03	0.01	0.00	0.87
	625			2	5	131	0.00	0.00	0.03	0.00	0.00	0.10	0.72
RÖII	470			2	4	80	0.00	0.00	0.37	0.06	0.01	0.00	0.61
	570			2	5	130	0.00	0.00	0.12	0.00	0.03	0.00	0.79
+*K*				—	4	9	0.00	0.00	0.00	0.00	0.00	0.00	0.05

[a] Isolated from generation 255 + 48 + 150.

[b] The six specific lethals are denoted as in Fig. 2; associations with inversions (see Table V) are labeled by I. Frequencies of specific and of "other" lethals are calculated relative to all chromosomes of a test, i.e., to lethal chromosomes included in complementation analysis (*N*) plus those not included plus nonlethal chromosomes (see Table II for numbers of chromosomes). Total frequency may exceed lethal frequency from Table II or even 1.0 because of >1 lethal per chromosome.

subpopulation $RÖI_0$, but also in *RÖII*, a population separated from the *RÖI* populations from the beginning of irradiation history. The surprising point is, however, that within the 100 generations between tests 4 and 5 in all populations (including $RÖI_{40}$), the frequency of lethals studied decreased quite drastically.

The unexpected loss from all populations of until-then frequent lethals required additional complementation studies with lethals from all populations (isolated in test 5) to look for replacement of the balanced lethals by others. The results are summarized in Table VII. The treatment of data corresponds to that in Table IV. So do the results: In general, the number of different lethals per chromosome is Poisson distributed with similar means of 1.2–1.5 in the continuously irradiated populations *RÖI*, $RÖI_{48}$, and *RÖII* and of 0.16–0.30 in the unirradiated populations $RÖI_{40}$ and $RÖI_0$. Mean occurrence of identical lethals in different chromosomes is about 3 in all populations. Recurrence of lethals is Poisson distributed. Rates of allelism are equal in the continuously irradiated populations *RÖII* (5.5%), *RÖI* (4.9%), and $RÖI_{48}$ (4.3%). They correspond to the 5.2% obtained for second-chromosome lethals of $RÖI_4$. Rates of allelism are higher in the unirradiated populations $RÖI_0$ (14.3%) and $RÖI_{40}$ (39.1), and in the latter case these are the remainders of the formerly frequent lethals.

These results clearly indicate that the balanced system of third chromosomes disappeared from all *RÖ* populations between generations 540 and 620 (*RÖI* count) and that it was not replaced by another system.

Function and Suppression of Third-Chromosome Lethals

Two questions remain: What was the advantage of the balanced system of third-chromosome inversions and lethals, and why did it finally break down? Studies on functional properties may help to answer them. Results of some tests are summarized in Table VIII.

In the first part of Table VIII, survival to adulthood is compared for eggs deposited by $RÖI_{40}$ females that were either fertilized by $RÖI_{40}$ males or by males of a marker strain. Since all other experimental conditions were carefully controlled and as equivalent as possible, the difference in developmental rates can be attributed to recessive lethals of $RÖI_{40}$ origin that become homozygous only in the "*inter se*" cross. The total frequency of casualities due to such homozygous lethals is calculated to affect 4.33% of eggs deposited. Compared to this figure, the expected frequencies of homozygous lethals as calculated alone from the frequencies of the six third-chromosome lethals dealt with in the previous sections (see Table VI for frequencies of these lethals in tests 3 and 4) amounts to 15.7% (test 3) or 10.8% (test 4)—

TABLE VII. Number of Lethals per Chromosome and Frequency of Specific Lethals According to Complementation Tests Performed with Third-Chromosome Lethals from Various *RÖ* Populations[a]

	Results for chromosomes from given population (at generation 620–630 of *RÖI*)				
	RÖII	*RÖI*	$RÖI_{48}$	$RÖI_{40}$	$RÖI_0$
Lethal chromosomes tested for complementation	60	60	80	24	28
Total chromosomes "tested"[b]	64	71	104	150	169
Different lethal factors detected	32	31	47	7	15
Mean number of different lethals per chromosome (see N_1 below)	97/64 = 1.52	86/71 = 1.21	135/104 = 1.30	24/150 = 0.16	50/169 = 0.30
Mean occurrence of identical lethals in different chromosomes	97/32 = 3.03	86/31 = 2.77	135/47 = 2.87	24/7 = 3.43	50/15 = 3.33
N_1					
0	4	11	24	126	141
1	33	42	49	24	13
2	18	12	19	0	9
3	8	4	5	0	5
4	1	2	4	0	1
5	0	0	1	0	0
6	0	0	2	0	0
Sum	64	71	104	150	169

Difference from Poisson distribution, a	0.0027	0.0007	0.03	0.25	10^{-7}
N_2					
0	7	10	12	4	5
1	12	7	17	1	0
2	4	5	6	0	3
3	3	6	4	1	3
4	2	0	3	0	0
5	1	1	3	0	4
6	1	0	1	0	0
7	1	1	0	0	0
8	0	1	0	0	0
10	1	0	0	0	0
13	0	0	0	1	0
14	0	0	1	0	0
Sum	32	31	47	7	15
Sum ($N \times$ value)	65	55	88	17	35
Sum (total number of lethals)	97	86	135	24	50
Difference from Poisson distribution, a	0.06	0.16	0.01	—	—

[a] See Table II, test 5, for lethal frequencies. N_1 is the number of different lethals per chromosome; 0 = chromosomes without lethals = total chromosomes − lethal chromosomes tested. N_2 is the recurrence of identical (allelic) lethals in different chromosomes; 0 = no allelism, i.e., one lethal; 1 = one recurrence, i.e., occurrence in two chromosomes; etc.

[b] Total chromosomes = (total chromosomes tested for complementation)/(frequency of lethals in test).

TABLE VIII. Experiments on Functional Superiority of Six Heterozygous Third-Chromosome Lethals from $RÖI_{40}$[a]

Lethality expressed in $RÖI_{40}$ (test 4) and expected from specific lethals

Matings females × males	Observed egg-to-adult survival: Eggs tested	Percent survival	Percent death by $RÖI_{40}$ lethals	Percent death expected by homozygosity of frequent $RÖI_{40}$ lethals
$RÖI_{40} \times RÖI_{40}$	6517	84.90		15.74 (test 3)
$RÖI_{40}$ × Marker	8302	88.74	4.33[b]	10.82 (test 4)

Mating behavior, oviposition, and egg-to-adult survival of specific lethals from $RÖI_{40}$ balanced over *TM3* ($RÖI_{40}/TM3$, test 4)

Crosses (four pairs per vial): $RÖI_{40}$ chromosomes in $RÖI_{40}/TM3$[c] females × males	Number of females tested	Copulations: percent of females copulated within given time (min): 30	45	60	75	Oviposition: number of eggs per day and female	Development: adults in percent of eggs tested: *TM3*	+
Q × S	72	8	33	70	90	30.54	40.70	0.00
S × Q	72	20	33	80	All	50.32	41.62	0.00
$E_1 \times E_2$	72	20	42	All	—	49.71	34.23	0.00
$E_2 \times E_1$	72	4	8	8	12	30.92	39.62	0.00
$Q \times E_1$	67	0	8	16	25	30.78	36.47	20.85

$S \times E_2$	72	25	38	50	All	58.07	40.28	20.14
$E_1 \times S$	71	20	40	90	All	46.42	27.46	13.26
$E_2 \times Q$	72	12	12	20	25	32.57	39.02	20.68
$Q + E_1 \times S + E_2$	132					39.55	35.89	9.21
$S + E_2 \times Q + E_1$	132					45.23	40.58	12.58
$Q + E_2 \times S + E_1$	129					35.71	42.18	9.77
$S + E_1 \times Q + E_2$	132					42.75	35.57	7.76

Mating preferences of double heterozygotes for lethal chromosomes (test 5)

Chromosomes: A, E, G, and N, respectively, heterozygous with F, M, Q, and S, respectively
Number of hybrids tested: $4^2 - 1 = 15$ (E/F are lethal)
Number of mating combinations: $15^2 = 225$
Hybrid types per observation chamber: 4 per sex (marked by wing cuts)
Number of flies per chamber: 3 per type and sex (a total of $3 \times 2 \times 4 = 24$)
Number of chambers used: 28 with a total of 672 flies
Records: Successful matings within 60 min
ANOVA: No different effects of chromosomes in males on mating success
No different effects of chromosomes in females on mating success
No interactions between chromosomes in either sex on mating success
No interactions between chromosomes in various competitions

[a] Chromosomes according to Fig. 3, test generations according to Table II.
[b] Determined as $100(1 - 0.8490/0.8874) = 4.33\%$.
[c] See Fig. 3 for lethals carried by these chromosomes; note that E_1 and E_2 are two different chromosomes of the same constitution but different origin.

both these comparisons are given, because the experiments on egg-to-adult survival were performed between tests 3 and 4. This result demonstrates unambiguously that at least part of the lethals constituting the balanced system of third chromosomes is not expressed, in the $RÖI_{40}$ genome, in homozygous condition.

In the remainder of Table VIII, a closer look is attempted at features of the lethal chromosomes themselves that may prevent lethal expression in the $RÖI_{40}$ genome. In the middle part of the table, matings were set up between heterozygotes for the balancer *TM3* and one or the other of types E, Q, and S chromosomes (see Fig. 2; the two type E chromosomes, E_1 and E_2, are of different origin). The rationale for taking heterozygotes was that any presumed factor should be dominant because homozygotes are lethal. Different combinations were tested for mating success, oviposition rate, and egg-to-adult survival. It is apparent that copulations are delayed with E_2 females and E_1 males, respectively, and that this effect is additive in $E_2 \times E_1$ matings. Oviposition rates solely depend on females; they are high in *TM3/S* and $TM3/E_1$ (50 eggs per day per female) and much lower (30) in $TM3/E_2$ and *TM3/Q*. Egg-to-adult survival apparently depends solely on homozygous lethality: In crosses where the *RÖ* chromosomes carry allelic lethals, about 40% of the eggs develop to *TM3* adults (50% is the theoretical expectation with normal segregation and without casualities due to other factors than the lethals considered here). The same rate is observed when *RÖ* chromosomes carry nonallelic lethals, and here the proportion of eggs surviving to normal adults is half that surviving to *TM3* adults, as is theoretically expected. It has no effect on these findings if competition is allowed, in both sexes, between heterozygotes for *TM3* and different *RÖ* chromosomes.

In the bottom part of Table VIII, one remaining possibility was looked at, namely preferential matings between flies without allelic lethals. These would reduce the occurrence of homozygous lethals. This was tested in various combinations of males and females that were double heterozygous for various lethals. The scheme allowed for ample mating competition. But there was no indication for any preferential matings.

Taken together, the data summarized in Table VIII do not give any hint that the balanced system of third-chromosome lethals itself reduces the amount of expressed lethality.

Therefore, in a further search for the underlying mechanisms, the genetic background was taken into account. Unfortunately, experiments could only be performed after the system of third-chromosome lethals had practically been lost from the populations. Hence, suppression of recessive lethals by other chromosomes was tested in general. The population studied was *RÖX* after about 800 generations of irradiation history (*RÖI* count). Results are summarized in Table IX. They include suppression of X-chromosome

TABLE IX. Suppression of Recessive Lethals in *RÖX*

Generations irradiated (*RÖI* count)	Generations unirradiated prior to test	Number of chromosomes tested	Percent nonsuppressed lethals		Percent suppressed lethals
			Expected[a]	Observed	
Y suppression of sex-linked lethals					
764[b] + test	0	1168	10.66	11.04	6.16
764	0	2666	5.63	5.49	4.09
764	1	2686	3.12	3.06	7.16
764	2	4191	1.86	2.36	4.10
764	3	1592	1.21	1.26	7.54
Chromosome 2 suppression of third-chromosome lethals					
800[c]	1	309		93.2	2.9

[a] X-chromosome lethals expected with mutation and selection: mutation rates are 0.30% in females and 0.19% in males at 0 R (Table I), 0.28%/kR in females (stage 7; see Nöthel, 1990); selection acts solely on hemizygous males, with reduction of lethal frequency by one-half in any generation.
[b] Females exposed to 17.6 kR (see Fig. 1).
[c] Females exposed to 22.0 kR (see Fig. 1).

lethals by Y chromosomes, and suppression of third-chromosome lethals by second chromosomes. The difference between the two sets of results is that, in the former, general effects of the heterochromatic Y are considered (it is not of *RÖ* origin) whereas, in the latter, interactions between *RÖ* second and third chromosomes on third-chromosome lethality were studied. The amount of Y-suppressed lethals is considerable (5.8% of the X chromosomes tested). It is apparently not newly induced: the frequency remained constant over several generations without further exposures, during which the rate of orthodox lethals decreased as expected with selection acting via hemizygous males. The amount of chromosome 2-suppressed third-chromosome lethals is 2.9% among the third chromosomes tested. These data demonstrate, at least, that part of the lethals carried by *RÖ* populations is suppressed within an *RÖ* genome and is not expressed as load.

DISCUSSION

The data on lethal frequencies and viabilities given in the present chapter were sampled over 30 years from several *RÖ* populations. During this time, public interest in the subject diminished because, as Wallace (1989) has recently put it, "this question dominated much of the research of the 1950s." Also in my own interests, other aspects dominated later. One is the relative radioresistance that evolved in the irradiated populations. But lethals still tell interesting stories with respect to long-lasting elevated mutagen exposures. Population histories of 800 generations at high radiation exposures are unique. Therefore, data are presented on genetic load in the course of this long history.

Population size (Fig. 1) as well as mutational yield per generation (Table I) were large enough to expect considerable frequencies of recessive lethals. These were observed (Table II). In the continuously irradiated populations, they were slightly higher than expected. This is irrespective of assuming any purifying selection (Table III). In spite of this touch of "heterozygous advantage of lethals," Poisson distributions of lethals per chromosome and of recurrence of specific lethals among different chromosomes (Fig. 2, Tables IV and VII) both point to merely mutational occurrence of lethals. They are apparently in mutation–random drift equilibria as expected for rare and completely recessive lethals. There is no hint of any heterozygous disadvantage of lethals. This whole picture corresponds to what is known from the literature (Sankaranarayanan, 1964, 1965; Salceda, 1967). There are, however, two points that merit special attention:

1. The immense increase in lethal frequencies is not correlated with a decrease in fitness, i.e., there is practically no measurable load arising from recessive lethals. On the contrary: After any increase in per generation exposures, progeny numbers recovered until carrying capacity of the system was regained (Nöthel, 1990). This was due to two components: (a) Evolution of relative radioresistance that enabled a tenfold increase in per-generation exposures. (b) Elevated general viability that is indicated by increases in oviposition rates (more than doubled) and in tolerances to desiccation, to higher temperatures, to various toxic compounds (alcohol, caffeine, DDT), and to infections. Recovery of population sizes to carrying capacity following any increase in per-generation exposures demonstrates that recessive detrimentals occurring in the course of this irradiation history are not hazardous to the gene pool. They are rather "helpful" in that they are part of the newly arising genetic variability that enables adaptive processes. It is indicative for the superior fitness of the *RÖ* populations that none of them was lost during 30 years (*RÖI*$_4$ was discarded after *RÖI*$_{48}$ had been safely established). In the meantime, not only had several marker stocks been lost from our lab, but also all *Berlin wild* lines (Fig. 1). These had been kept independently by several researchers and technicians and yet were lost: In the original *Berlin wild* (+) a second chromosome detrimental increased in frequency to 100% (Bochnig *et al.*, 1965), and *+60* could not stand the failure of cooling facilities during the hot summer of 1969 (Nöthel, 1973).

2. A balanced system of third-chromosome lethals was present in *RÖI*$_4$. It was first observed in *RÖI*$_{40}$ after 150 generations without exposures (Fig. 3). It had frequencies of specific lethals of >20%. It was connected, in part, with inversions (Table V). It persisted for an additional 70 generations of *RÖI*$_{40}$. It was likewise observed in all other *RÖ* populations, but not in *Berlin wild K* (Table VI). In *RÖI*$_{40}$ it was not fully expressed as genetic load: egg-to-adult survival was not reduced as expected with statistically occurring homozygosity of these lethals (Table VIII). The frequency of this system drastically decreased after 1980 (Tables VI and VII). Two questions have to be answered, namely (a) which mechanism(s) prevented the expression of the lethals and (b) what was the adaptive advantage of the system and why did this advantage suddenly disappear in 1980–1981?

(a) Expression of lethality cannot be prevented by any type of reproductive superiority of heterozygotes or by segregation distortion in the *RÖ* genome, because these would equally affect detection (or frequency) of lethal chromosomes and expression of load. Mating disadvantages between carriers of allelic lethals might explain the nonexpression of load, but they were not observed (Table VIII). The remaining explanation is a suppression of lethals in the *RÖ* genome so that part of the lethals recovered in the various test schemes are "synthetic" in that they have been separated from their suppressors. The general occurrence of lethal suppressions in an *RÖ* genome has been shown (Table IX). Due to the loss of the balanced system of third-chromosome lethals from the *RÖ* populations, it could not be tested whether or not these lethals are especially suppressed.

(b) The adaptive advantage of the balanced system of third-chromosome lethals and inversions has to be seen within the time borders of its existence. The system was found in all *RÖ* populations, but not in the control population *Berlin wild K* (Table VI) [see also Wienberg (1983) for inversions].

It might therefore be considered an adaptive response to the irradiation conditions. But genes controlling relative resistance against the induction of genetic radiation damage (*rar* factors), are completely different in the two main branches of the phylogenetic tree, namely *RÖI* and *RÖII* (Nöthel, 1987). Moreover, only the factor *rar-3* is located on chromosome 3. It was a new acquisition of *RÖI*$_4$. It quickly became homozygous in this population. It is located genetically at 3-50.1 and cytologically between 86E2-4 and 87C6-7 (Nöthel, 1990). In chromosomes of type S from *RÖI*$_{40}$ (Table V), this locus is positioned between two inversions that are apparently never separated by recombination. It has recently been shown that *rar-3* originates from a transposon (Nöthel, 1990). This may explain why it occurs between the two inversions. In any case, homozygosity and this location of *rar-3* indicate that the lethal system of chromosome 3 cannot be part of the adaptive resistance, but that the latter evolved in spite of this system.

If the balanced system is present in all *RÖ* populations and is not part of the adaptive resistance, it may have existed from the very beginning of the irradiation history, i.e., already in the basic population *Berlin wild* (+). To the best of my knowledge, this has never been looked at. It may therefore be assumed that the balanced system already existed in the 1950s or even at the time of the work of Timoféeff-Ressovsky, when *Berlin wild* was caught outdoors. Why, then, should it so suddenly break down in 1980–1981? Going through old protocols, there never was any change in culture conditions that was reflected in altered experimental results. There is only one exception: As far back as records go (1936), all *Drosophila* cultures of our lab have suffered severe mite infections. Besides large white food mites, these were especially *Histiostoma* sp. whose hypopi at times tremendously crowded the genitalia of flies. These infections were less severe in cultures crowded by *Drosophila* larvae and therefore moist compared to dry, undercrowded vials. This is somewhat contradictory to the descriptions by Spencer (1950) or Ashburner and Thompson (1978). In any case, during the winter of 1980–1981 the mites disappeared from all our cultures after kitchen facilities had been greatly improved. They never came back. I consider this not an incidental but a causal correlation with the existence and breakdown of the third-chromosome lethal system: apparently, the balanced system enabled the flies to withstand somewhat easier the severe mite infections. This system was the more necessary, the less crowded the fly cultures were, as was the case, for example, with an *RÖ* population after any increase in per-generation irradiation level.

I consider +*K* as the *experimentum crucis* to this hypothesis. This population is only a subline of the long-lost basic populations (see Fig. 1). It was sent in 1958 to the lab of Prof. H. Traut, then in Karlsruhe. It returned to Berlin 5 years later after loss of the original *Berlin wild* (+) line. This traveling meant some bottlenecks in population size and different culture conditions for a total of about 130 generations. Those bottlenecks were mild, however, compared to the two $RÖI_{40}$ sublines maintained at <50 pairs for >180 generations and yet preserving part of the balanced system of third-chromosome lethals (see Table VI). *But this system was lost from +K!* I therefore asked Horst Traut, during the final preparation of this chapter, whether his lab in Karlsruhe suffered from mite infections. The answer was a clear-cut "no" (H. Traut, personal communication). This supports the "mite hypothesis."

Combinations between specific alleles of various genes are well known that are locked together in gene complexes or inversions. They are thus protected against separation by recombination. Their advantage is most often expressed in the various heterozygotes that, in general, are optimally adapted to fluctuating environmental conditions. This has apparently been the case with the third-chromosome system in our populations. However, the general disadvantage of heterotic inversion systems is that they will accumulate recessive lethals in considerable frequencies. Hitchhiking effects of lethals can be expected to be especially pronounced with elevated mutational input of lethals. This happened with the third-chromosome system, too. It is an interesting perspective that part of such lethals might be suppressed, perhaps in a way similar to Y-suppressed lethals, so that accumulated detrimentals are not expressed, in the segregating homozygotes, as genetic load. It is fascinating that an apparently old balanced system of this type cannot be overturned within more than 500 generations with extreme mutagen exposures and only breaks down after environmental changes render the system superfluous. This means as a conclusion that even extreme inputs of mutations alone will neither change the basic genetic composition of a population nor reduce its viability.

ACKNOWLEDGMENTS

It is a pleasure to acknowledge gratefully the generous help of Dr. V. Wieczorek, who started *RÖI* and *RÖII* and performed lethal tests of the first period. In the course of further irradiation history, technical assistance in the laborious work reported here was carefully contributed (in chronological order ranging over 25 years) by Almuth Aberle, Birgit Heinrich, Marianne Weber, Dieter Lotze, Meike Werner, Amei Bergholz, Ingrid Blochmann, Stephanie Pelz, Andrea Albrecht, and Ruth Franz. The final parts of this work were supported by grant No. 150/2-1 from the Deutsche Forschungsgemeinschaft.

REFERENCES

Ashburner, M., and Thompson, J. N., Jr., 1978, The laboratory culture of *Drosophila*, in: *The Genetics and Biology of Drosophila* (M. Ashburner and T. R. F. Wright, eds.), Vol. 2a, pp. 1–109, Academic Press, London.

Bochnig, V., Belitz, H. J., and Nöthel, H., 1965, Increase in frequency of a detrimental in a laboratory wild stock, *Drosophila Information Service* **40**:45.

Bonnier, G., Jonsson, U. B., and Ramel, C., 1958, Selection pressure on irradiated populations of *Drosophila melanogaster*, *Hereditas* **44**:378–406.

Bonnier, G., Jonsson, U. B., and Ramel, C., 1959, Experiments on the effects of homozygosity and heterozygosity on the rate of development in *Drosophila melanogaster*, *Genetics* **44**:679–704.

Carson, H. L., 1964, Population size and genetic load in irradiated populations of *Drosophila melanogaster, Genetics* **49:**521–528.

Cordeiro, A. R., Marques, E. K., and Veiga-Neto, A. J., 1973, Radioresistance of a natural population of *Drosophila willistoni* living in a radioactive environment, *Mutat. Res.* **19:**325–329.

Eiche, A., 1972, Effects of sublethal X-ray doses on the number of ovarioles in *Drosophila melanogaster* populations, *Hereditas* **71:**253–258.

Falk, R., 1967, Viability of heterozygotes for induced mutations in *Drosophila melanogaster.* III. Mutations in spermatogonia, *Mutat. Res.* **4:**59–72.

Green, E. L., 1968, Genetic effects of radiation on mammalian populations, *Annu. Rev. Genet.* **2:**87–120.

Hartl, D. L., 1980, *Principles of Population Genetics,* Sinauer Associates, Sunderland, Massachusetts.

King, R. C., 1970, *Ovarian Development in Drosophila melanogaster,* Academic Press, London.

Kratz, F. L., 1975, Radioresistance in natural populations of *Drosophila nebulosa* from a Brazilian area of high background radiation, *Mutat. Res.* **27:**347–355.

Lindsley, D. L., and Grell, E. H., 1968, *Genetic Variations of Drosophila melanogaster,* Carnegie Institution of Washington, publication No. 627.

Marques, E. K., 1973, The development of radioresistance in irradiated *Drosophila nebulosa* populations, *Mutat. Res.* **17:**59–72.

Morton, N. E., Crow, J. F., and Muller, H. J., 1956, An estimate of mutational damage in man from data on consanguineous marriages, *Proc. Natl. Acad. Sci. USA* **42:**855–863.

Muller, H. J., 1927, Artificial transmutation of the gene, *Science* **66:**84–87.

Muller, H. J., 1928*a*, The production of mutations by X-rays, *Proc. Natl. Acad. Sci. USA* **14:**714–718.

Muller, H. J., 1928*b*, The problem of genetic modification, in: *Verhandlung 5 International Kongress Vererbsforschung Berlin,* Vol. 1.

Muller, H. J., 1950, Our load of mutations, *Am. J. Hum. Genet.* **2:**111–176.

Nöthel, H., 1967, Der Einfluss von Röntgenstrahlen auf Vitalitätsmerkmale von *Drosophila melanogaster.* II. Untersuchungen über die Fekundität, *Strahlentherapie* **134:**607–624.

Nöthel, H., 1973, Investigations on radiosensitive and radioresistant populations of *Drosophila melanogaster.* III. Persistence of the lower radiosensitivity in stage-7 oocytes of *RÖI* females after treatment with oxygen and nitrogen pre- and post-irradiation, *Mutat. Res.* **19:**187–197.

Nöthel, H., 1987, Adaptation of *Drosophila melanogaster* populations to high mutation pressure: Evolutionary adjustment of mutation rates, *Proc. Natl. Acad. Sci. USA* **84:**1045–1049.

Nöthel, H., 1990, Mutagen–mutation equilibria in evolution, *Adv. Mutagen. Res.* **1:**70–88.

Poodry, C., Hal, L., and Suzuki, D. T., 1973, Developmental properties of the *shibire*^{ts1}, a pleiotropic mutation producing larval and adult paralysis, *Devl. Biol.* **32:**373–386.

Salceda, V. M., 1967, Recessive lethals in second chromosomes of *Drosophila melanogaster* with radiation histories, *Genetics* **57:**691–699.

Sankaranarayanan, K., 1964, Genetic loads in irradiated experimental populations of *Drosophila melanogaster, Genetics* **50:**131–150.

Sankaranarayanan, K., 1965, Further data on the genetic loads in irradiated populations of *Drosophila melanogaster, Genetics* **52:**153–154.

Sankaranarayanan, K., and Sobels, F. H., 1976, Radiation genetics, in: *The Genetics and Biology of Drosophila* (M. Ashburner and E. Novitski, eds.), Vol. 1c, pp. 1089–1250, Academic Press, New York.

Spencer, W. P., 1950, Collection and laboratory culture, in: *Biology of Drosophila* (M. Demerec, ed.), pp. 535–590, Wiley, New York.

Timoféeff-Ressovsky, N. W., and Zimmer, K. G., 1947, *Das Trefferprinzip in der Biologie-Biophysik,* Vol. 1, S. Hirzel Verlag, Leipzig.

Wallace, B., 1956, Studies on irradiated populations of *Drosophila melanogaster, J. Genet.* **54:**280–293.

Wallace, B., 1958, The average effect of radiation-induced mutations on viability in *Drosophila melanogaster, Evolution* **12:**532–552.

Wallace, B., 1963, Further data on the overdominance of induced mutations, *Genetics* **48:**633–651.

Wallace, B., 1970, *Genetic Load: Its Biological and Conceptual Aspects,* Prentice-Hall, Englewood Cliffs, New Jersey.

Wallace, B., 1989, Analyzing variation in egg-to-adult viability in experimental populations of *Drosophila melanogaster, Proc. Natl. Acad. Sci. USA* **86:**2117–2121.

Wallace, B., and Blohowiak, C. E., 1985*a,* Rank-order selection and the analysis of data obtained by ClB-like procedures, *Evol. Biol.* **19:**99–146.

Wallace, B., and Blohowiak, C. E., 1985*b,* Rank-order selection and the interpretation of data obtained by ClB-like procedures, *Biol. Zentralbl.* **104:**683–700.

Wienberg, J., 1983, Cytologische Untersuchungen an bestrahlten Populationen von *D. melanogaster,* Thesis, Freie Universität Berlin.

5

Evolutionary Adaptation and Stress

The Fitness Gradient

P. A. PARSONS

THE "SUBSIDY–STRESS" GRADIENT

Following Hoffmann and Parsons (1991), stress is defined to be an environmental factor that causes a potentially injurious change in a biological system. Under this definition, stress has a major impact on many evolutionary and ecological processes because of irreversible damage irrespective of the density or frequency of organisms. Stresses at this level of severity tend to be physical. Quantitative resistance traits that can be related to the physical features of the environment and are important in determining the distribution and abundance of organisms (Andrewartha and Birch, 1954) are therefore emphasized. In many widespread species such ecological phenotypes vary clinally in a manner predictable from *a priori* considerations of climatic variation, whereby periods of short-term extreme temperature and desiccation stress are effective in defining climatic races (Parsons, 1987; Hoffmann and Parsons, 1991). Under these circumstances competition is usually unimportant, but may transiently interact with stress resistance by narrowing tolerance limits (C. H. Peterson and Black, 1988).

Odum *et al.* (1979) view stress as an agent placing an organism or an ecosystem at a disadvantage requiring continued expenditure of excess energy ultimately threatening survival. This is likely at species boundaries,

P. A. PARSONS • Waite Institute, University of Adelaide, Glen Osmond, S.A. 5064, Australia.

Evolutionary Biology, Volume 26, edited by Max K. Hecht *et al.* Plenum Press, New York, 1992.

where environmental stress levels normally exceed those of more central areas (Parsons, 1991*a*). A general early warning sign of the stressed state may be increased respiration, since repairing damage caused by stress requires increased energy expenditure. In this way, physical stresses pose costs for the maintenance of normal protoplasmic homeostasis and functioning of membranes and enzymes. A widely applicable biochemical approach to stress assessment is via measures such as the adenylate energy charge (AEC), which quantifies the amount of metabolically available energy stored in the adenine nucleotide pool (Atkinson, 1977). The AEC can be used to express the deviation of an organism from a stready metabolic state, and is reduced under stress when metabolic energy resources are drained, so increasing the vulnerability of the organism to further stress. The AEC falls from around 0.8–0.9 in optimal and nonstressed organisms to 0.5 where conditions are so severely stressful that a return to normal conditions is associated with unrestored fitness losses (Ivanovici and Weibe, 1981).

This discussion of stress is largely restricted to natural selection acting on genetic variants differing in metabolic efficiency which may occur when there is differential impairment or mortality of genotypes after exposure to stress. Shorter-term effects discussed more extensively elsewhere (Hoffmann and Parsons, 1991) include stress evasion by behavioral means or the development of quiescence, the induction of heat-shock proteins, and acclimation.

In contrast, environmental perturbations at levels occurring regularly in natural habitats may lead to a favorable deflection, or "subsidy." For example, plant productivity may be maximized under moderate flooding compared with reductions under high flooding. Field investigations on the relationship between drought stress on bush beans, *Phaseolus vulgaris,* and outbreaks of the two-spotted spidermite, *Tetranychus urticae,* showed that mite populations were highest on well-watered and on severely stressed bean plants, and lowest on slightly-to-moderately-stressed plants (English-Loeb, 1990). Such nonlinear relationships may be a feature of the interaction between plant stress and insect performance, especially for bark beetles and sucking insects (Larsson, 1989).

Plotting the consequences of a stressor from zero to lethality can give a "subsidy–stress" gradient (Fig. 1) which is relatable to Darwinian fitness. This fitness gradient applies to many stressors. Some toxic substances are regarded as totally deleterious, but show a "subsidy" at the low levels in which they occur in natural habitats. This is an effect known for many decades from the field of pharmacology as hormesis.

The main aim of this chapter is to compare the evolutionary situation at the stressful region of the "subsidy–stress" gradient with the "subsidy" region. The stressful region corresponds to conditions likely to pertain at spe-

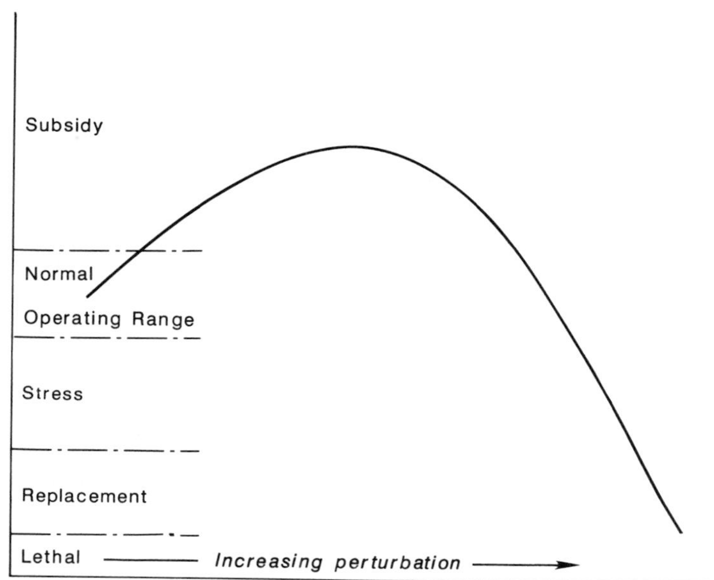

FIG. 1. Performance curve for a perturbed ecosystem indicating the "subsidy–stress" or fitness gradient. [Modified in Hoffmann and Parsons (1991) from Odum *et al.* (1979).]

cies borders as defined ecologically, while the "subsidy" region relates to benign and equable habitats. Comparisons between the stressful and "subsidy" regions are important in the light of recent difficulties in extrapolating from (1) the effects of chemicals that are demonstrably carcinogenic at high levels to the concentrations existing in background habitats (Ames and Gold, 1990*a,b*), and (2) the known deleterious effects of ionizing radiation at high doses and rates to the low doses and rates of natural habitats (Parsons, 1990*a*). Even if a "subsidy" is not apparent or detectable, there remains the general issue of extrapolating from high stress levels to equable levels. Finally, the analysis of the stressful region and the question of possible adaptation of organisms to more extreme stresses is important in interpreting responses to global climatic changes of the type predicted in the near future. Even if these generalized predictions are found to be wanting, rapid deforestation as in the Amazonian region is having an effect in terms of significant increases in surface temperature, a decrease in evapotranspiration and precipitation, and a longer dry season (Shukla *et al.*, 1990). Evolutionary change under stress is therefore a problem of general importance of consequence for both species distributions and biodiversity generally (Parsons, 1991*b*).

HABITATS AND FITNESS

Ionizing Radiation

Radiation protection criteria are largely based on the linear extrapolation of the harmful effects of radiation at high doses and rates to low doses and rates. This leads to a dilemma: efforts to reduce exposures at low levels are based on risks derived from radiation levels that are much higher. Any threshold below which harmful effects are reduced would greatly modify risk expectations. As much of this debate concerns quite small changes in radiation exposure and rare events such as genetic defects and carcinomas, any approach based on extrapolation presents an almost intractable assessment problem.

Organisms usually become increasingly adapted to those habitats where they live, and where Darwinian fitness is highest (Fisher, 1930). Highest fitness might then be expected at background radiation levels, or around 2–2.5 millisieverts per year (mSv/y) in most parts of the world (Eisenbud, 1987; Clarke and Southwood, 1989). Fitness at or close to zero radiation should be lower than at around background (Parsons, 1989*a*, 1990*a*). The estimation of fitness is not easy, since it is a measure of the capacity to survive, reproduce, and contribute to future generations (Wallace, 1981). However, fitness estimates involve all individuals in a population rather than a small minority as do studies of genetic defects and carcinomas. To replicate the situation in nature, radiation experiments should cover the whole life span. Radiation shielding devices are also needed to obtain radiation levels close to zero.

It is not surprising that the best experiments are from microorganisms. In the protozoan *Paramecium tetraurelia* and the cyanobacterium *Synechococcus lividus,* cell proliferation experiments indicate a lower fitness close to zero radiation in a shielding device than around background levels (Planel *et al.,* 1987). Furthermore, fitness increased when radionuclides were added to bring the γ-radiation level back to background and somewhat above—a result consistent with higher background radiation in certain parts of the world and in earlier geological epochs (Eisenbud, 1987; Moore and Sastry, 1982).

Fitness levels may therefore be highest at the low levels of radiation found in natural habitats. This is radiation hormesis, for which there are many claims (Luckey, 1982), mostly involving short and intense exposures, so that connections with natural levels are difficult; however, the evidence covers microorganisms, plants, invertebrates, and vertebrates, including humans (Hickey *et al.,* 1981; Nambi and Soman, 1987), where interpretations of epidemiological surveys are confounded by correlated variables.

Most claims for hormesis involve γ-radiation and X-radiation, with fewer reports dealing with β-radiation (Luckey, 1982). The α-emitter radon occurs in "man-made" habitats, such as houses and mines, at substantially higher levels than in nature (Clarke and Southwood, 1989), which implies an artificial habitat in terms of the evolutionary argument. Although lung cancer rates are below expectation in several parts of the world where radon levels are high (Cohen, 1987), other data show a linear relationship between risk and exposure (Clarke and Southwood, 1989). Henshaw *et al.* (1990) have shown that the incidence of various cancers in different countries is significantly correlated with radon levels in the home, but cohort studies with suitable controls are needed.

The radiation exposure giving maximum fitness should be similar in all taxa since it depends upon the background level; however, local variations could lead to minor differences which could be detectable in whole-of-life experiments in microorganisms. Such investigations should cover a range of natural and experimentally-manipulated exposures to ascertain the upper exposure threshold at which fitness falls below the "control" level close to zero radiation. For a range of traits, hormesis claims involve fitness increases above the control not normally exceeding 10% (Luckey, 1982), which means that the upper threshold may be difficult to estimate accurately.

Mechanisms leading to deviations from the linear extrapolation model following low levels of radiation include the enhancement of DNA repair, increase of free radical scavengers, and the stimulation of antigen production (Sagan, 1989; Wolff, 1989; Sagan and Cohen, 1990). These deviations would give a threshold below which harmful effects are reduced, and hormesis could occur as a consequence of any of these deviations either singly or in combination. At the population level, it is noteworthy that the most probable gametic doubling dose for chronic radiation is about one half of that for acute radiation for the children of parents exposed to atomic bombs (Neel *et al.*, 1990), indicating the difficulty of extrapolating across dose intensities.

A recent recommended International Commission for Radiation Protection (ICRP) exposure limit for members of the public is 5 mSv/y, with a move to reduce this to 1 mSv/y (Clarke, 1990). However, the evolutionary argument predicts that there is an effective zero in terms of environmentally deleterious radiation effects upon populations in the background range of radiation which exceeds 5 mSv/y in many parts of the world. This shows that there is a real need to introduce evolutionary thinking and hypotheses into this field, which has been dominated by models based mainly upon physics and statistics, since these models may overestimate the hazardous effects of low levels of radiation. Fitness estimates in populations of microorganisms exposed to high background levels of radiation are needed with some urgency to provide additional tests of the evolutionary hypothesis.

An Organic Metabolite

The underlying biochemical mechanisms are more readily understood for some organic metabolites than radiation and heavy metals for which there are many claims for hormesis. An example is acetaldehyde, an intermediary in the metabolism of ethanol to acetic acid, which is normally regarded as highly toxic in insects, rodents, and humans. However, it is a metabolite of high intrinsic activity (Moxon *et al.*, 1985) and low concentrations occur in nature, even if transiently (Holmes *et al.*, 1986). Hence the expectation is for higher fitness at low concentrations than at zero concentration. This is shown in Fig. 2, which gives adult survival expressed as mean LT_{50}, the mean number of hours at which 50% of flies had died at various

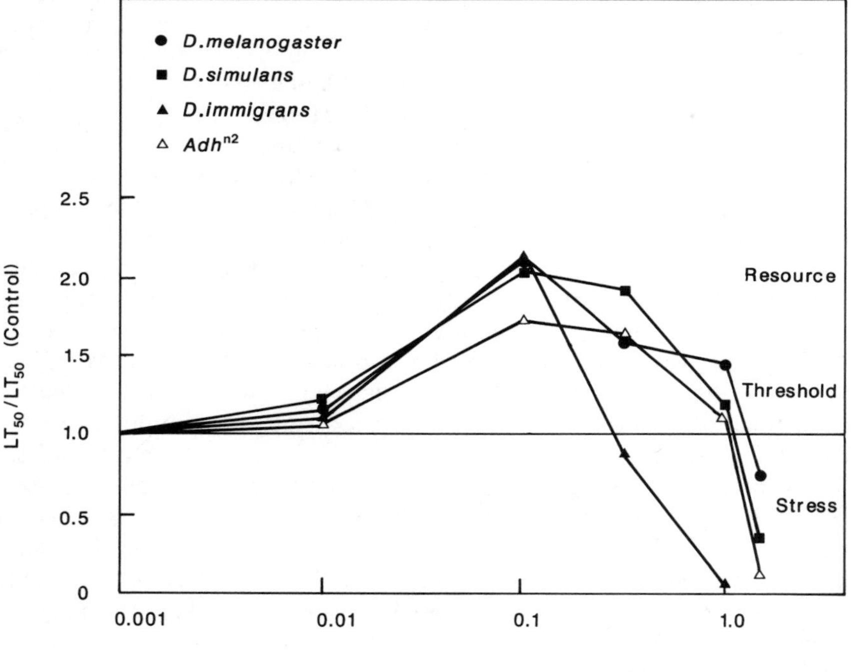

FIG. 2. Adult survivorship expressed as the ratio LT_{50}/LT_{50}(control) for various gaseous acetaldehyde concentrations for three *Drosophila* species and the Adh^{n2} mutant of *D. melanogaster.* The intersection of the plots with the horizontal straight line gives the threshold concentration between acetaldehyde as a resource and a stress. [Simplified from Parsons and Spence (1981*a*).]

gaseous acetaldehyde concentrations, divided by LT_{50} for water vapor for three *Drosophila* species and an alcohol dehydrogenase null mutant in *D. melanogaster,* Adh^{n2}. All four curves indicate enhanced survival under acetaldehyde concentrations up to around 1%, and at higher concentrations acetaldehyde is exceedingly toxic, giving a "subsidy–stress" gradient incorporating hormesis (Parsons, 1989*a*).

Adults and larvae should be attracted to low concentrations of acetaldehyde as found in nature, and this has been demonstrated in the laboratory (Parsons and Spence, 1981*a*). For larvae, attraction and repulsion occurred at similar concentrations to those enhancing adult survival (Fig. 1). In the case of adults, acetaldehyde above 1% repelled flies as expected, but the minimum concentrations eliciting responses (Hoffmann and Parsons, 1984) were far lower than those at which resources are demonstrably utilized, implying that acetaldehyde has an additional role as a "resource recognition compound" (Parsons and Spence, 1981*b*). The association with habitat is further emphasized by the observation of enhanced survival following exposure to 0.1% acetaldehyde vapor for *Asobara persimilis,* a braconid parasite of larvae of some *Drosophila* species (Owen, 1985).

Such habitat-related acetaldehyde hormesis has not been found in *Drosophila inornata* and *Drosophila hibisci,* which are not attracted to fermented-fruit baits, are electrophoretically *Adh* null, have very low levels of ADH activity, and survive poorly in the presence of ethanol (Holmes *et al.,* 1980; Moxon *et al.,* 1982). This indicates that for flies obtaining energy from alternative pathways, acetaldehyde is of little significance in terms of fitness and habitat-related tests.

Concluding Comments

Hormesis has been observed in plants and animals for many chemicals representing diverse chemical classes and measured by many biological and/or toxicological endpoints (Stebbing *et al.,* 1984; Calabrese *et al.,* 1987). For example, in animal studies, the range includes antibiotics, polychlorinated biphenyls, ethanol, polycyclic aromatic hydrocarbons, heavy metals, essential trace elements, pesticides, and various chemotherapeutic agents. The range of biological assessments includes weight, growth, hatching, tumor inhibition, longevity, reproductive span, and behavioral tests. Overall, these assessments incorporate substantial fitness components, so that the observation of hormesis implies a higher fitness in the presence of a chemical agent, or an environmental perturbation, at low levels than at the zero level, while at higher levels fitness falls progressively as the agent or perturbation becomes increasingly stressful.

Another example comes from laboratory population cage experiments with *Drosophila,* which tend to be carried out at constant temperatures. Provided that temperature fluctuations are not at levels exceeding those regularly found in nature which would be highly stressful, the evolutionary expectation is for higher fitness under fluctuating temperatures than under a constant temperature derived from the average of the fluctuating temperatures. In agreement, Oshima (1969) found that the fitness of *D. melanogaster* heterozygotes for lethal, semilethal, and quasinormal genes (those which are not lethal or semilethal or as homozygotes) was greater in a fluctuating environment of 25 ± 5°C with a mean of 25°C compared with a constant 25°C. This extends the "subsidy–stress" gradient to the characteristics of the physical habitat generally.

STRESS EXTREMES

Significance and Variability

Under extreme stress, a major difficulty concerns the selection of the elusive stress level where there are fertile survivors, yet lethality is close. These are conditions exemplified by short bursts of extreme stress occurring at climatic and ecological margins of species where just a small increased environmental perturbation would mean extinction. Experiments at this critical boundary pose logistic problems not apparent in most studies in experimental evolutionary biology where more optimal environments are chosen. The evidence for the importance of this boundary derives from a variety of sources discussed in detail by Parsons (1987) and Hoffmann and Parsons (1991), and include:

1. The biogeographic importance of extremes of heat and cold in determining the boundaries of the distributions of many species, together with supportive laboratory experiments on stress resistance.
2. Associations between extreme stresses and certain genotypes, including those determined electrophoretically in natural populations of various taxa, including *Drosophila, Dacus,* and *Avena.*
3. Observations of rapid and major morphological changes in natural populations under conditions of severe climatic stress, such as found in the Galapagos finch, *Geospiza fortis.*
4. The development of heat-shock proteins in response to extreme temperatures and stresses in general that on continuous exposure would be lethal; not surprisingly, since most natural environments are

scarce in nutrients, there are now reports of starvation and nutrient limitation proteins, some of which are induced by heat shocks (Matin *et al.*, 1989).

5. The importance of the adaptive response of enzymes to temperature stresses, especially in fish and higher plants.

These and additional examples cover integration levels ranging from the biogeographic to the molecular, and the stresses are typically of such intensity that a quite minor escalation could wipe out populations.

In considering survival under extreme stress, it is important to ascertain whether survivors are fertile (Stanley *et al.*, 1980). Table I gives the percent fertilities after exposure to high temperatures for 6 hr at 0% RH for six members of the *melanogaster* subgroup of species of *Drosophila*. Deviations from a sample size of 50 provide a measure of mortality from stress before carrying out the fertility test. Even when mortality from stress is exceedingly high, some flies still remain fertile, indicating that survival of extreme stress is a good guide to population continuity. More experiments would be useful incorporating changes in energy carriers such as measured by AEC.

In many organisms phenotypic and genotypic variability tend to be high under severe environmental stress, provided that the stress is not so severe as to cause complete lethality. However, much relevant data derive from experiments not primarily concerned with stress and variability. It is not then surprising that there are exceptions to this generalization. Indeed, in many

TABLE I. Percent Fertilities of *Drosophila* Species after Exposure to High Temperatures for 6 hr at 0% RH[a]

Species and sex		22°C	24°C	26°C	28°C	30°C	32°C
D. melanogaster	♂	—	98.0 (50)	100.0 (50)	74.4 (39)	56.9 (30)	60.0 (5)
	♀	—	100.0 (50)	100.0 (50)	75.0 (40)	51.2 (43)	72.2 (36)
D. simulans	♂	—	95.6 (46)	90.2 (41)	— (1)	33.3 (3)	— (0)
	♀	—	98.0 (49)	95.4 (43)	65.0 (20)	50.0 (6)	— (0)
D. mauritiana	♂	—	83.3 (42)	80.0 (15)	— (1)	— (1)	— (0)
	♀	—	84.4 (45)	88.9 (36)	90.3 (31)	64.7 (17)	— (0)
D. teissieri	♂	—	48.3 (31)	30.0 (30)	— (0)	— (0)	— (0)
	♀	—	90.7 (43)	100.0 (44)	35.0 (20)	42.9 (7)	— (0)
D. yakuba	♂	—	20.0 (25)	21.4 (14)	— (0)	— (0)	— (0)
	♀	—	79.6 (44)	30.0 (30)	— (0)	— (0)	— (0)
D. erecta	♂	72.7 (11)	66.7 (9)	— (0)	— (0)	— (0)	— (0)
	♀	92.0 (50)	77.8 (9)	— (0)	— (0)	— (0)	— (0)

[a] The numbers in parentheses are the sample sizes, and the magnitudes of deviations below 50 provide a measure of mortality from stress before carrying out the fertility test. After Stanley *et al.* (1980).

agricultural plants, heritability has been found to change with stress levels, but without consistent trends. To some extent this may be a reflection of the complication of phenotypic plasticity, which tends to be high in plant species and in populations experiencing nonextreme environmental fluctuations (Hoffmann, 1990; Hoffmann and Parsons, 1991). Given the need for additional work, especially extrapolations to field conditions, there is now, however, sufficient evidence to accept the association between stress and variability as a working hypothesis, at least in populations not previously subjected to natural selection for the stress in question. There are supportive data for the association on recombination, temperature-dependent catalytic properties of enzymes, and morphology assessed by fluctuating asymmetry (Parsons, 1988, 1990*b;* Karvountzi *et al.,* 1989). Parallel arguments appear in Holloway *et al.* (1990), who considered evolution in toxin-stressed environments, and argued for an increase in additive genetic variance when the population is transferred to a new (and stressful) toxin-rich environment.

Following from high genetic variability under severe stress, generalizations appear which are difficult to perceive under more optimal conditions. Correlations between resistances to diverse stresses at this level of intensity often occur because of an underlying genetic mechanism associated with reduced resting metabolic rate. Animal examples come from *Drosophila,* mole rats, and beef cattle (Nevo and Shkolnik, 1974; Frisch, 1981; Hoffmann and Parsons, 1989*a,b,* 1991). Simplistically, the availability of metabolic energy is expected to be positively correlated with increased fitness of life-history traits, which gives an expectation of positive correlations with metabolic rate. This is consistent with reduced fecundity, behavioral activity, and metabolic rate of stress-resistant strains of *D. melanogaster,* implying that fitness is reduced (Service and Rose, 1985; Hoffmann and Parsons, 1989*a,b;* Sierra and Comendador, 1989). Indeed, mechanisms increasing stress resistance may divert energy and other resources from growth and reproduction, so that resistance to environmental stresses may be negatively correlated with measures of many life-history traits, including fecundity, development time, and behavioral activity. Evidence for such associations is accumulating especially in *D. melanogaster.* The exception is longevity and associated traits, since stress-resistant individuals should live longer than more sensitive individuals because of their lowered metabolic rate. Experimental data from the hymenopterous parasite *Aphytis lingnanensis* and *D. melanogaster* tend to agree with this expectation (White *et al.,* 1970; Rose, 1984; Luckinbill *et al.,* 1988; Service *et al.,* 1988).

Associations of this nature depend upon severe stress to be realized fully, since under these conditions a high level of additivity in terms of genetic architecture is the expectation (Parsons, 1974), as shown by the study of longevity in *D. melanogaster* under a range of exposures to ^{60}Co γ-

radiation (Westerman and Parsons, 1973*a*). In other words, severe stress can be regarded as an "environmental probe" capable of revealing generalizations difficult to perceive under more optimal conditions, especially with regard to associations between life-history traits.

Energetic Limitations

The association of stress and reduced fitness is consistent with the view that species borders can be regions where the metabolic cost under stress may be restrictive for range expansions (Root, 1988*a,b;* Bozinovic and Rosenmann, 1989; Novoa *et al.,* 1990; Parsons, 1990*c,* 1991*a*). Energetic limitations at species borders as determined by maximum attainable metabolic rate would restrict normal physiological and behavioral processes, some of which are exceedingly expensive energetically, requiring almost all of the energy intake of an animal.

In rain beetles, *Pleocoma* spp., the energetic cost of walking is small relative to the cost of maintaining a temperature excess during sustained endothermy (Morgan, 1987) which releases the beetles from physiological constraints on activity that would otherwise be imposed on them by a low ambient temperature. This has the effect of enhancing terrestrial locomotion, the efficiency of searching for females, and so ultimately mating success. In view of the high energetic cost of such activities and the fact that adult *Pleocoma* do not eat, it is hardly surprising that a male remains active for only a few days and that the activity of males and females is closely synchronized.

Energetically expensive behaviors are features of many species at critical phases in their life cycle. The energetic cost of web construction in the web-building spider, *Angelena limbata,* ranges from 9 to 19 times the daily maintenance energy requirements (Tanaka, 1989). Not surprisingly, the daily rate of web relocation was below 1%, indicating high web-site tenacity. A comparison of various spiders shows that web relocation is more frequent when the energetic cost of web building is lower, indicating that the cost of web construction is important in determining the frequency of web relocation.

The energetic cost of flight is approximately 11 times the basal metabolic rate for many avian species (Goldstein, 1988); and other activities, such as locomotion, perching, preening, eating, singing, and calling, while substantially above the basal metabolic rate, tend to be less expensive energetically. Energetic constraints are now appearing in the literature to explain limits to avian distributions and abundances. Based upon the premise that external environmental factors such as temperature are primary in shaping biogeographic ranges, a continent-wide analysis of avian (passerine) distributions shows that the location of winter distribution and abundance patterns of several species are directly related to their physiological demands (Root, 1988*a,b*). An association of temperature extremes and physiological mechanisms was found at species borders, whereby birds are limited to regions where they do not need to raise their metabolic rates beyond about 2.5 times the basal level in order to stay warm. This means that on a broad scale, the winter ranges of a large number of passerines are limited by the energy expenditure needed to compensate for cooler ambient temperatures.

Acoustic signals often mediate the mating process through the action of female choice. The amount of energy involved in calling has been shown to exceed resting levels by up to 20 times the resting rate in frogs (Ryan, 1988). Hence, the mating process can be extremely expensive energetically, but needs to be assessed together with the costs of other activities, including nest-building, predatory, and escape behaviors (Bucher *et al.*, 1982). In a number of frog species, the mating success of males increases with increasing chorus tenure (Pough, 1989), whereby the total time a male spends in the breeding chorus explained more of the variability in mating success than any other variable. In other words, fitness assessed by mating success is relatable to available metabolic energy. In great tits, *Parus major,* and male pied flycatchers, *Ficedula hypoleuca,* in breeding condition, resting metabolic rate was found to be correlated significantly with dominance rank, whereby individuals with the highest metabolic rate were the most dominant (Røskoft *et al.,* 1986). A substantial degree of intraspecific variation in oxygen consumption rate was noted by these and other authors, and social status explained much of this variation. Hence, the social status of an individual is correlated with the oxygen consumption rate, whereby dominant individuals have the highest energy requirement. Somewhat parallel results were obtained in the fish *Betta splendens,* since winners and dominant individuals in a hierarchy produced more energy per unit time than losers and submissives (Haller and Wittenberger, 1988). In male sage grouse, *Centrocercus urophasianus,* daily energy expenditure increased significantly with increasing display rate and time spent on the lek (Vehrencamp *et al.,* 1989). Daily energy expenditure for the most vigorously displaying males was two times higher than for a nondisplaying male and four times higher than the basal metabolic rate. Furthermore, estimates of the instantaneous rate of energy expenditure during display ranged from 13.9 to 17.4 times basal metabolic rate.

These and other examples, especially from insects, where the amount of energy involved in calling may exceed the resting rate by up to 30 times (Ryan, 1988), show the substantial energetic costs in mating and in related behaviors in the broadest sense. At species boundaries, the energetic cost of stress would restrict successful mating as well as other energetically expensive but critical behaviors (endothermy, web construction, flight, etc.). Necessarily species continuity would then be threatened.

So far the restrictive environmental stress has been assumed to be climatic. However, disease is an important stress which interacts with other stresses. In mice, the effect of isolation or cold stress was investigated with West Nile virus (WNV). Exposure of mice for 5 min/day to cold water (1 ± 0.5°C) for 8–10 days showed 92% mortality compared with 47% in control mice, and mice stressed in isolation in individual cages showed 88% mortality compared with 50% in control mice (Ben-Nathan and Feuerstein, 1990). The data suggest that both physical and nonphysical stresses enhance WNV encephalitis by accelerating virus proliferation and so increasing mortality. Such data lead to the possibility of multiple-stress experiments, especially as there are a number of reports (Ben-Nathan and Feuerstein, 1990) of enhanced susceptibility to mice exposed to a variety of stress situations, including cold, high-intensity sound, and avoidance-learning stress. In addition, diseases can increase metabolic costs in mammals through higher body temperatures and metabolic rates, although reduced energy intake can also occur (Yuill, 1987). Similarly, threespine sticklebacks, *Gasterosteus aculeatus*, parasitized by the cestode *Schistocephalus solidus* have a greater need for energy than do unparasitized fish (Godin and Sproul, 1988).

More generally, increases in respiration have been recorded as an immediate response to stress in the environment, such as when larvae of the crustacean *Cancer irroratus* were exposed to Cu and Cd, and sea otters, *Enhydra lutris*, to crude oil (Johns and Miller, 1982; Davis *et al.*, 1988). The allocation of energy to countering stress of all kinds therefore increases immediate survival, but the energy available for reproduction, growth, and other processes is reduced, which threatens continuity across generations.

C. C. Peterson *et al.* (1990) analyzed sustained metabolic rates (SusMR), which are time-averaged metabolic rates that are measured in free-ranging animals maintaining constant body mass over periods long enough that metabolism is fueled by food intake rather than by depletion of energy reserves. In a survey of 37 species (humans, 31 other endodermic vertebrates, and 5 ectothermic vertebrates), they concluded that SusMR is between 1.5 and 7 times the resting metabolic rate, with most values being 5 and lower. When these limits are consistently exceeded, the existance of the species is threatened. In this way, the effect of stress of any type is to move the position of the population or species under test along the "subsidy–stress" or fitness gradient toward reduced fitness, and possibly to lethality and hence extinction. Energetically expensive behaviors at critical life-cycle stages would be particularly vulnerable in this fitness shift.

RELATIONSHIPS ACROSS THE GRADIENT

A dilemma arising from the previous section is that in stressful environments metabolic energy becomes restrictive under conditions when genetic variability is likely to be highest. Conversely, in benign habitats, especially in the "subsidy" region of the fitness gradient, the expection is for adequate metabolic energy and low genetic variability. In becoming adapted to equable environments, the additive genetic variance of fitness is expected to approach zero as fitness becomes maximized (Fisher, 1930); however, in natural populations this limit is not normally achieved. In a major survey of published data, Mousseau and Roff (1987) found that life-history traits (which have major fitness components) measured under nonstressful conditions generally have lower heritabilities than do morphological traits, while those for behavioral and physiological traits fall between these extremes (Table II). In *Drosophila,* heritabilities of life-history and behavioral traits are relatively low, but may exceed 0.20; those for morphological and physiological traits are higher (Roff and Mousseau, 1987).

While more data are needed, heritabilities for direct stress resistance traits may be substantially higher than those summarized in Table II; for example, >0.60 for desiccation resistance in *D. melanogaster* (Hoffmann and Parsons, 1989*a*). This suggests that under stress, any life-history or fitness trait associated with desiccation resistance should show increased heritability and additive genetic variance compared with less stressful conditions. In the cotton stainer bug, *Dysdercus bimaculatus,* the heritability for the timing of the first clutch increased from close to 0 under benign conditions to 0.3–0.4 under conditions of desiccation stress approximating levels in nature (Derr, 1980). Other examples in Hoffmann and Parsons (1991) mainly fit the expectation of increased additive genetic variance when organisms are placed in stressful habitats, and furthermore correlations between fitness

TABLE II. Means, Standard Errors, and Medians for Heritability Estimates from Data on Wild and Outbred Populations Collected by Mousseau and Roff (1987) from the Published Record for Life-History, Physiological, Behavioral, and Morphological Traits

	Life history	Physiology	Behavior	Morphology
n	341	104	105	570
Mean	0.262	0.330	0.302	0.461
SE	0.012	0.027	0.023	0.004
Median	0.250	0.262	0.280	0.428

traits tend in some circumstances to become more positive under stress (Service and Rose, 1985; Holloway *et al.*, 1990; Hoffmann and Parsons, 1991).

Assuming that some stress is a normal component of most habitats, a consequence would be higher heritabilities of fitness traits compared with the ultimate theoretical expectation of zero. Mousseau and Roff (1987) point out that many explanations have been advanced to explain the maintenance of the genetic variance of fitness traits above zero. These include heterozygote advantage, frequency-dependent selection, variable selection in heterogeneous environments, diversifying selection, migration, and antagonistic pleiotropy. These need to be considered in relation to the effects of stressful environments, which may have more predictable consequences.

These qualifications do not negate the conclusion that the "subsidy" and stress regions of the fitness gradient are extremes of a continuum for both genetic variability and fitness which vary inversely with each other. A testable prediction is that in situations where there is a significant "subsidy" the zero level is expected to be more stressful, and should show higher additive genetic variability and lowered fitness compared with the "subsidy" region.

The difficulty of extrapolating from the harmful effects of radiation at high doses and rates to low doses and rates has been discussed mainly in terms of experimental data. Neel *et al.* (1990) drew attention to this problem when considering the results of acute gonadal doses (mainly to males) of radiation in mouse experiments and their relationship to data from human exposures. From the human perspective, the mouse doses are unrealistically large. Therefore it appears likely that the exposures are in different regions of the fitness gradient. However, because of differences in the maturity of mice and humans at birth, plus radiation-induced litter-size effects that strongly influence pre- and postnatal survival, it is difficult to compare much of the mouse and human data. Based upon a linear relationship, a genetic doubling dose from human data is around 2 Sv for acute and 4 Sv for chronic radiation, and the human and mouse doubling-dose estimates for chronic radiation are very similar (Neel *et al.*, 1990; Neel and Lewis, 1990). Interestingly, the genetic risk for humans of low-dose-rate, low-LET radiation is considered to be less than has been generally assumed during the last 30 years. The dilemma is the large sums that are being used to reduce exposure to low levels of radiation based upon health risks obtained by linear extrapolation from radiation exposure levels much higher than those at around background levels, and implying comparisons across different regions of the fitness gradient.

Similar but clearer extrapolation problems are emerging in animal cancer tests, which are characteristically conducted at near toxic doses (the maximum tolerated dose, MTD) of the test chemical for long periods of time. Such a procedure can cause induced cell division, or mitogenesis, which plays a dominant role in carcinogenesis. Chronic dosing at the MTD can be thought of as a chronic wounding, which is known to be both a promoter of carcinogenesis in animals and a risk factor for cancer in humans. This means that a high percentage of all chemicals might be expected to be carcinogenic at chronic, near toxic doses, which is what is found. Indeed, about half of all chemicals tested chronically at the MTD are carcinogenic (Ames and Gold, 1990*a,b*). In other words, the tests are being carried out in a region of the fitness gradient that is distant from real environments. Ames and Gold (1990*a*) consider evolutionary implications:

> In the evolutionary war between plants and animals, animals have developed layers of general defenses, almost all inducible, against toxic chemicals. This means that humans are well buffered against toxicity at low doses from both man-made and natural chemicals. Given the high proportion of carcinogens among those natural chemicals tested, human exposure to rodent carcinogens is far more common than generally thought; however, at the low doses of most human exposures (where cell-killing and mitogenesis do not occur), the hazards may be much lower than is commonly assumed and often will be zero. Thus, without studies of the mechanism of carcinogenesis, the fact that a chemical is a carcinogen at the MTD in rodents provides no information about low-dose risk to humans. [Ames and Gold (1990*a*), p. 971.]

While unwilling to downplay the risks from chemicals in the environment, it is clear that an evolutionary perspective requires tests at levels actually present in natural habitats (Ames *et al.*, 1990), since these have most validity in predicting risk. Fitness experiments at around background levels in microorganisms in a parallel manner to the radiation experiments suggestive of hormesis should be informative, especially as there is some overlap between chemicals claimed as carcinogenic and/or clastogenic (ability to break chromosomes) at high doses in vertebrates with those for which hormesis has been found at low doses (Calabrese *et al.*, 1987).

MULTIPLE STRESSES: METABOLIC COSTS

Interactions and Metabolic Costs

Many chemicals are increasing in natural habitats due to human activities. Such pollutants, especially those in the atmosphere, may underlie present and predicted future climatic trends (Schneider, 1989). Global average temperature increases of 2–6°C may occur in the next century, but there are many uncertainties, which are considerably greater for variables such as soil moisture and precipitation patterns. To detect biological consequences, the arguments presented in this chapter suggest that a reasonable approach is to concentrate initially upon stressed marginal habitats close to the limits of distributions of species. For example, during the Little Ice Age, areas of marginal farmland in the uplands of northern Europe rapidly showed the effects of climatic shifts by the abandonment of cultivation (Parry and Carter, 1985).

An example from a marginal habitat in Canada suggestive of recent climatic change comes from Schindler *et al.* (1990), who analyzed 20 years of climatic, hydrologic, and ecological records for the Experimental Lakes Area of northwestern Ontario, and found that air and lake temperatures increased by 2°C and the length of the ice-free season by 3 weeks. Higher than normal evaporation and lower than average precipitation have decreased rates of water renewal in lakes. Concentrations of most chemicals have increased in lakes and streams because of decreased water renewal and forest fires in the catchments. Increased wind velocities, increased transparency, and increased exposure to wind of lakes in burned catchments caused thermoclines to deepen. As a result, summer habitats for cold stenothermic organisms like lake trout and opossum shrimp have decreased. A preview is therefore provided of the effects of greenhouse warming on such lakes.

It is in these ecologically extreme habitats that simplifications appear which are difficult to unravel in more benign regions. Under these circumstances (1) extreme events occur relatively often, (2) the frequency of such events may change dramatically following minor climatic changes, and (3) the chance of two successive extremes is even more sensitive to changes in the mean (Parry, 1978). Hence, in marginal regions, minor temperature shifts could have major biological (and agricultural) implications.

Greenhouse predictions are often expressed as temperature increases; however, the above example indicates that this stressor should not be considered in isolation, since desiccation stress is also important. Considering individual atmospheric pollutants, substantial additive genetic variability has been documented for SO_2 tolerance and ozone damage in various trees and herbs (Pitelka, 1988). In humans, on days of air pollution with high SO_2 levels, mortality is somewhat increased in heavily polluted cities such as Athens (Katsouyanni *et al.*, 1990). Reports of synergistic effects (Hutchinson, 1984) indicate the importance of interacting combinations superimposed upon direct effects. For example, forest decline in Europe depends upon several primary and a multitude of secondary stress factors (Hinrichsen, 1986). It is important to appreciate that interactions are most likely under conditions of severe stress, so that many negative and conflicting reports need to be assessed in terms of overall stress levels. Oppenheimer (1989) examined the interaction among several environmental variables—climatic change, UV-B enhancement, air pollution, and increasing nutrient fluxes. He makes the point that an important chemical outcome of atmospheric change is an increase in oxidant levels throughout the lower atmosphere and hydrosphere. These oxidants are phytotoxic, and contribute directly and indirectly along with other stresses on ecosystems to the acceleration of the S, N, and C cycles. With particular reference to climatic change in temperate forests, Oppenheimer emphasizes the likelihood of synergistic interactions between ozone, sulfur dioxide, and nitrogen oxides which would be exacerbated if temperature increases. Considering UV-B and ozone, Feder and Shrier (1990) have shown that combinations of these two stresses reduce pollen tube growth in *Nicotiana tabacum* and *Petunia hybrida* more than either stress does alone (Table III).

There is a lack of information on the evolution of resistance to multiple environmental stresses which might be expected when there are underlying metabolic costs. The possibility that responses to diverse stresses involve common molecular mechanisms should facilitate the evolution of increased resistance to multiple stresses. One suggested molecular mechanism concerns specific components of membrane lipids (Hochachka and Somero, 1984). Experiments in microorganisms might be useful in elucidating mechanisms; for example, in nondifferentiating *Escherichia coli,* starvation proteins were induced by both heat-shock and oxidative stresses (Matin *et al.*, 1989). Genetic changes in response to multiple stresses, particularly on the interaction of high-temperature stress with an array of more specific stresses such as desiccation, are needed, especially where hypotheses can be put for-

TABLE III. Comparison of Percent Reduction in Mean Tube Length and Range of Tube Length of Pollen Incubated for 24 hr after Treatment with O_3, UV-B, and a Combination of O_3 and UV-B[a]

Plant	Treatment	Mean tube length[b] (mm × 150)	SD	Percent reduction
Nicotiana tabacum	Control	194.8	47.3	0.0
Bel-W3	O_3	75.6	30.4	62.0
	UV-B	94.3	31.3	44.0
	UV-B + O_3	39.6	14.8	79.0
Petunia hybrida	Control	191.1	48.5	0.0
White	O_3	137.4	38.0	34.0
Cascade	UV-B	83.0	33.2	59.0
	UV-B + O_3	55.3	22.1	75.0

[a] Data from Feder and Shrier (1990).
[b] Mean tube lengths are significantly different from each other; $P < 0.001$, Wilcoxon–Mann–Whitney nonparametric test.

ward on energetic considerations. Cold temperature may involve rather different mechanisms, since in desiccation-resistance-selection experiments in *D. melanogaster* (Hoffmann and Parsons, 1989*a,b*) there was no association with cold shock, although cold shock elicits the expression of heat-shock proteins in *D. melanogaster* and the flesh fly, *Sarcophaga crassipalpis* (Burton *et al.*, 1988; Joplin *et al.*, 1990). Furthermore, there are reports on parasitic wasps and copepods that successful selection for resistance to cold may be associated with increased resistance to heat (Huey and Kingsolver, 1989). In addition, cold interacts with atmospheric gases, since winter stress is enhanced by exposure to ozone in peas, a result perhaps applicable to herbaceous species and trees generally (Barnes *et al.*, 1988).

The urgent need for looking at the direct effects and interactions among multiple pollutants is shown by a marine example. Coral species in a polluted estuary in Hong Kong are becoming locally extinct, and coral cover is being restricted to an increasingly smaller area of shallow coast. Growth rates are declining and are significantly lower in the more polluted sites. Heavy metal contaminants (Al, Cd, Cu, Pb, U, V, Y) are present in the skeletons of corals at high concentrations, and total acid-insoluble material measured by skeletal dissociation is high (Scott, 1990). As discussed earlier, exposure to heavy metals and oil pollution may have substantial metabolic costs, providing models for the effects of pollution on coral species.

Monitoring Atmospheric Pollution

The presence of atmospheric pollution brings up the problem of biological assessment. Stress-resistant strains of *D. melanogaster* developed by selection have a lowered metabolic rate and increased longevity (Hoffmann and Parsons, 1991). The reverse possibility is to develop high-metabolic-rate strains that should be stress sensitive with reduced longevity. On energetic grounds they should be very vulnerable to further environmental changes (Ivanovici and Weibe, 1981), and could have a role as sensitive monitors of the effects of relatively small environmental perturbations.

However, responses to directional selection for sensitivity appear difficult to achieve. An artificial stress showing associations with metabolic rate in parallel with desiccation resistance is the stress of extremely high doses of ^{60}Co γ-radiation. Using a sib-selection procedure, a five-generation selection program (Westerman and Parsons, 1973*b*) for resistance to 1.1 kGy of ^{60}Co γ-radiation assessed by mortality after 24 hr gave a realized heritability of 0.68 ± 0.21 for increased resistance and -0.02 ± 0.23 for increased sensitivity. In other words, obtaining stress-sensitive strains based upon natural populations was unsuccessful.

An alternative approach is via major mutants (Parsons, 1991*c*). A common consequence of mutants that do not inactivate an enzyme is heat sensitivity of the enzyme. Among many examples are recessive alleles at the β-galactosidase locus in *E. coli* (Langridge, 1968). In *D. melanogaster* numerous temperature-sensitive genes are known (Suzuki, 1970) as well as in other organisms. Temperature and stress sensitivity (including desiccation) are a feature of a number of deleterious behavioral mutants in *D. melanogaster* (Homyk *et al.,* 1980). These include "shakers" which shake their legs vigorously while etherized, as well as being abnormally active and showing an enhanced sensitivity to environmental stimuli and to other flies. Furthermore, A. R. Barros, L. M. Sierra and M. A. Comendador (personal communication) have shown that some of these mutants are unusually sensitive to acrolein stress.

Based upon the data of Trout and Kaplan (1970), various shaker mutant females are compared with standard Canton-S females in Table IV for metabolic rate and activity level, which are strongly correlated as shown by the rankings (0–5). The increase in metabolic rate is large for some mutants, especially Hk^1, when compared with the selection lines of Hoffmann and Parsons (1989*a*) for desiccation resistance where a 54% increase in LT_{50} for resistance was associated with a 20% decrease in metabolic rate. An approach via major mutant genes may therefore provide substantially more extreme phenotypes for metabolic rate than a reliance upon variation within natural populations (Parsons, 1991*c*). Finally, the rankings in Table IV show that high metabolic rate is associated with reduced longevity. Any increased stress should reduce longevity further by imposing an additional energetic cost, especially in high-metabolic-rate mutants such as Hk^1.

Trout and Hanson (1971) carried out a preliminary experiment on the effect of Los Angeles smog and found that at room temperature (20°C) smog

TABLE IV. The Metabolic Rate, Activity Level, and Longevity of Shaker and Canton-S Females of *Drosophila melanogaster,* with Ranks[a]

Strain	Metabolic rate (μl O_2/mg per hr)	Rank	Activity level (buzzes/fly per min)	Rank	Mean longevity (days after eclosion)	Rank
Hk^1	5.57	5	6.67	5	39.9	−5
Sh^5	5.01	4	4.87	3	40.5	−4
Hk^1Sh^5	4.75	3	5.94	4	59.5	−3
Hk^2	3.73	2	1.35	2	65.9	−1
Eag	3.44	1	.79	1	63.2	−2
Canton-S	3.39	0	0.56	0	76.2	0

[a] Adapted from Trout and Kaplan (1970).

reduced longevity by 9% in the Canton-S stock and 18% for Hk^1. In vials outdoors in the shade experiencing a temperature range of 3–30°C and averaging 15°C, the decrease in longevity was 9% for Canton-S and 33% for Hk^1. Therefore, both experiments show that Hk^1 is more sensitive to smog than is Canton-S. Furthermore, this sensitivity is greatest in the vials outdoors, where superimposed upon smog stress is stress from temperature extremes. Trout and Hanson (1971) conclude by commenting that, "A larger experiment with ozone, artificially generated at Los Angeles summer level, has shown that this major component of photochemical smog has a similar effect on the lifespan."

Positive correlations between fitness and metabolic rate have been discussed earlier in this chapter. However, the high-metabolic-rate mutants analyzed in Table IV are not found in natural populations, implying reduced fitness, as shown directly by Burnet *et al.* (1974), who found that Hk^{1P} males tend to switch rapidly from one behavior to another and have reduced mating speed caused by frequent loss of females during courtship. Extremes of high and low metabolic rates are therefore associated with reduced fitness as would be expected under stabilizing selection. More generally, while experimental data suggest a positive relationship between metabolic flux and fitness, Beaumont (1988) has shown that intrinsic selective constraints occur against the production of extreme phenotypic variants.

In summary, the high activity levels of some behavioral mutants in *Drosophila* form a major drain on metabolic energy, giving them the potential to be sensitive biological monitors of stress perturbations. Furthermore, the assessment period is short, enabling seasonal and locality differences to be monitored. Longevities following exposure to specific components of smog and mixtures of these components could be measured. For those stresses affecting metabolic rate, amplifications of separate effects are likely as well as interactions with stresses from climatic extremes.

The association between metabolic activity and sensitivity has been suggested in wider contexts, such as in explaining causes of forest damage in Europe (Prinz, 1987). Similarly, metabolically expensive behaviors in fish, such as swimming, are turning out to be useful indicators of sublethal toxicity (Little and Finger, 1990) covering a wide variety of species exposed to different chemicals under various conditions.

The use of stress-sensitive mutants as monitors of air pollution has been suggested for plants such as *Nicotiana* (Ashmore *et al.*, 1980). Meaningful results depend upon careful control and measurement of environmental conditions, and furthermore many other factors can produce symptoms similar to those from pollutants. There is also the problem of disentangling the two primary mechanisms of stress resistance and stress avoidance. Considering avoidance, one important mechanism is the exclusion of the pollutant from the plant by increased stomatal resistance (Koziol *et al.*, 1986). Consequently, it is not uncommon to find plants regarded as being of somewhat restricted use as pollution monitors (Moriarty, 1988). However, lichens are useful as indicators of atmospheric levels of SO_2 (Hawksworth and Rose, 1976). This is a pollutant that may modify resistance levels genetically within plant populations (Horsman *et al.*, 1979), and is of particular interest because of reports (Bennett *et al.*, 1990) that SO_2 and NO_2 together amplify their separate effects on plants. Indeed, both of these pollutants affect ATP levels and hence bioenergetic processes (Mansfield and Freer-Smith, 1981).

THE FITNESS GRADIENT

Very small undirected changes in either organism or environment approach a 50% chance of being advantageous, while very great changes will normally be maladaptive (Fisher, 1930). This means that as the magnitude of change increases, the probability of improvement will approach zero. Any perturbation not normally experienced is likely to move organisms to a region of reduced fitness on the fitness gradient. Such perturbations can be genetic and/or environmental, so that maximum fitness is expected to be associated with a minimum of genomic and environmental stress.

Domesticated organisms live in environments contrived by humans which are normally more uniform with respect to those physical and biotic factors determining distribution and abundance than is the situation in the wild. For example, captive animals are often housed in temperature-regulated facilities, where after some generations the capacity to adapt to extreme temperatures is reduced (Kohane and Parsons, 1988) in favor of maximizing fitness in a contrived, rather invariant environment. Such animals would then be vulnerable to climatic variation if released to the wild, implying a shift toward the stressed end of the fitness gradient. Similar shifts may occur in the wild under the rapid climatic change predicted under the greenhouse scenario, so that conservation strategies should be directed toward the preservation of the maximum potential of organisms to respond to directional selection for specified climate-related environmental factors. It suggests that conserving populations from more benign habitats may be a strategy of restricted use when the primary aim is the maintenance of high levels of stress resistance to ensure survival against future environmental change (Hoffmann and Parsons, 1991; Parsons, 1990*c*), since populations from benign habitats would need to be shifted greater distances toward the stressed end of the fitness gradient than populations already exposed to more stressful habitats. These comments imply that approaches to conservation genetics that depend mainly upon maintaining the overall level of genetic variability as measured electrophoretically may be somewhat limited, except under conditions of little or no climatic change (Parsons, 1989*b*). Conservation genetics therefore needs to incorporate the phenotypic and genetic analysis of specific stress-response traits demonstrably important in determining distribution and abundance, using populations from the varying habitats covering the range of a species emphasizing species boundaries. In management programs, genetic variation for stress-resistance traits need to be incorporated by incorporating heterogeneous conditions.

At the genetic level, the fall in fitness in the development of desiccation-resistant selection lines of *D. melanogaster* is an example of the general observation of a fall in fitness following genetic changes consequent upon periods of directional selection. Another example is the ease whereby an increase in resistance to the stress of heavy metals in *Agrostis tenuis* can be obtained, but at the cost of reduced competitive ability of genetically resistant plants compared with normal plants (Bradshaw, 1984).

An artificial form of genetic perturbation is the introduction of extra macromolecules in the formation of transgenic organisms. This implies (1) an energetic cost because of increased synthetic activity and (2) the disruption of normal physiological processes by genes promoting destabilizing metabolic excesses and unforeseen phenotypic effects. Lenski and Nguyen (1988) give many examples compatible with a fitness cost expressed as an energetic burden in transgenic organisms. In other words, the development of transgenic organisms is likely to be associated with a fall in fitness, as shown in transgenic pigs with high growth rate, which simultaneously have a high incidence of gastric ulcers, arthritis, cardiomegaly, dermatitis, and renal disease (Pursel *et al.*, 1989). Such fitness reductions may not be of concern under the relatively invariant environments contrived by humans, since transgenic organisms are likely to be selected for adaptation and survival to such conditions. However, exposing transgenic organisms to the variable and stressful environments of nature implies a fitness cost due to environmental perturbations superimposed upon the costs of the genetic changes involved in their formation (Parsons, 1990*d*). Therefore transgenic organisms may not normally pose a threat when released into natural habitats, although this does not obviate the need for proceeding cautiously. However, it does emphasize that evaluation and regulation depend upon the biological properties of transgenic organisms in their environments rather than the genetic techniques used to produce them (Tiedje *et al.*, 1989).

One way of assessing the consequences of stress is via fluctuating asymmetry (FA), since increased FA represents a deterioration in developmental homeostasis at the molecular, chromosomal, and epigenetic levels that is normally measured in adult morphology (Parsons, 1990*b*). Perturbations increasing FA can be genetic, including intense directional selection, and certain specific genes, or environmental, including temperature extremes, protein deprivation, audiogenic stress, and exposure to pollutants. Assuming that transgenic organisms are in a state of genomic stress, it follows that FA assessments may be a useful way of assessing their overall fitness. This suggestion is based upon an expectation of an inverse association of FA levels with fitness, as discussed in detail in Parsons (1990*b*).

STRESS AND REDUCTIONISM

Williams (1985) defined reductionism as, "the seeking of explanations for complex systems entirely in what is known of their component parts and processes." It is reductionist to regard gene frequency change as the essence of evolution, since, "such change is a bookkeeping process that records past reproductive success and predicts what is likely to happen in the future." This approach occurs in the "selfish-gene" view of natural selection and evolutionary change (Dawkins, 1976, 1982). On the other hand, Mayr (1984) emphasizes adaptive complexes which are favored by natural selection as the units of evolution. He stresses the unity of the genotype and the importance of gene interactions and balanced gene complexes. Controversies on the unit of selection have engaged many authors covering many perspectives (see, for example, Lewontin, 1970; Brandon and Burian, 1984), ranging from those who argue for the primary importance of contributions at the gene level to those who argue for the importance of epistatic interactions.

In any environment, organisms will accumulate over successive generations those variations fitting them best to their surroundings based upon *phenotypic variation* manifested in various ways, including morphological, physiological, and behavioral. As the environment changes, new variants will become advantageous and will tend to supplant less well-adapted variants. In other words, there is *differential fitness* whereby different phenotypes vary in survival and reproductive rates in different environments. Charles Darwin emphasized the importance of phenotypic variation and differential fitness in leading to natural selection via the organism–environment interaction regarded by many as central for an understanding of evolution. The effectiveness of this process depends upon the existence and modification of underlying *genotypic variation,* and there is continuing debate concerning the relative importance of the various components of this variation underlying evolutionary change.

Stress leads to an organism–environment interaction with the following consequences:

1. Phenotypic and genotypic variation is high.
2. There are major fitness differences among phenotypes.
3. Genetic variation underlying fitness and associated traits is high.

When considering stress resistance in terms of physiological change, its common association with reduced metabolic rate indicates a physiological variable of underlying fundamental significance. Suggestive associations between habitat, life-history characteristics, and stress resistance emerge which are difficult to perceive under more optimal conditions. Taking *D. melanogaster* as a case study, genetic correlations between stress resistance, low metabolic rate, and life-history traits such as fecundity, development time, and behavioral activity tend to be negative on the one hand, while the reverse is true for traits involved with longevity on the other. Stress can therefore be regarded as an "environmental probe" capable of revealing associations among an array of traits because of an underlying dependence upon metabolic rate.

The efficacy of stress as an environmental probe follows from the high heritability of stress resistance, and the substantial increase of the heritability of life-history traits when estimated under stressful conditions. Under these circumstances of high additive genetic variance, contributions at the gene level are emphasized, which is a reductionist paradigm. Unfortunately, our understanding of specific genes for stress-resistance traits is not great, a point applying even more to associated physiological traits. The gap from genes to metabolic rate, stress resistance, and fitness (under stressful conditions) needs to be explored, although it is likely that additive genes of large effect are involved.

The role of metabolic rate in this reductionist paradigm follows from the point that variations in metabolic rate can be reduced to the adenine nucleotide level, which is important in many reactions of intermediary metabolism. An involvement at this level incorporating stress has been suggested by Kohane (1988) in studies in larval fitness in *D. melanogaster.* Indeed, the AEC, which monitors changes in stress levels, is an expression of changes in concentrations of energy carriers (Atkinson, 1977). The likelihood of generalizations via substances at the adenine nucleotide level has been emphasized by Morowitz (1989) in his attempts to simplify the "matrix of biological information," and in descriptions of biological systems in terms of energetic costs (Hoffmann and Parsons, 1991).

Under stress, there is therefore the potential for reductionism to the gene level especially in environments approaching lethality on the fitness gradient; these occur in nature, as shown by studies at integration levels from the molecular to the biogeographic. In other words, the reductionist paradigm depends upon regular exposure to severe stress in natural habitats showing that its applicability is environment dependent. Extrapolations across stress levels can only be carried out with the utmost of caution.

Periods of environmental stress have a major role in the determination of species distributions and characteristics (Hoffmann and Parsons, 1991). This suggests that the reductionist paradigm could be close to universal, implying a convergence with those who regard the gene as the unit of natural selection. Rather than viewing the reductionism of G. C. Williams and R. Dawkins on the one hand and the adaptive complexes of E. Mayr on the other as alternatives, it may be more appropriate to consider these approaches as opposite ends of a continuum. This is because there are some organisms living in habitats where physical variations appear low, for example, arthropods restricted to cave habitats (Ahearn and Howarth, 1982; Humphreys and Collis, 1990). However, exposure to stress may be exacerbated by climatic changes in the near future, so it becomes difficult not to argue for the predominance of the reductionist approach. With special reference to life-history and fitness traits, the model is one of stress as an environmental probe lowering metabolic rate and magnifying additive genetic variability whereby genotype–phenotype relationships become direct and in many cases potentially reducible to the gene level. In other words, the organism can remain as the unit of selection, but following Williams' (1966) principle that adaptation should be attributed to no higher level of organization than is demanded by the evidence, reductionism is the expectation *if* extreme stress is a predominant environmental force underlying evolutionary processes.

SUMMARY

1. Impacts of environmental perturbations from zero to extreme change can be described in terms of a "subsidy–stress" gradient. This is also a gradient of fitness which is inversely associated with stress level.

2. A "subsidy" may appear under environmental conditions to which organisms are most commonly exposed; for example, highest fitness at around background levels of ionizing radiation is a possibility. This phenomenon of hormesis is expected on evolutionary grounds and occurs for many physical and chemical agents (including acetaldehyde) that are stressful at high levels.

3. The extreme end of the fitness gradient close to lethality is characterized by high additive genetic variability. Extrapolation across regions of the fitness gradient is therefore problematic, which is of direct importance in attempts (a) to use the known harmful effects of radiation at high doses and rates as indicative of the situation at background levels, and (b) to apply the results of animal cancer tests at near toxic levels to background exposure levels.

4. Any environmental or genetic perturbation will shift populations toward the stressful end of the fitness gradient, which would happen under climatic stress especially at species borders. Energetically expensive behaviors at critical life-cycle stages would be particularly vulnerable in this fitness shift.

5. The association of stress resistance with reduced metabolic rate means that metabolic limitations are likely to be restrictive in adapting to additional stresses due to climatic change or increases in pollution levels. The metabolic cost incurred in high-activity behavioral mutants in *Drosophila* makes them vulnerable to environmental perturbations, giving them the potential to be sensitive biological monitors of atmospheric pollution.

6. Energetic arguments suggest that the successful release of transgenic organisms should be inversely proportional to the magnitude of both their phenotypic effects and of the environmental perturbations of natural habitats. Fluctuating asymmetry of morphological traits may be a useful way of assessing the fitness of transgenic organisms.

7. Stress is an environmental probe which lowers metabolic rate, increases additive genetic variability, and reveals associations among life-history traits. Furthermore, genotype–phenotype relationships become direct and potentially reducible to the gene level. Since stress appears to be the norm in most natural populations, this reductionist approach to evolutionary change may have some generality, although caution is needed.

REFERENCES

Ahearn, G. A., and Howarth, F. G., 1982, Physiology of cave arthropods in Hawaii, *J. Exp. Zool.* **222**:227–238.

Ames, B. N., and Gold, L. S., 1990*a*, Too many rodent carcinogens: Mitogenesis increases mutagenesis, *Science* **249**:970–971.

Ames, B. N., and Gold, L. S., 1990*b*, Chemical carcinogenesis: Too many rodent carcinogens, *Proc. Natl. Acad. Sci. USA* **87**:7772–7776.

Ames, B. N., Profet, M., and Gold, L. S., 1990, Nature's chemicals and synthetic chemicals: Comparative toxicology, *Proc. Natl. Acad. Sci. USA* **87**:7782–7786.

Andrewartha, H. G., and Birch, L. C., 1954, *The Distribution and Abundance of Animals,* University of Chicago Press, Chicago, Illinois.

Ashmore, M. R., Bell, J. N. B., and Reily, C. L., 1980, The distribution of phytotoxic ozone in the British Isles, *Environ. Pollut. B* **1**:195–216.

Atkinson, D. E., 1977, *Cellular Energy Metabolism and its Regulation,* Academic Press, New York.

Barnes, J. D., Reiling, K., Davison, A. W., and Renner, C. J., 1988, Interaction between ozone and winter stress, *Environ. Pollut.* **53**:235–254.

Beaumont, M. A., 1988, Stabilizing selection and metabolism, *Heredity* **61**:433–438.

Ben-Nathan, D., and Feuerstein, G., 1990, The influence of cold or isolation stress on resistance of mice to West Nile virus encephalitis, *Experientia* **46**:285–290.

Bennett, J. H., Lee, E. H., and Heggestad, H. E., 1990, Inhibition of photosynthesis and leaf conductance interactions induced by SO_2, NO_2 and $SO_2 + NO_2$, *Atmos. Environ.* **24A**:557–562.

Bozinovic, F., and Rosenmann, M., 1989, Maximum metabolic rate of rodents: Physiological and ecological consequences on distributional limits, *Funct. Ecol.* **3**:173–181.

Bradshaw, A. D., 1984, Adaptation of plants to soils containing toxic metals—A test for conceit, in: *Origin and Development of Adaptation* (CIBA Foundation Symposium 102), pp. 4–19, Pitman, London.

Brandon, R. N., and Burian, R. M. (eds.), 1984, *Genes, Organisms and Populations: Controversies over the Units of Selection,* MIT Press, Cambridge, Massachusetts.

Bucher, T. L., Ryan, M. J., and Bartholomew, G. A., 1982, Oxygen consumption during resting, calling, and nest building in the frog *Physalaemus pustulosus, Physiol. Zool.* **55**:10–22.

Burnet, B., Connolly, K., and Mallinson, M., 1974, Activity and sexual behavior of neurological mutants in *Drosophila melanogaster, Behav. Genet.* **4**:227–235.

Burton, V., Mitchell, H. K., Young, P., and Petersen, N. S., 1988, Heat shock protection against cold stress of *Drosophila melanogaster, Mol. Cell. Biol.* **8**:3550–3552.

Calabrese, E. J., McCarthy, M. E., and Kenyon, E., 1987, The occurrence of chemically induced hormesis, *Health Phys.* **52**:531–541.

Clarke, R. H., 1990, A summary of the draft recommendations of ICRP 1990, *J. Radiat. Protection* **10**:143–145.

Clarke, R. H., and Southwood, T. R. E., 1989, Risks from ionizing radiation, *Nature* **338**:197–198.

Cohen, B. L., 1987, Tests of the linear, no-threshold dose–response relationship for high-LET radiation, *Health Phys.* **52**:629–636.

Davis, R. W., Williams, T. M., Thomas, J. A., Kastelein, R. A., and Cornell, L. H., 1988, The effects of oil contamination and cleaning on sea otters (*Enhydra lutris*) II Metabolism, thermoregulation, and behavior, *Can. J. Zool.* **66**:2782–2790.

Dawkins, R., 1976, *The Selfish Gene,* Oxford University Press, Oxford.

Dawkins, R., 1982, *The Extended Phenotype,* W. H. Freeman, Oxford.

Derr, J. A., 1980, The nature of variation in life history characters of *Dysdercus bimaculatus* (Heteroptera: Pyrrhocoridae), a colonizing species, *Evolution* **34**:348–357.

Eisenbud, M., 1987, *Environmental Radioactivity,* 3rd ed., Academic Press, Orlando, Florida.

English-Loeb, G. M., 1990, Plant drought stress and outbreaks of spider mites: A field test, *Ecology* **71**:1401–1411.

Feder, W. A., and Shrier, R., 1990, Combination of U.V.-B and ozone reduces pollen tube growth more than either stress alone, *Environ. Exp. Bot.* **30**:451–454.

Fisher, R. A., 1930, *The Genetical Theory of Natural Selection,* Clarendon Press, Oxford.

Frisch, J. E., 1981, Changes occurring in cattle as a consequence of selection for growth rate in a stressful environment, *J. Agric. Sci.* **96**:23–38.

Godin, J.-G. J., and Sproul, C. D., 1988, Risk taking in parasitized sticklebacks under threat of predation: Effects of energetic need and food availability, *Can. J. Zool.* **66:**2360–2367.

Goldstein, D. L., 1988, Estimates of daily energy expenditure in birds: The time-energy budget as an integrator of laboratory and field studies, *Am. Zool.* **28:**829–844.

Haller, J., and Wittenberger, C., 1988, Biochemical energetics of hierarchy formation in *Betta splendens, Physiol. Behav.* **43:**447–450.

Hawksworth, D. L., and Rose, F., 1976, *Lichens as Pollution Monitors,* Edward Arnold, London.

Henshaw, D. L., Eatough, J. P., and Richardson, R. B., 1990, Radon as a causative factor in induction of myeloid leukemia and other cancers, *Lancet* **335:**1008–1012.

Hickey, R. J., Bowers, E. J., Spence, D. E., Zemel, B. S., Clelland, A. B., and Clelland, R. C., 1981, Low level ionizing radiation and human mortality: Multi-regional epidemiological studies, *Health Phys.* **40:**625–641.

Hinrichsen, D., 1986, Multiple pollutants and forest decline, *Ambio* **15:**258–265.

Hochachka, P. W., and Somero, G. N., 1984, *Biochemical Adaptation,* Princeton University Press, Princeton, New Jersey.

Hoffmann, A. A., 1990, Acclimation for desiccation resistance in *Drosophila melanogaster* and the association between acclimation responses and genetic variation, *J. Insect Physiol.* **36:**885–891.

Hoffmann, A. A., and Parsons, P. A., 1984, Olfactory response and resource utilization in *Drosophila:* Interspecific comparisons, *Biol. J. Linn. Soc.* **22:**43–53.

Hoffmann, A. A., and Parsons, P. A., 1989*a,* An integrated approach to environmental stress tolerance and life-history variation. Desiccation tolerance in *Drosophila, Biol. J. Linn. Soc.* **37:**117–136.

Hoffmann, A. A., and Parsons, P. A., 1989*b,* Selection for increased desiccation resistance in *Drosophila melanogaster:* Additive genetic control and correlated responses for other stresses, *Genetics* **122:**837–845.

Hoffmann, A. A., and Parsons, P. A., 1991, *Evolutionary Genetics and Environmental Stress,* Oxford University Press, Oxford.

Holloway, G. J., Sibly, R. M., and Povey, S. R., 1990, Evolution in toxin-stressed environments, *Funct. Ecol.* **4:**289–294.

Holmes, R. S., Duley, J. A., Algar, E. M., Mather, P. B., and Rout, U. K., 1986, Biochemical and genetic studies on enzymes of alcohol metabolism: The mouse as a model organism for human studies, *Alcohol and Alcoholism* **21:**41–56.

Holmes, R. S., Moxon, L. N., and Parsons, P. A., 1980, Genetic variability of alcohol dehydrogenase among Australian *Drosophila* species: A correlation of ADH biochemical phenotype with ethanol resource utilization, *J. Exp. Zool.* **214:**199–204.

Homyk, T., Szidonya, J., and Suzuki, D. T., 1980, Behavioral mutants of *Drosophila melanogaster* III Isolation and mapping of mutations by direct visual observations of behavioral phenotypes, *Mol. Gen. Genet.* **177:**553–565.

Horsman, D. C., Roberts, T. M., and Bradshaw, A. D., 1979, Studies on the effect of sulphur dioxide on perennial ryegrass (*Lolium perenne* L.), *J. Exp. Bot.* **30:**495–501.

Huey, R. B., and Kingsolver, J. G., 1989, Evolution of thermal sensitivity of ectotherm preference, *Trends Ecol. Evol.* **4:**131–135.

Humphreys, W. F., and Collis, G., 1990, Water loss and respiration of cave arthropods from Cape Range, Western Australia, *Comp. Biochem. Physiol.* **95A:**101–107.

Hutchinson, T. C., 1984, Adaptation of plants to atmospheric pollutants, in: *Origins and Development of Adaptation* (CIBA Foundation Symposium 102), pp. 52–72, Pitman Books, London.

Ivanovici, A. M., and Weibe, W. J., 1981, Towards a working "definition" of "stress": A review and critique, in: *Stress Effects on Natural Ecosystems* (G. W. Barrett and R. Rosenberg, eds.), pp. 13–17, Wiley, New York.

Johns, D. M., and Miller, D. C., 1982, The use of bioenergetics to investigate the mechanisms of pollutant toxicity in crustacean larvae, in: *Physiological Mechanisms of Marine Pollutant Toxicity* (W. B. Vernberg, A. Calabrese, F. P. Thurberg, and F. J. Vernberg, eds.), pp. 261–288, Academic Press, New York.

Joplin, K. H., Yocum, G. D., and Denlinger, D. L., 1990, Cold shock elicits expression of heat shock proteins in the flesh fly, *Sarcophaga crassipalpis, J. Insect. Physiol.* **36**:825–834.

Karvountzi, E., Goulielmos, G., Kalpaxis, D. L., and Alahiotis, S. N., 1989, Adaptation of *Drosophila* enzymes to temperature—VI. Acclimation studies using the malate dehydrogenase (MDH) and lactate dehydrogenase (LDH) systems, *J. Therm. Biol.* **14**:55–61.

Katsouyanni, K., Karakatsani, A., Messari, I., Touloumi, G., Hatzakis, A., Kalandidi, A., and Trichopolous, D., 1990, Air pollution and cause specific mortality in Athens, *J. Epidemiol. Community Health* **44**:321–324.

Kohane, M. J., 1988, Stress, altered energy availability and larval fitness in *Drosophila melanogaster, Heredity* **60**:273–281.

Kohane, M. J., and Parsons, P. A., 1988, Domestication: Evolutionary change under stress, *Evol. Biol.* **23**:31–48.

Koziol, M. J., Shelvey, J. D., Lockyer, D. R., and Whatley, F. R., 1986, Response of SO_2-sensitive and resistant genotypes of ryegrass (*Lolium perenne* L.) to prolonged exposure to SO_2, *New Phytol.* **102**:345–357.

Langridge, J., 1968, Thermal responses of mutant enzymes and temperature limits to growth, *Mol. Gen. Genet.* **103**:116–126.

Larsson, S., 1989, Stressful times for the plant stress—Insect performance hypothesis, *Oikos* **56**:277–283.

Lenski, R. E., and Nguyen, T. T., 1988, Stability of recombinant DNA and its effect on fitness, *Trends Ecol. Evol.* **3**:518–520.

Lewontin, R. C., 1970, The units of selection, *Annu. Rev. Ecol. Syst.* **1**:1–18.

Little, E. E., and Finger, S. E., 1990, Swimming behavior as an indicator of sublethal toxicity in fish, *Environ. Toxicol. Chem.* **9**:13–19.

Luckey, T. D., 1982, Physiological benefits from low levels of ionizing radiation, *Health Phys.* **43**:771–789.

Luckinbill, L. S., Graves, J. L., Tomkiw, A., and Sowirka, O., 1988, A qualitative analysis of some life-history correlates of longevity in *Drosophila melanogaster, Evol. Ecol.* **2**:85–94.

Mansfield, T. A., and Freer-Smith, P. H., 1981, Effects of urban air pollution on plant growth, *Biol. Rev.* **56**:343–368.

Matin, A., Auger, E. A., Blum, P. H., and Schultz, J. E., 1989, Genetic basis of starvation survival in nondifferentiating bacteria, *Annu. Rev. Microbiol.* **43**:293–316.

Mayr, E., 1984, The unity of the genotype in: *Genes, Organisms, Populations: Controversies over the Units of Selection* (R. N. Brandon and R. M. Burian, eds.), pp 69–84, MIT Press, Cambridge, Massachusetts.

Moore, F. D., and Sastry, K. S. R., 1982, Intracellular potassium: ^{40}K as a primordial gene irradiator, *Proc. Natl. Acad. Sci. USA* **79**:3556–3559.

Morgan, K. R., 1987, Temperature regulation, energy metabolism and mate-searching in rain beetles (*Pleocoma* spp.), winter active, endothermic scarabs (*Coleoptera*), *J. Exp. Biol.* **128**:107–122.

Moriarty, F., 1988, *Ecotoxicology: The Study of Pollutants in Ecosystems,* 2nd ed., Academic Press, London.

Morowitz, H. J., 1989, Models, theory and the matrix of biological knowledge, *BioScience* **39**:177–179.

Mousseau, T. A., and Roff, D. A., 1987, Natural selection and the heritability of fitness components, *Heredity* **59:**181–197.

Moxon, L. N., Holmes, R. S., and Parsons, P. A., 1982, Comparative studies of aldehyde oxidase, alcohol dehydrogenase and aldehyde resource utilization among Australian *Drosophila* species, *Comp. Biochem. Physiol.* **71B:**387–395.

Moxon, L. N., Holmes, R. S., Parsons, P. A., Irving, M. G., and Doddrell, D. M., 1985, Purification and molecular properties of alcohol dehydrogenase from *Drosophila melanogaster:* Evidence from NMR and kinetic studies for function as an aldehyde dehydrogenase, *Comp. Physiol. Biochem.* **80B:**523–525.

Nambi, K. S. V., and Soman, S. D., 1987, Environmental radiation and cancer in India, *Health Phys.* **52:**653–657.

Neel, J. V., and Lewis, S. E., 1990, The comparative radiation genetics of humans and mice, *Annu. Rev. Genet.* **24:**327–362.

Neel, J. V., Schull, W. J., Awa, A. A., Satoh, C., Kato, H., Otake, M., and Yoshimoto, Y., 1990, The children of parents exposed to atomic bombs: Estimates of the genetic doubling dose of radiation in humans, *Am. J. Hum. Genet.* **46:**1053–1072.

Nevo, E., and Shkolnik, A., 1974, Adaptive metabolic variation of chromosomal forms in mole rats *Spalax, Experientia* **30:**724–726.

Novoa, F. F., Bozinovic, F., and Rosenmann, M., 1990, Maximum metabolic rate and temperature regulation in the rufous-collared sparrow, *Zonotrichia capensis,* from Central Chile, *Comp. Biochem. Physiol.* **95A:**181–183.

Odum, E. P., Finn, J. T., and Franz, E. H., 1979, Perturbation theory and the subsidy stress gradient, *BioScience* **29:**349–352.

Oppenheimer, M., 1989, Climate change and environmental pollution: Physical and biological interactions, *Climatic Change* **13:**255–270.

Oshima, C., 1969, Persistance of some recessive lethal genes in natural populations of *Drosophila melanogaster, Jpn. J. Genet.* **44**(Suppl. 1):209–216.

Owen, R. E., 1985, Utilization and tolerance of ethanol, acetic acid and acetaldehyde vapour by *Asobara persimilis* a parasitoid of *Drosophila, Entomol. Exp. Appl.* **39:**143–147.

Parry, M. L., 1978, *Climatic Change, Agriculture and Settlement,* Dawson, Folkeston, England.

Parry, M. L., and Carter, T. R., 1985, The effect of climatic variations on agricultural risk, *Climatic Change* **7:**95–110.

Parsons, P. A., 1974, Genetics of resistance to environmental stresses in *Drosophila* populations, *Annu. Rev. Genet.* **7:**239–265.

Parsons, P. A., 1987, Evolutionary rates under environmental stress, *Evol. Biol.* **21:**311–347.

Parsons, P. A., 1988, Evolutionary rates: Effects of stress upon recombination, *Biol. J. Linn. Soc.* **35:**49–68.

Parsons, P. A., 1989*a,* Acetaldehyde utilization in *Drosophila:* An example of hormesis, *Biol. J. Linn. Soc.* **37:**183–189.

Parsons, P. A., 1989*b,* Environmental stresses and conservation of natural populations, *Annu. Rev. Ecol. Syst.* **20:**29–49.

Parsons, P. A., 1990*a,* Radiation hormesis: An evolutionary expectation and the evidence, *Appl. Radiat. Isot.* **41:**857–860.

Parsons, P. A., 1990*b,* Fluctuating asymmetry: An epigenetic measure of stress, *Biol. Rev.* **65:**131–145.

Parsons, P. A., 1990*c,* The metabolic cost of multiple environmental stresses: Implications for climatic change and conservation, *Trends Ecol. Evol.* **5:**315–317.

Parsons, P. A., 1990*d,* Risks from genetically engineered organisms: Energetics and environmental stress, *Funct. Ecol.* **4:**265–271.

Parsons, P. A., 1991*a*, Evolutionary rates: Stress and species boundaries, *Annu. Rev. Ecol. Syst.* **22:**1–18.

Parsons, P. A., 1991*b*, Biodiversity conservation under global climatic change: The insect *Drosophila* as a biological indicator. *Global Ecol. Biogeog. Letters* **1:**77–83.

Parsons, P. A., 1991*c*, Can atmospheric pollution be monitored from the longevity of stress sensitive behavioural mutants in *Drosophila?, Funct. Ecol.* **5:**713–715.

Parsons, P. A., and Spence, G. E., 1981*a*, Acetaldehyde: A low-concentration resource and larval attractant in 3 *Drosophila* species, *Experientia* **37:**576–577.

Parsons, P. A., and Spence, G. E., 1981*b*, Longevity, resource utilization and larval preferences in *Drosophila:* Inter- and intraspecific variation, *Aust. J. Zool.* **29:**671–678.

Peterson, C. C., Nagy, K. A., and Diamond, J., 1990, Sustained metabolic scope, *Proc. Natl. Acad. Sci. USA* **87:**2324–2328.

Peterson, C. H., and Black, R., 1988, Density-dependent mortality caused by physical stress interacting with biotic history, *Am. Nat.* **131:**257–270.

Pitelka, L. F., 1988, Evolutionary responses of plants to anthropogenic pollutants, *Trends Ecol. Evol.* **3:**233–236.

Planel, H., Soleilhavoup, R., Tixador, G., Richoilley, G., Conter, A., Croute, F., Caratero, C., and Gawbin, Y., 1987, Influence on cell proliferation of background radiation on exposure to very low, chronic γ-irradiation, *Health Phys.* **52:**571–578.

Pough, F. H., 1989, Organismal performance and Darwinian fitness: Approaches and interpretations, *Physiol. Zool.* **62:**199–236.

Prinz, B., 1987, Causes of forest damage in Europe, *Environment* **29:**11–15.

Pursel, V. G., Pinkert, C. A., Miller, K. F., Bolt, D. J., Campbell, R. G., Palmiter, R. D., Brinster, R. L., and Hammer, R. E., 1989, Genetic engineering of livestock, *Science* **224:**1281–1288.

Roff, D. A., and Mousseau, T. A., 1987, Quantitative genetics and fitness: Lessons from *Drosophila, Heredity* **58:**103–118.

Root, T., 1988*a*, Energy constraints on avian distributions and abundances, *Ecology* **69:**330–339.

Root, T., 1988*b*, Environmental factors associated with avian distributional limits, *J. Biogeog.* **15:**489–505.

Rose, M. R., 1984, Genetic covariation in *Drosophila* life history: Untangling the data, *Am. Nat.* **123:**565–569.

Røskaft, E., Jarvi, T., Bakken, M., Bech, C., and Reinertsen, R. E., 1986, The relationship between social status and metabolic rate in great tits (*Parus major*) and pied flycatchers (*Ficedula hypoleuca*), *Anim. Behav.* **34:**838–842.

Ryan, M. J., 1988, Energy, calling, and selection, *Am. Zool.* **28:**885–898.

Sagan, L. A., 1989, On radiation, paradigms, and hormesis, *Science* **245:**574.

Sagan, L. A., and Cohen, J. J., 1990, Biological effects of low-dose radiation: Overview and perspective, *Health Phys.* **59:**11–13.

Schindler, D. W., Beaty, K. G., Fee, E. J., Cruikshank, D. R., DeBruyn, E. R., Findlay, D. L., Linsey, G. A., Shearer, J. A., Stainton, M. P., and Turner, M. A., 1990, Effects of climatic warming on lakes of the central boreal forest, *Science* **250:**967–970.

Schneider, S. H., 1989, The greenhouse effect: Science and policy, *Science* **243:**771–781.

Scott, P. J. B., 1990, Chronic pollution recorded in coral skeletons in Hong Kong, *J. Exp. Mar. Biol. Ecol.* **139:**51–64.

Service, P. M., and Rose, M. R., 1985, Genetic covariation among life-history components: The effect of novel environments, *Evolution* **39:**943–945.

Service, P. M., Hutchinson, E. W., and Rose, M. R., 1988, Multiple genetic mechanisms for the evolution of senescence in *Drosophila melanogaster, Evolution* **42:**708–716.

Shukla, J., Nobre, C., and Sellers, P., 1990, Amazonian deforestation and climate change, *Science* **247**:1322–1325.

Sierra, L. M., and Comendador, M. A., 1989, Selection for acrolein tolerance in *Drosophila melanogaster, Genet. Select. Evol.* **21**:253–267.

Stanley, S. M., Parsons, P. A., Spence, G. E., and Weber, L., 1980, Resistance of species of the *Drosophila melanogaster* subgroup to environmental extremes, *Aust. J. Zool.* **28**:413–421.

Stebbing, A. R. D., Norton, J. P., and Brinsley, M. D., 1984, Dynamics of growth control in a marine yeast subjected to perturbation, *J. Gen. Microbiol.* **130**:1799–1808.

Suzuki, D. T., 1970, Temperature-sensitive mutations in *Drosophila melanogaster, Science* **170**:695–706.

Tanaka, K., 1989, Energetic cost of web construction and its effect on web relocation in the web-building spider *Angelena limbata, Oecologia* **81**:459–464.

Tiedje, J. M., Colwell, R. K., Grossman, Y. L., Hodson, R. E., Lenski, R. E., Mack, R. N., and Regal, R. J., 1989, The planned introduction of genetically engineered organisms: Ecological considerations and recommendations, *Ecology* **70**:298–315.

Trout, W. E., and Hanson, G. P., 1971, The effect of Los Angeles smog on the longevity of normal and hyperkinetic *Drosophila melanogaster, Genetics* **68**:S69.

Trout, W. E., and Kaplan, W. D., 1970, A relation between longevity, metabolic rate, and activity in shaker mutants of *Drosophila melanogaster, Exp. Gerontol.* **5**:83–92.

Vehrencamp, S. L., Bradbury, J. W., and Gibson, R. M., 1989, The energetic cost of display in male sage grouse, *Anim. Behav.* **38**:885–896.

Wallace, B., 1981, *Basic Population Genetics,* Columbia University Press, New York.

Westerman, J. M., and Parsons, P. A., 1973*a,* Variation in genetic architecture at different doses of γ-radiation as measured by longevity in *Drosophila melanogaster, Can. J. Genet. Cytol.* **15**:289–298.

Westerman, J. M., and Parsons, P. A., 1973*b,* Asymmetrical responses to directional selection for radiation resistance and sensitivity in *Drosophila, Experientia* **29**:722.

White, E. B., Debach, P., and Garber, M. J., 1970, Artificial selection for genetic adaptation to temperature extremes in *Aphytis lingnanensis* (Hymenoptera: Aphelinidae), *Hilgardia* **40**:161–192.

Williams, G. C., 1966, *Adaptation and Natural Selection,* Princeton University Press, Princeton, New Jersey.

Williams, G. C., 1985, A defence of reductionism in evolutionary biology, *Oxf. Surv. Evol. Biol.* **2**:1–27.

Wolff, S., 1989, Are radiation-induced effects hormetic?, *Science* **245**:575.

Yuill, T. M., 1987, Diseases as components of mammalian ecosystems: Mayhem and subtlety, *Can J. Zool.* **65**:1061–1066.

6

Molecular Clones within Organismal Clones

Mitochondrial DNA Phylogenies and the Evolutionary Histories of Unisexual Vertebrates

JOHN C. AVISE, JOSEPH M. QUATTRO, and ROBERT C. VRIJENHOEK

INTRODUCTION

Some vertebrate "species"* exist predominantly or exclusively as females, exhibiting asexual or semisexual reproduction (Dawley and Bogart, 1989). Examples occur among the fishes, amphibians, and squamate reptiles. Essentially all known unisexual vertebrates carry the nuclear genomes of two or

* The usual definitions of biological species do not apply to unisexual forms. "Biotype" will be used in this chapter, though its meaning also may be unclear, unless the genomic constitution of a hybrid unisexual form is specified (Dawley, 1989). For sake of continuity with the literature, we will employ traditional Latin binomials where they have been assigned to unisexual biotypes, and use hybrid genomic designations (Schultz, 1969) where they have been applied, as summarized in Vrijenhoek *et al.* (1989).

JOHN C. AVISE • Department of Genetics, University of Georgia, Athens, Georgia 30602. JOSEPH M. QUATTRO • Center for Theoretical and Applied Genetics, Rutgers University, New Brunswick, New Jersey 08903-0231; *present address:* Hopkins Marine Station, Stanford University, Pacific Grove, California 93950. ROBERT C. VRIJENHOEK • Center for Theoretical and Applied Genetics, Rutgers University, New Brunswick, New Jersey 08903-0231.

Evolutionary Biology, Volume 26, edited by Max K. Hecht *et al.* Plenum Press, New York, 1992.

more bisexual species, and thus arose via interspecific hybridization. These all-female biotypes reproduce without genetic recombination, by one of three modes (Fig. 1): (a) parthenogenesis, in which the female's nuclear genome is transmitted intact to the egg, which then develops into an offspring genetically identical to the mother; (b) gynogenesis, in which the process is the same except that sperm from a related bisexual species is required to stimulate egg development; and (c) hybridogenesis, in which an ancestral genome from the maternal line is transmitted to the egg without recombination, while paternally-derived chromosomes are discarded only to be replaced in each generation through fertilization by sperm from a related sexual species.

Conventional wisdom holds that the rarity of unisexual reproduction in higher animals stems from both proximate and evolutionary factors. The window of opportunity for production of unisexual biotypes may be quite narrow—presumably, genetic differences between the hybridizing taxa must be sufficient to disrupt recombinant processes during gametogenesis, but not so great as to severely impair viability, fecundity, or other fitness components (Moritz *et al.,* 1989*a;* Vrijenhoek, 1989). Longer-term evolutionary constraints presumably involve a paucity of genetic variation by which unisexuals might adapt to changing environments, and the accumulation of deleterious mutations and gene combinations that cannot be purged in the absence of genetic recombination (Felsenstein, 1974; Kondrashov, 1988; Muller, 1964). Nonetheless, approximately 70 unisexual vertebrate biotypes have been identified (Vrijenhoek *et al.,* 1989), and they are common in some groups, such as *Cnemidophorus* lizards. Some unisexuals also have large

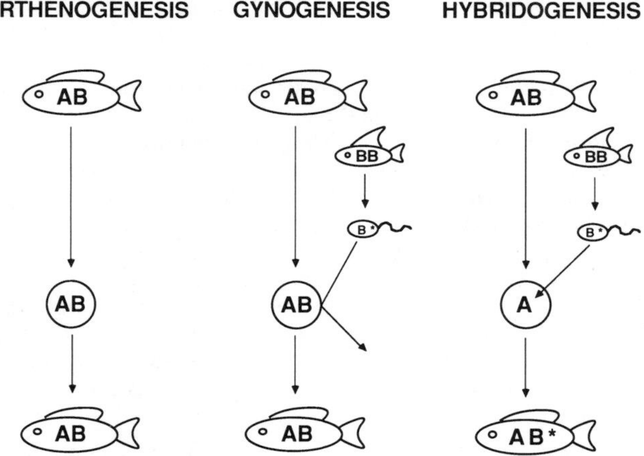

FIG. 1. Schematic summary of three modes of unisexual reproduction. [Redrawn from Dawley (1989).]

populations and occupy extensive ranges, suggesting that asexuality can be a successful evolutionary strategy at least in the short term (Vrijenhoek, 1990). How frequently do unisexual biotypes arise? What are the mechanics of their origin? Which bisexual species provided the male and female parents in the original hybridizations? Finally, how long do unisexual lineages survive? Answers to these and related questions are of interest in their own right, and may offer broader insights into the significance of sexual reproduction (Maynard Smith, 1978; Michod and Levin, 1988; Williams, 1975).

Here we review the recent literature that addresses these problems from the perspective of mitochondrial (mt) DNA lineages. The mtDNA molecule is uniquely favorable for analyses of unisexual complexes because it evolves rapidly in nucleotide sequence (Brown *et al.*, 1979) and exhibits maternal, nonrecombining transmission through organismal pedigrees (Avise and Vrijenhoek, 1987). Thus, mtDNA provides a common genetic yardstick by which to compare the magnitudes and patterns of maternal lineage separation in populations both of unisexual biotypes and their sexual progenitors. Furthermore, unlike the situation in sexually reproducing species, where each gene genealogy represents only a minute component of the organismal phylogeny (Avise, 1989), the transmission pathway of mtDNA within a unisexual biotype is in principle one-and-the-same as the organismal pedigree.

METHODS

Data Bases

Genetic analyses of mtDNA have thus far been conducted on more than 25 unisexual vertebrate biotypes and their sexual relatives (Table I). Most studies involved assay of 50–200 restriction sites per individual, revealed as digestion profiles produced either by many endonucleases with five- and six-base recognition sequences, or by a smaller number of four-base cutting enzymes. Typically, the authors calculated sequence divergence values (p) between mtDNA haplotypes using conventional formulas (e.g., Nei and Li, 1979), and generated estimates of phylogeny from raw data (presence–absence of restriction sites or fragments) or distance matrices.

Estimation of Genetic Diversity

For comparative purposes, we used the published estimates of sequence divergence in conjunction with sample sizes to calculate genotypic and nucleotide diversities according to the following formulas (Nei, 1987):

TABLE I. Directions of Cross Involved in Production of Unisexual Vertebrates[a]

Unisexual biotype	Ploidy level	Reproduction mode[b]	Bisexual parental species: Male	Bisexual parental species: Female	Ref.[c]
Cnemidophorus lizards					
C. uniparens	$3n$	P	*C. burti*	*C. inornatus* (2)	1
C. flagellicaudus	$3n$	P	*C. burti* (2)	*C. inornatus*	1
C. opatae	$3n$	P?	*C. burti* or *C. costatus* (2)	*C. inornatus* (2)	1
C. sonorae	$3n$	P	*C. burti* (2)	*C. inornatus*	1
Undescribed species "C," "N," "O," and "S"	$3n$	P?	*C. burti* or *C. costatus* (2)	*C. inornatus*	1
Undescribed species "P"	$2n$	P?	*C. burti* or *C. costatus*	*C. inornatus*	1
C. tesselatus	$2n$	P	*C. septemvittatus*	*C. marmoratus*[d]	2, 3
	$3n$	P	*C. sexlineatus, C. septemvittatus*	*C. marmoratus*[d]	2, 3
C. neomexicanus	$2n$	P	*C. inornatus*	*C. marmoratus*	2, 3
C. velox	$3n$	P	*C. inornatus* (2)	*C. burti* or *C. costatus*	4
C. exsanguis	$3n$	P	*C. inornatus, C. septemvittatus*	*C. burti* or *C. costatus*	4
C. laredoensis	$2n$	P	*C. sexlineatus*	*C. gularis*	5
C. lemniscatus	$2n$, $3n$?	P	Cryptic sp. in *C. lemniscatus*	Cytotype D in *C. lemniscatus*	6
Heteronotia lizards					
H. binoei (widespread form)	$3n$	P	*H. binoei*	Undescribed sp. "CA6"	4, 7

Ambystoma salamanders					
A. laterale-texanum forms	$2n$, $3n$	H?	*A. laterale, A. texanum, A. tigrinum*	*A. texanum*	8
A. 2-laterale-jeffersonianum	$3n$	H?	*A. laterale, A. jeffersonianum*	*A. texanum*	9
Rana frogs					
R. esculenta	$2n$	H	*R. lessonae, R. ridibunda*[e]	*R. ridibunda, R. lessonae*[e]	14
Menidia fish					
M. clarkhubbsi complex	$2n$	G	*M. beryllina*	*M. peninsulae*	11
Phoxinus fish					
P. eos-neogaeus	$2n$, $3n$	G	*M. eos*	*M. neogaeus*	12
Poecilia fish					
P. formosa	$2n$	G	*P. latipinna*	*P. mexicana*	13
Poeciliopsis fish					
P. monacha-lucida	$2n$	H	*P. lucida*	*P. monacha*	14, 15
P. monacha-occidentalis	$2n$	H	*P. occidentalis*	*P. monacha*	16
P. 2-monacha-lucida	$3n$	G	*P. lucida*	*P. monacha* (2)	17
P. monacha-2 lucida	$3n$	G	*P. lucida* (2)	*P. monacha*	17

[a] The bisexual parental species were identified previously from comparisons of morphology, karyotype, allozymes, geographic ranges, or other information. The female parent was determined from mtDNA genotypes. The ploidy level is that of assayed specimens.
[b] P, parthenogenetic; G, gynogenetic; H, hybridogenetic (see Fig. 1).
[c] References: 1, Densmore *et al.*, 1989*a*; 2, Brown and Wright, 1979; 3, Densmore *et al.*, 1989*b*; 4, Moritz *et al.*, 1989*a*; 5, Wright *et al.*, 1983; 6, Vyas *et al.*, 1990; 7, Moritz, 1991; 8, Kraus, 1989; 9, Kraus and Miyamoto, 1990; 10, Spolsky and Uzzell, 1986; 11, Echelle *et al.*, 1989; 12, Goddard *et al.*, 1989; 13, Avise *et al.*, 1991; 14, Avise and Vrijenhoek, 1987; 15, Quattro *et al.*, 1991; 16, Quattro *et al.*, 1992*a*; 17, Quattro *et al.*, 1992*b*.
[d] Treated as a subspecies of *C. tigris* by some authors.
[e] See text.

$$\text{genotypic diversity} = \frac{n}{n-1}(1 - \sum f_i^2)$$

where f_i is the frequency of the ith haplotype in a sample of size n, and

$$\text{nucleotide diversity} = \frac{n}{n-1} \sum f_i f_j p_{ij}$$

where f_i and f_j are the frequencies of the ith and jth haplotypes, respectively, and p_{ij} is the estimated sequence divergence between the ith and jth sequences. We also present UPGMA phenograms (Sneath and Sokal, 1973) for selected data sets. While these are not necessarily the best estimates of phylogeny, they offer the advantage of simple visual comparison of the depths and patterns of mtDNA divergence across studies. Furthermore, the phenograms in all cases considered faithfully reflected the qualitative patterns of phylogenetic relationships between unisexuals and their bisexual ancestors as gauged by alternative appraisals such as maximum parsimony.

The following sections summarize evolutionary information derived from mtDNA assays of unisexual vertebrates and their sexual relatives.

INHERITANCE OF mtDNA IN UNISEXUAL VERTEBRATES

Verification of Maternal Inheritance in Hybridogens

A unisexual fish, *Poeciliopsis monacha-lucida,* provided an unusually critical test of the possibility of paternal leakage of mtDNA in a hybridogenetic system. In the laboratory, crosses of *P. monacha* females × *P. lucida* males sometimes produce viable hybridogenetic lineages, whereas the reciprocal matings do not (Schultz, 1973). Naturally-occurring strains of all-female *P. monacha-lucida* presumably arose in this same fashion, and are perpetuated by hybridogenetic reproduction involving sperm from *P. lucida* males (Fig. 1). Thus, in effect, the unisexual biotype is maintained by continued backcrosses to *P. lucida,* although the *P. lucida* nuclear genome is lost prior to each hybrid meiosis. If even a minute fraction of zygote mtDNA derives from sperm in each generation, natural hybridogenetic strains should have accumulated considerable *P. lucida* mtDNA over time. Nonetheless, restriction assays failed to detect *P. lucida* mtDNA within natural or synthetic strains of *P. monacha-lucida* (Avise and Vrijenhoek, 1987). Results are consistent with strict maternal inheritance of mtDNA in these fishes.

Gynogens and Parthenogens

Unisexual vertebrates that reproduce by gynogenesis are probably less susceptible to sperm-mediated mtDNA input, since fusion of egg and sperm nuclei does not take place (Fig. 1). No evidence for paternal leakage of mtDNA was found in gynogenetic triploid forms of *Poeciliopsis* (Quattro *et al.*, 1992*b*). The role of sperm during fertilization of these fish is unknown. Does it just contribute a mechanical signal to divide, or does it provide an essential chemical input? Parthenogens do not require sperm at all, and thus are not susceptible to paternal leakage of mtDNA. For the remainder of this report, we assume strict maternal inheritance of mtDNA for all unisexual taxa considered.

ORIGINS OF UNISEXUAL VERTEBRATES

Direction of Hybridization Events Producing Unisexuals

Although the bisexual progenitors of most unisexual vertebrates were known or suspected from earlier comparisons of morphology, karyotype, allozymes, geographic range, or other information, the direction(s) of crosses had remained conjectural in essentially all cases. Mitochondrial assays have closed this information gap. For example, *Poecilia latipinna* and *P. mexicana,* sexual progenitors of the gynogenetic fish *P. formosa,* differ in mtDNA digestion profiles for 12 endonucleases, whereas the mtDNA of *P. formosa* is essentially indistinguishable from that in *P. mexicana* (Avise *et al.*, 1991). Thus, *P. mexicana* was the female parent of the assayed gynogens.

To date, similar mitochondrial inspections have allowed unambiguous determination of the female progenitor for more than 25 unisexual biotypes (Table I). Typically, the bisexual relatives of the unisexuals have proved highly distinct in mtDNA genotype, while mtDNAs of the unisexuals are closely related or indistinguishable from those of only one of the sexual progenitors. Thus, an emerging generalization is that most extant unisexual biotypes originated through asymmetrical hybridization events, occurring in one direction only (e.g., A♀ × B♂ versus B♀ × A♂). In many cases, mtDNA analyses have further pinpointed the geographic and genetic maternal source of the unisexuals. For example, nine unisexual biotypes in the *sexlineatus* group of *Cnemidophorus* lizards all appear to stem from females within one of the four nominate geographic subspecies of *C. inornatus*–*C. i. arizonae* (Densmore *et al.*, 1989*a*); and five triploid unisexual strains of *Poeciliopsis* in the *monacha–lucida* complex trace phylogenetically to mtDNA haplotype

"Mt-type M.2" observed in extant bisexual *P. monacha* from the Rio Fuerte in northwestern Mexico (Quattro *et al.*, 1992*b*).

One exception to such straightforward hybrid origins involves the hybridogenetic frog *Rana esculenta,* in which individuals exhibit mtDNA genotypes normally characteristic of either *R. lessonae* or *R. ridibunda* (Spolsky and Uzzell, 1986). *Rana esculenta* is unique among the assayed "asexual" biotypes in consisting of high frequencies of both males and females. From behavioral considerations, the initial hybridizations producing *esculenta* have been postulated to involve male *R. lessonae* × female *R. ridibunda.* Once the hybridogen was formed, occasional matings of male *R. esculenta* with female *R. lessonae* secondarily may have introduced *R. lessonae*-type mtDNA into *R. esculenta* (Spolsky and Uzzell, 1986). Furthermore, females belonging to such *R. esculenta* lineages appear to have served as a natural bridge for interspecies transfer of *R. lessonae* mtDNA into certain *R. ridibunda* populations via matings with *R. ridibunda* males (Spolsky and Uzzell, 1984). Such crosses apparently produced "*ridibunda*" frogs with normal nuclear genomes (because the *R. lessonae* chromosomes are excluded during meiosis), but *R. lessonae*-type mtDNA.

Another complex scenario surrounds the hypothesized maternal ancestry of the triploid salamander *Ambystoma* 2-*laterale-jeffersonianum,* which by allozyme evidence contains nuclear genomes of the bisexual species *A. laterale* and *A. jeffersonianum,* but reportedly carries mtDNA from *A. texanum* (Kraus and Miyamoto, 1990). Kraus and Miyamoto (1990) favor an explanation in which an original *A. laterale-texanum* hybrid female produced an ovum with primarily *A. laterale* nuclear chromosomes, but the female-determining sex chromosome (W) and the mtDNA of *A. texanum.* When fertilized by a male *A. laterale,* female progeny with two *A. laterale* nuclear genomes and the mtDNA of *A. texanum* would result. Subsequent hybridization with male *A. jeffersonianum* could then produce the observed *A.* 2-*laterale-jeffersonianum* biotypes carrying *A. texanum* mtDNA. While this scenario remains speculative, its mere feasibility suggests that distinct reticulate histories could characterize different genomic elements in some hybridogens.

All other unisexual biotypes in Table I appear to lack the sorts of peculiarities associated with hybridogenetic reproduction in the *Ambystoma* and *Rana* amphibians, and interpretations of maternal ancestry have been more straightforward.

Formation of Polyploids

About 64% of the known unisexual biotypes are polyploid (Vrijenhoek *et al.,* 1989). Schultz (1969) suggested that normal meiotic processes are disrupted in interspecific hybrids, such that triploids (for example) might have arisen through a hybrid intermediate that produced unreduced diploid eggs subsequently fertilized by haploid sperm (henceforth the "primary hybrid origin" hypothesis). Alternatively, under the "spontaneous origin" hypothesis advanced by Cuellar (1974, 1977), parthenogenetic triploids might have arisen when unreduced oocytes from a diploid nonhybrid were fertilized by sperm from a second bisexual species.

As diagrammed in Fig. 2, mtDNA assays permit a test of these competing hypotheses. If a unisexual biotype arose spontaneously from sexual ancestors and hybridization was involved only secondarily, the paired homospecific genomes should derive from the maternal parent, and thus should be coupled with mtDNA from the same species (i.e., the AA genome of AAB should be coupled with mtDNA type a, and vice versa for ABB). Conversely, under a primary hybrid origin, the paired homospecific genomes could be coupled either with mtDNA type a or b (depending on the details by which a nuclear genome was duplicated or added—see below and Fig. 2).

Such genetic analyses have provided strong support the "primary hybrid origin" hypothesis for 8 of 10 parthenogenetic *Cnemidophorus* lizards (Densmore *et al.*, 1989*a;* Moritz *et al.*, 1989*b*). For example, the triploid parthenogen *C. flagellicaudus* possesses the mtDNA of *C. inornatus* but two

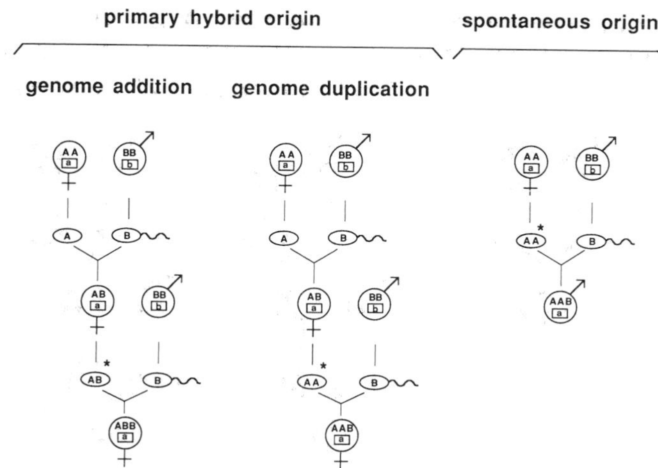

FIG. 2. Schematic representation of competing scenarios for the origin of triploid parthenogens: "hybrid origin" and "spontaneous origin" hypotheses. Each upper-case letter represents one nuclear gene set (A or B) from the respective parent species, and the lower-case letters in boxes similarly refer to the maternally-transmitted mtDNA genomes. Smaller ovals indicate sperm and eggs, the latter being unreduced where indicated by stars. In the genome duplication scenario, this suppression of reduction occurs during an equational division such that the AB hybrid produces AA (or BB) ova.

homospecific nuclear genomes from *C. burti* (Table I). Similarly, the triploid gynogenetic fish *Poeciliopsis monacha*-2 *lucida* possesses the mtDNA of *P. monacha* but two nuclear genomes from *P. lucida* (Quattro *et al.*, 1992*b*). In these cases, the genetic data clearly refute the "spontaneous origin" scenario, and support Schultz's (1969) postulated relationship between hybridization, unisexuality, and polyploidy in vertebrates.

Assuming correctness of the primary hybrid origin scenario, two further cytogenetic pathways to triploidy can be distinguished. Under the "genomic addition" scenario (Schultz, 1969), interspecific F_1 hybrids produce unreduced ova (AB) that upon backcrossing to one of the sexual ancestors leads to allotriploid biotypes AAB or ABB. Under the "genomic duplication" scenario (Cimino, 1972), suppression of an equational division in an F_1 hybrid could produce unreduced AA or BB ova, which upon backcrossing to species A or B would produce AAB or ABB offspring (Fig. 2) (autopolyploid AAA or BBB progeny could also result from this process, but no self-sustaining populations of autopolyploid unisexual vertebrates have been found). An important distinction between these pathways involves the predicted level of heterozygosity at loci in the homospecific nuclear genomes. Heterozygosity should be extremely low under the genome duplication pathway (the only variation being derived from postformational mutations), whereas normal heterozygosity is predicted under genomic addition. In triploid *Poeciliopsis* gynogens, all assayed strains proved to be heterozygous for homospecific nuclear markers at one or more allozyme loci, a result which effectively excludes the genome duplication hypothesis for these fishes (Quattro *et al.*, 1992*b*).

GENETIC DIVERSITY WITHIN UNISEXUAL AND BISEXUAL TAXA

Table II summarizes estimates of mtDNA genotypic and nucleotide diversity within unisexual taxa and the extant sexual descendants of their sexual maternal progenitors. Two major points are evident. First, mtDNA variability has been detected within most unisexual taxa. For example, several such "species" show genotypic diversity values greater than 0.50, indicating that random pairs of individuals are distinguishable with high probability, even under these limited genetic assays, which typically surveyed about 2% of the mtDNA genomic sequence. Second, as evidenced by a comparison of nucleotide diversities (Table II), mtDNA variation within the unisexuals is normally much lower than that within sexual relatives. Of 13 unisexual taxa studied (excluding *Rana esculenta* because its mtDNA derives from two ancestral species), 12 showed mtDNA diversities less than or

TABLE II. Estimates of mtDNA Diversity in Unisexual Vertebrates (U) and Their Maternal Bisexual Progenitors (B)[a]

Unisexual biotype	Sample size		Number of mtDNA clones		Genotypic diversity		Nucleotide diversity		Maximum p within species		Phylogenetic category[b] (minimum number of hybrid origins)
	U	B	U	B	U	B	U	B	U	B	
C. uniparens	18	25	8	8	0.75	0.40	0.002	0.024	0.006	0.068	Paraphyly (1)
C. tesselatus	55	21	3	3	0.17	0.50	0.001	0.001	0.013	0.020	Polyphyly (2)
C. velox[c]	8	5	5	2	0.89	0.27	0.003	0.007	0.007	0.012	Paraphyly (1)
C. exsanguis[c]	12	5	8	2	0.94	0.27	0.002	0.007	0.006	0.012	Paraphyly (1)
C. lemniscatus	45	19	4	13	~0.75[d]	~0.91	~0.001	~0.064	0.001	0.019	Paraphyly (1)
H. binoei	78	12	15	12	0.82	1.00	0.001	0.027	0.007	0.054	Paraphyly (1)
A. laterale-texanum	36	14	2	4	~0.51	~0.79	0.001	~0.038	0.003	0.088	Paraphyly (1)
A. 2-laterale-jeffersonianum	17	14	1	4	0.00	~0.79	0.000	~0.038	0.000	0.088	Paraphyly (1)
R. esculenta[e]	74	38	4	2	0.37	0.50	0.016	0.041	0.085	0.082	Paraphyly (1)
		40		2		0.33		0.011		0.034	Polyphyly (4)
M. clarkhubbsi	17	9	3	9	0.22	1.00	0.001	0.004	0.004	0.008	Polyphyly (2)
P. eos-neogaeus	7	4	3	2	0.67	0.50	0.007	0.007	0.013	0.010	Polyphyly (2)
P. formosa	15	13	1	4	0.00	0.52	0.000	0.001	0.000	0.004	Paraphyly (1)
P. monacha-lucida	20	24	8	9	0.85	0.78	0.005	0.003	0.009	0.011	Polyphyly (5)
P. monacha-occidentalis	38	24	5	9	0.44[f]	0.78	0.001[f]	0.003	0.003[f]	0.011	Paraphyly (1)

[a] Information is presented only from studies that involved seven or more assayed unisexuals, and only for those specimens assayed with most of the endonucleases employed. The value p is estimated sequence divergence, and nucleotide diversity is the mean p among individuals. The maternal parent species and references are given in Table I. Also shown are the mtDNA phylogenetic relationships of each unisexual/sexual pair and the *minimum* number of unisexual hybrid origins as inferred strictly from the mtDNA phylogeny.

[b] See text and Fig. 4.

[c] Values presented for the bisexual ancestors are means for the two candidate species, *C. burti* and *C. costatus.*

[d] This and some other indicated values in the table are approximate only, because sample sizes were not reported completely in the original citation, and different enzymes were employed on different specimens.

[e] The two sets of values displayed for the bisexual ancestors refer to *R. ridibunda* and *R. lessonae,* respectively (see text).

[f] Within the major clade of single-hybridization origin (see text).

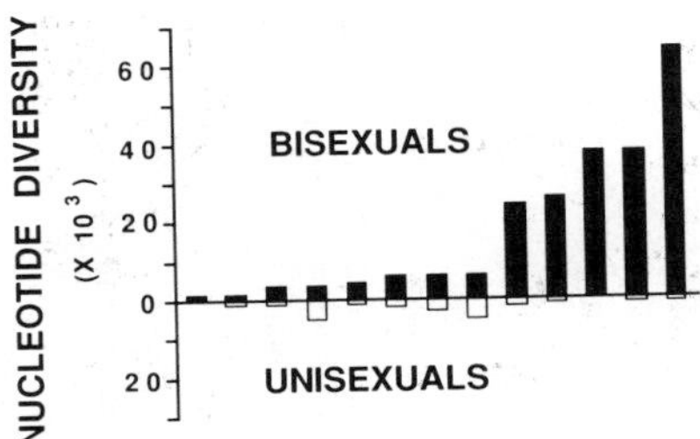

FIG. 3. MtDNA nucleotide diversities in bisexual species (above) and their respective unisexual derivatives (below), arranged in rank order from left to right by the magnitude of variation within the bisexuals. Data are from Table II.

equal to those of their maternal sexual parents (Table II; Fig. 3). Some comparisons were dramatic. For example, nucleotide diversities in the unisexuals *Cnemidophorus uniparens, Heteronotia binoei,* and *Ambystoma laterale-texanum* were more than an order of magnitude lower than such values in their sexual counterparts. Only in *P. monacha-lucida* did nucleotide diversity in a unisexual exceed (slightly) that of its sexual cognate.

Nevertheless, such *prima facie* comparisons are of limited value for the following reasons. First, genotypic and nucleotide diversity estimates can be strongly influenced by the sampling design and geographic distribution of genotypes, both of which varied greatly among studies. Second, taxonomic assignments can exert an overriding influence on interpretations of genetic diversity estimates. For example, the unisexual *Cnemidophorus tesselatus* exhibits nucleotide diversity essentially identical to that of its sexual cognate *C. marmoratus* (Table II), yet *C. marmoratus* is sometimes treated as a subspecies of the more widespread and variable *C. tigris.* If *C. tigris* as a whole were considered the bisexual ancestor of *C. tesselatus,* the estimate of mtDNA variability within the sexual form would be vastly greater than that within the unisexual (Densmore *et al.*, 1989*b*). Taxonomic conventions applied to unisexual biotypes can similarly influence perceptions of genetic diversity. For example, nine unisexual "species" of *Cnemidophorus* lizards that had *C. inornatus* mothers and *C. burti* (or *C. costatus*) fathers (Table I) are given taxonomic recognition on the basis of morphological or karyotypic distinctions (Densmore *et al.,* 1989*a*), whereas all diploid *Poeciliopsis* unisexuals with *P. monacha* mothers and *P. lucida* fathers are conventionally considered a single taxon, *P. monacha-lucida,* despite differences among strains in morphology, ecology, and behavior (Vrijenhoek, 1978, 1984*a*).

PHYLOGENETIC RELATIONSHIPS OF UNISEXUAL AND BISEXUAL TAXA

In assessing the significance of mtDNA variability to questions of unisexual origins and ages, an important step is to distinguish between unisexual lineages that arose only once from those that had multiple hybrid origins. Toward that end, relationships in the matriarchal phylogenies of unisexual–bisexual complexes have been assessed. With respect to maternal phylogeny, three categories of relationship are possible between a unisexual biotype and its maternal sexual cognate (Fig. 4): (a) *Reciprocal monophyly,* in which all mtDNA lineages within the sexual species are more closely related to one another than to any lineages within the unisexual, and *vice versa;* (b) *paraphyly,* in which all mtDNA lineages within the unisexual are more closely related to one another than to any bisexual, but some lineages in the sexual species are more closely related to those in unisexuals than to one another (the converse of this direction of paraphyly is also conceivable); and (c) *polyphyly,* in which extant lineages of neither the unisexual nor the bisexual form a distinct clade.

These phylogenetic categories should be a function both of the mode of origin of the unisexuals and of demography-based processes of maternal lineage sorting in the unisexual and sexual taxa (Neigel and Avise, 1986). For example, a bisexual that gave rise to a unisexual arising through a single hybridization event would initially exhibit a paraphyletic mtDNA relationship to the unisexual (or in other words, the unisexual would constitute a monophyletic lineage within the broader matriarchal phylogeny of the sexual); through time, maternal lineage sorting within the sexual species might then convert the relationship to one of reciprocal monophyly. However, a unisexual that arose through multiple hybridization events involving unrelated females would initially appear polyphyletic in mtDNA ancestry. Maternal lineage turnover within the sexual and unisexual forms could subsequently lead to the appearance of paraphyly and then reciprocal monophyly in the mtDNA tree.

Table II summarizes the phylogenetic status of mtDNA in the 14 unisexual–bisexual complexes studied. Five such complexes exhibit mtDNA polyphyly, and these provide *prima facie* evidence for multiple, independent

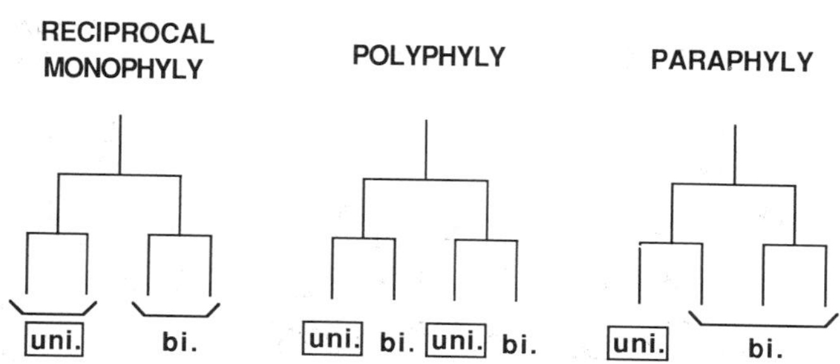

FIG. 4. Schematic representation of three classes of mtDNA phylogenetic relationship possible within a bisexual–unisexual complex.

hybrid origins of unisexuals from crosses involving distantly related female ancestors. For example, at least five independent origins for diploid *Poeciliopsis monacha-lucida* are inferred from the mtDNA phylogeny of extant lineages; that is, genotypes in the unisexuals trace to at least five distinct nodes in the mtDNA phylogeny of the sexual ancestor *P. monacha* (Quattro *et al.*, 1991). Similar documentations of multiple hybrid origins occur in *Cnemidophorus tesselatus* (Densmore *et al.*, 1989*b*), *Menidia clarkhubbsi* (Echelle *et al.*, 1989), *Phoxinus eos-neogaeus* (Goddard *et al.*, 1989), and, for the biological reasons discussed earlier, *Rana esculenta* (Spolsky and Uzzell, 1986). Examples of polyphyly are presented in Fig. 5.

The remaining unisexual–bisexual complexes exhibit a paraphyletic status (Table II), in which the unisexuals form a mtDNA clade within the

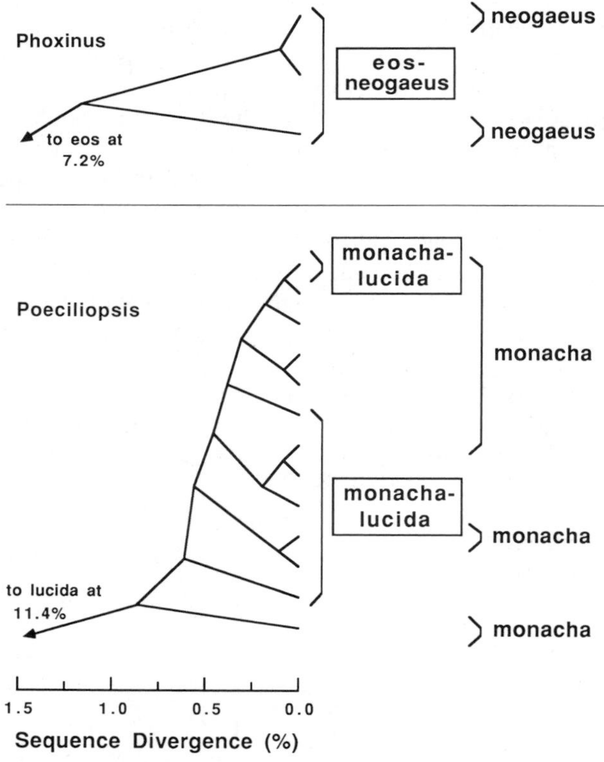

FIG. 5. Examples of unisexual–bisexual complexes exhibiting a polyphyletic relationship in mtDNA (references given in Table I). Both UPGMA phenograms are plotted on the same scale of genetic distance.

broader matriarchal phylogeny of the sexual ancestor (examples in Fig. 6). Often, the evolutionary depths in the bisexual matriarchal phylogeny were far greater than those within the unisexual derivatives, as indicated by the mtDNA phylogenies themselves and the maximum observed sequence divergences within the sexual versus unisexual taxa (Table II). As discussed below, such instances of bisexual–unisexual paraphyly are consistent with but do not prove that single hybridization events were involved in the formation of these unisexuals.

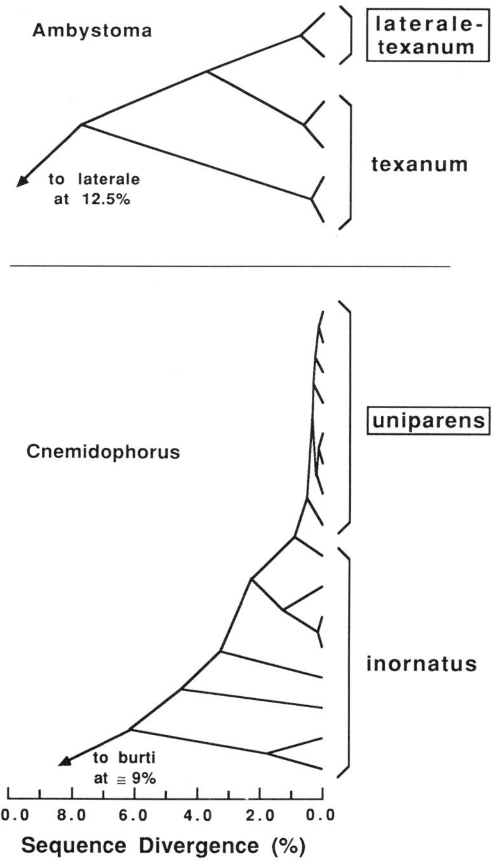

FIG. 6. Examples of unisexual–bisexual complexes exhibiting a paraphyletic relationship in mtDNA (references given in Table I). Both UPGMA phenograms are plotted on the same scale of genetic distance.

THE NUMBER OF FORMATIONAL HYBRID EVENTS

Each particular mtDNA clade within a polyphyletic or paraphyletic phylogeny discussed above appears to have arisen from hybridization event(s) involving one or more closely related bisexual females. Nonetheless, an important distinction should be drawn between unique mtDNA origin and individual hybridization event, and this can be seen most forcefully by noting that hybridization of even a single female with more than one male could produce multiple unisexual lineages differing in nuclear genotype but identical in mtDNA. In some cases, different nuclear genotypes, as identified by allozymes, morphology, or results of tissue grafts, have indeed been observed within a mtDNA clone or clade in a unisexual. For example, about 12 different diploid allozyme genotypes were found (Parker, 1979; Parker and Selander, 1976) in populations of *Cnemidophorus tesselatus* that from mtDNA evidence appear to have only two or three distinct origins within the matriarchal phylogeny of the sexual progenitor *C. marmoratus* (Densmore *et al.*, 1989*b*); and each of several mtDNA clones within *Poeciliopsis monacha-lucida* could be decomposed into distinct allozyme classes (Quattro *et al.*, 1991). Conversely, several allozymically-defined clones of *P. monacha-lucida* could be subdivided into distinguishable mtDNA genotypes with the restriction assays employed.

A serious complication in interpreting variability in nuclear or mitochondrial genomes concerns distinguishing postformational mutations from genetic differences frozen during separate hybrid origins. Ancillary information may help in determining how many hybridizations were involved. For example, unique alleles that are rare and localized in unisexuals probably arose through postformational mutations, while distinct genotypic combinations that are common in unisexuals and shared with extant sexual populations probably derive from independent hybridizations. Parker and Selander (1976) used such reasoning to suggest that 4 of 12 observed allozyme genotypes in *Cnemidophorus tesselatus* originated through separate hybrid events. Similarly, most clonal diversity in hybridogenetic *Poeciliopsis* also arose via multiple hybrid events, but in these fish postformational mutations resulting in silent enzymes, deleterious recessives, and mtDNA variants marked some clones (Leslie and Vrijenhoek, 1978, 1980; Quattro *et al.*, 1992*a;* Spinella and Vrijenhoek, 1982; Vrijenhoek, 1984*b*). In most cases where the potential sexual progenitors of unisexual vertebrates are more poorly known and incompletely sampled, it is difficult to discriminate between postformational mutations and frozen variation. In general, although it is clear that multiple hybridizations were involved in the formation of several of the unisexual taxa, it is not yet possible to specify the precise number of such events for any biotype. The *minimum* numbers of independent hybridization origins inferred strictly from the mtDNA phylogenies are presented in Table II.

EVOLUTIONARY AGES OF UNISEXUAL BIOTYPES

The relatively low mtDNA genetic variability within unisexual taxa (Fig. II) and the pattern of mtDNA paraphyly in the majority of unisexual–bisexual complexes examined (Table III; Fig. 6) suggest severe constraints on the origin and survival of most asexual vertebrate lineages. However, these observations alone do not establish the absolute evolutionary durations of surviving unisexuals.

Recent Origins for Most Unisexual Vertebrates

One approach to estimating the age of an asexual lineage involves genetic comparison to the closest sexual relative. Figure 7 presents estimates of

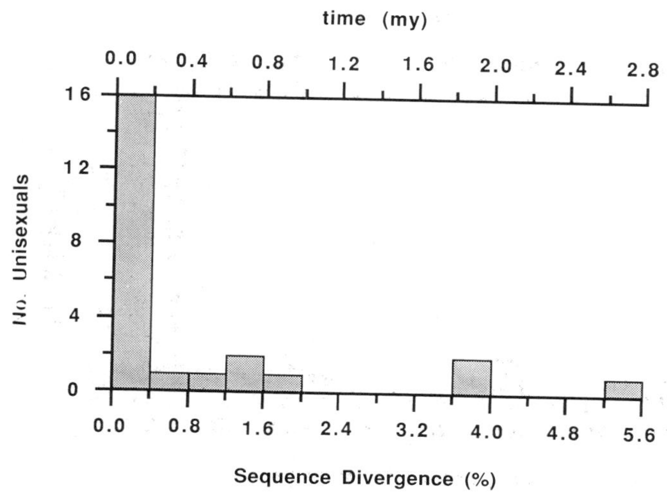

FIG. 7. Frequency distribution of the smallest mtDNA genetic distances observed between mtDNA clades in unisexuals and their closest assayed sexual relatives. Above is a scale of associated times of separation based on the conventional mtDNA clock calibration of 2% sequence divergence per million years (Brown *et al.*, 1979) (see text for reservations).

mtDNA sequence divergence between each putative unisexual clade and the genetically closest sexual haplotype assayed. Among the 24 unisexual clades identified by mtDNA comparisons, 13 were indistinguishable in current assays from an extant genotype in the sexual species, indicating a very recent evolutionary separation; and five additional unisexual clades differed from nearest assayed bisexual at sequence divergence estimates less than 1%, suggesting times of origin within the last 500,000 years (assuming a "conventional" vertebrate mtDNA clock calibration of 2% sequence divergence between a pair of lineages per million years) (Brown *et al.,* 1979; Shields and Wilson, 1987). However, a few unisexual haplotypes show greater sequence divergences, which translate into literal estimates of evolutionary origins as much as 2.75 million years ago (Table II; Fig. 7). However, a serious reservation about such estimates is that closer relatives within the sexual progenitor may have gone extinct after unisexual separation, or otherwise remained unsampled in the collections. If so, unisexual ages could be seriously overestimated. Indeed, because of the lower mtDNA diversity within unisexuals (Fig. 3), most authors have concluded that unisexuals arose very recently, even when close mtDNA lineages were not observed among the sexual relatives sampled (Vyas *et al.,* 1990).

A Relatively Ancient Clonal Lineage of *Poeciliopsis*

A second and more conservative approach to estimating evolutionary ages of unisexual taxa involves assessing the postformational spectrum of mtDNA mutations within unisexual clades that by *independent* criteria are thought to derive from a single hybridization event. This approach is less likely to produce artifactual overestimates of unisexual age, but on the other hand could seriously underestimate origination times because of postformational lineage extinctions within the unisexual clade. Few unisexual clades are well enough characterized to permit such mtDNA age assessments, but one case does stand out as exemplary of this method. Quattro *et al.* (1992*a*) described postformational mtDNA diversity within a monophyletic lineage of *Poeciliopsis monacha-occidentalis* that on the basis of independent zoogeographic evidence, protein-electrophoretic data, and tissue grafting analysis was of single-hybridization origin. From the observed nucleotide diversity within the clade ($\cong$0.0012), and using a conventional mtDNA clock calibration for vertebrates, this unisexual lineage was estimated to be roughly 100,000–200,000 generations (60,000 years) old. Although this estimate is provisional due to uncertainties of mtDNA evolutionary rate, an ancient origin is also indicated by the postformational mutations at allozyme and histocompatibility loci, and by the 550-km geographic range of the lineage

(Quattro *et al.*, 1992*a*). Clearly, at least some unisexual vertebrate lineages can achieve considerable ecological success and evolutionary longevity.

SUMMARY AND CONCLUSIONS

Genetic surveys of mitochondrial (mt) DNA diversity have provided a novel class of information on the evolutionary histories of more than 25 "species" of unisexual vertebrates and their bisexual progenitors. A review of this literature reveals that: (1) mtDNA inheritance in hybridogenetic fishes (and, by extension, other gynogenetic and parthenogenetic unisexuals) is indeed strictly maternal; (2) most unisexuals arose through nonreciprocal hybridization events between bisexual species in which the female parent has now been identified; (3) most polyploid unisexuals arose via fertilization of unreduced eggs in a diploid hybrid, rather than spontaneously via fertilization of unreduced eggs in a nonhybrid; (4) with few exceptions, overall mtDNA genetic diversity within an assayed unisexual taxon is considerably lower than that within its maternal bisexual cognate; (5) in terms of matriarchal phylogeny, a few bisexual–unisexual complexes exhibit a pattern of polyphyly (demonstrating independent hybridization origins of unisexuals from unrelated female ancestors), while most show paraphyly (in which cases the unisexual appears to have arisen only once or a few times from within a subset of the matriarchal genealogy of the sexual parent); (6) although many unisexuals are closely related to maternal lineages in the sexual relative, and thus appear evolutionarily young, in one well-characterized unisexual clade (involving *Poeciliopsis monacha-occidentalis*), numerous postformational mutations indicate an evolutionary age of at least 100,000 generations. Thus, contrary to conventional wisdom, some unisexual vertebrate clades can achieve considerable evolutionary longevity.

Unisexual biotypes provide exceptions to the norm of sexual reproduction in vertebrates. Mitochondrial DNA provides an exception to the norm of biparental Mendelian inheritance. These "aberrant" systems studied together have provided a synergistic boost to our understanding of the evolutionary genetics of clonal systems.

ACKNOWLEDGMENTS

Our mtDNA work on unisexual fishes has been supported by NSF grants BSR-8805360 (J.C.A.) and BSR-8805361 (R.C.V.), the Theodore

Roosevelt Memorial Fund, American Museum of Natural History (J.M.Q.), and the Leatham–Steinetz Stauber Graduate Research Fund of Rutgers University (J.M.Q.). We thank D. A. Hendrickson for help in the field and W. S. Nelson for assistance in the laboratory.

REFERENCES

Avise, J. C., 1989, Gene trees and organismal histories: A phylogenetic approach to population biology, *Evolution* **43:**1192–1208.

Avise, J. C., and Vrijenhoek, R. C., 1987, Mode of inheritance and variation of mitochondrial DNA in hybridogenetic fishes of the genus *Poeciliopsis, Mol. Biol. Evol.* **4:**514–525.

Avise, J. C., Trexler, J. C., Travis, J., and Nelson, W. S., 1991, *Poecilia mexicana* is the recent female parent of the unisexual fish *P. formosa, Evolution* **45:**1530–1533.

Brown, W. M., and Wright, J. W., 1979, Mitochondrial DNA analysis and the origin and relative age of parthenogenetic lizards (genus *Cnemidophorus*), *Science* **203:**1247–1249.

Brown, W. M., George, M., and Wilson, A. C., 1979, Rapid evolution of animal mitochondrial DNA, *Proc. Natl. Acad. Sci. USA* **76:**1967–1971.

Cimino, M. C., 1972, Egg production, polyploidization and evolution in a diploid all-female fish of the genus *Poeciliopsis, Evolution* **26:**294–306.

Cuellar, O., 1974, On the origin of parthenogenesis in vertebrates: The cytogenetic factors, *Am. Nat.* **108:**625–648.

Cuellar, O., 1977, Animal parthenogenesis, *Science* **197:**837–843.

Dawley, R. M., 1989, An introduction to unisexual vertebrates, in: *Evolution and Ecology of Unisexual Vertebrates* (R. M. Dawley and J. P. Bogart, eds.), pp. 1–18, New York State Museum, Albany, New York.

Dawley, R. M., and Bogart, J. P. (eds.), 1989, *Evolution and Ecology of Unisexual Vertebrates*, New York State Museum, Albany, New York.

Densmore, L. D., III., Moritz, C. C., Wright, J. W., and Brown, W. M., 1989*a*, Mitochondrial-DNA analyses and the origin and relative age of parthenogenetic lizards (genus *Cnemidophorus*). IV. Nine *sexlineatus*-group unisexuals, *Evolution* **43:**969–983.

Densmore, L. D., III., Wright, J. W., and Brown, W. M., 1989*b*, Mitochondrial-DNA analyses and the origin and relative age of parthenogenetic lizards (genus *Cnemidophorus*). II. *C. neomexicanus* and the *C. tesselatus* complex, *Evolution* **43:**943–957.

Echelle, A. A., Dowling, T. E., Moritz, C. C., and Brown, W. M., 1989, Mitochondrial-DNA diversity and the origin of the *Menidia clarkhubbsi* complex of unisexual fishes (Atherinidae), *Evolution* **43:**984–993.

Felsenstein, J., 1974, The evolutionary advantage of recombination, *Genetics* **78:**737–756.

Goddard, K. A., Dawley, R. M., and Dowling, T. E., 1989, Origin and genetic relationships of diploid, triploid, and diploid–triploid mosaic biotypes in the *Phoxinus eos–neogaeus* unisexual complex, in: *Evolution and Ecology of Unisexual Vertebrates* (R. Dawley and J. Bogart, eds.), pp. 268–280, New York State Museum, Albany, New York.

Kondrashov, A. S., 1988, Deleterious mutations and the evolution of sexual reproduction, *Nature* **336:**435–440.

Kraus, F., 1989, Constraints on the evolutionary history of the unisexual salamanders of the *Ambystoma laterale–texanum* complex as revealed by mitochondrial DNA analysis., in: *Evolution and Ecology of Unisexual Vertebrates* (R. M. Dawley and J. P. Bogart, eds.), pp. 218–227, New York State Museum, Albany, New York.

Kraus, F., and Miyamoto, M. M., 1990, Mitochondrial genotype of a unisexual salamander of hybrid origin is unrelated to either of its nuclear haplotypes, *Proc. Natl. Acad. Sci. USA* **87**:2235–2238.

Leslie, J. F., and Vrijenhoek, R. C., 1978, Genetic dissection of clonally inherited genomes of *Poeciliopsis* I. Linkage analysis and preliminary assessment of deleterious gene loads, *Genetics* **90**:801–811.

Leslie, J. F., and Vrijenhoek, R. C., 1980, Consideration of Muller's ratchet mechanism through studies of genetic linkage and genomic compatibilities in clonally reproducing *Poeciliopsis, Evolution* **34**:1105–1115.

Maynard Smith, J., 1978, *The Evolution of Sex,* Cambridge University Press, Cambridge.

Michod, R. M., and Levin, B. R., 1988, *The Evolution of Sex: An Examination of Current Ideas,* Sinauer Associates, Sunderland, Massachusetts.

Moritz, C. C., 1991, The origin and evolution of parthenogenesis in *Heteronotia binoei* (*Gekkonidae*): Evidence for recent and localized origins of widespread clones, *Genetics* **129**:211–219.

Moritz, C., Brown, W. M., Densmore, L. D., Wright, J. W., Vyas, D., Donnellan, S., Adams, M., and Baverstock, P., 1989*a,* Genetic diversity and the dynamics of hybrid parthenogenesis in *Cnemidophorus* (Teiidae) and *Heteronotia* (Gekkonidae), in: *Evolution and Ecology of Unisexual Vertebrates* (R. Dawley and J. Bogart, eds.), pp. 87–112, New York State Museum, Albany, New York.

Moritz, C. C., Wright, J. W., and Brown, W. M., 1989*b,* Mitochondrial-DNA analyses and the origin and relative age of parthenogenetic lizards (genus *Cnemidophorus*). III. *C. velox* and *C. exsanguis, Evolution* **43**:958–968.

Muller, H. J., 1964, The relation of mutation to mutational advance, *Mutat. Res.* **1**:2–9.

Nei, M., 1987, *Molecular Evolutionary Genetics,* Columbia University Press, New York.

Nei, M., and Li., W. H., 1979, Mathematical model for studying genetic variation in terms of restriction endonucleases, *Proc. Natl. Acad. Sci. USA* **76**:5269–5273.

Neigel, J. E., and Avise, J. C., 1986, Phylogenetic relationships of mitochondrial DNA under various demographic models of speciation, in: *Evolutionary Processes and Theory* (E. Nevo and S. Karlin, eds.), pp. 515–534, Academic Press, Orlando, Florida.

Parker, E. D., 1979, Ecological implications of clonal diversity in parthenogenetic morphospecies, *Am. Zool.* **19**:753–762.

Parker, E. D., and Selander, R. K., 1976, The organization of genetic diversity in the parthenogenetic lizard *Cnemidophorus tesselatus, Genetics* **84**:791–805.

Quattro, J. M., Avise, J. C., and Vrijenhoek, R. C., 1991, Molecular evidence for multiple origins of hybridogenetic fish clones (Poeciliidae: *Poeciliopsis*), *Genetics* **127**:391–398.

Quattro, J. M., Avise, J. C., and Vrijenhoek, R. C., 1992*a,* An ancient clonal lineage in the fish genus *Poeciliopsis* (Atheriniformes: Poeciliidae), *Proc. Natl. Acad. Sci. USA* **89**:348–352.

Quattro, J. M., Avise, J. C., and Vrijenhoek, R. C., 1992*b,* Mode of origin and sources of genotypic diversity in triploid fish clones (Poeciliopsis: *Poeciliidae*), *Genetics* **130.**

Schultz, R. J., 1969, Hybridization, unisexuality and polyploidy in the teleost *Poeciliopsis* (Poeciliidae) and other vertebrates, *Am. Nat.* **103**:605–619.

Schultz, R. J., 1973, Unisexual fish: Laboratory synthesis of a "species," *Science* **179**:180–181.

Shields, G. F., and Wilson, A. C., 1987, Calibration of mitochondrial DNA evolution in geese, *J. Mol. Evol.* **24**:212–217.

Sneath, P. H. A., and Sokal, R. R., 1973, *Numerical Taxonomy,* W. H. Freeman, San Francisco.

Spinella, D. G., and Vrijenhoek, R. C., 1982, Genetic dissection of clonally inherited genomes of *Poeciliopsis.* II. Investigation of a silent carboxylesterase allele, *Genetics* **100**:279–286.

Spolsky, C., and Uzzell, T., 1984, Natural interspecies transfer of mitochondrial DNA in amphibians, *Proc. Natl. Acad. Sci. USA* **81**:5802–5805.

Spolsky, C., and Uzzell, T., 1986, Evolutionary history of the hybridogenetic hybrid frog *Rana esculenta* as deduced from mtDNA analyses, *Mol. Biol. Evol.* **3**:44–56.

Vrijenhoek, R. C., 1978, Coexistence of clones in a heterogeneous environment, *Science* **199**:549–552.

Vrijenhoek, R. C., 1984*a,* Ecological differentiation among clones: The frozen niche variation model, in: *Population Biology and Evolution* (K. Wöhrmann and V. Loeschcke, eds.), pp. 217–231, Springer, Heidelberg, Germany.

Vrijenhoek, R. C., 1984*b,* The evolution of clonal diversity in *Poeciliopsis,* in: *Evolutionary Genetics of Fishes* (B. J. Turner, eds.), pp. 399–429, Plenum Press, New York.

Vrijenhoek, R. C., 1989, Genetic and ecological constraints on the origins and establishment of unisexual vertebrates, in: *Evolution and Ecology of Unisexual Vertebrates* (R. Dawley and J. Bogart, eds.), pp. 24–31, New York State Museum, Albany, New York.

Vrijenhoek, R. C., 1990, Genetic diversity and the ecology of asexual populations, in: *Population Biology and Evolution* (K. Wöhrmann and S. Jain, eds.), pp. 175–197, Springer-Verlag, Berlin.

Vrijenhoek, R. C., Dawley, R. M., Cole, C. J., and Bogart, J. P., 1989, A list of known unisexual vertebrates, in: *Evolution and Ecology of Unisexual Vertebrates* (R. Dawley and J. Bogart, eds.), pp. 19–23, New York State Museum, Albany, New York.

Vyas, D. K., Moritz, C., Peccinini-Seale, D. M., Wright, J. W., and Brown, W. M., 1990, The evolutionary history of parthenogenetic *Cnemidophorous lemniscatus* (Sauria: Teiidae). II. Maternal origin and age inferred from mitochondrial DNA analyses, *Evolution* **44**:922–932.

Williams, G. C., 1975, *Sex and Evolution,* Princeton University Press, Princeton, New Jersey.

Wright, J. W., Spolsky, C., and Brown, W. M., 1983, The origin of the parthenogenetic lizard *Cnemidophorus laredoensis* inferred from mitochondrial DNA analysis, *Herpetologica* **39**:410–416.

7

Crop Domestication and the Evolutionary Ecology of Cocona (*Solanum sessiliflorum* Dunal)

JAN SALICK

CROP DOMESTICATION

Of approximately 200,000 species of flowering plants, perhaps 3000 are eaten as food. About 200 species have been domesticated as crops, and of these, only 15–20 are now of major importance (Heiser, 1973; NAS, 1972). Despite this statistical minority of domesticated crops, few activities have so changed the history of the human race as has crop domestication.

A review of the literature on crop domestication evidences a bias toward a historical perspective based on the major cultigens.* Crops such as wheat, maize, and potatoes were domesticated thousands of years ago, have differentiated greatly from their nearest relatives, and, now, are affected little by the processes of domestication. Studies of crop domestication based on these major cultigens are limited by the historical nature of the event. Since vir-

* For example, Ucko and Dimbleby (1969), in the introduction to their compendium *The Domestication and Exploitation of Plants and Animals,* state, "The domestication of plants and animals was one of the greatest steps forward taken by mankind, and although it was first achieved so long ago we still need to know what led to it and how, and even when, it took place."

JAN SALICK • Department of Botany, Ohio University, Athens, Ohio 45701.

Evolutionary Biology, Volume 26, edited by Max K. Hecht *et al.* Plenum Press, New York, 1992.

tually all of the major crop plants cultivated today were fully domesticated in prehistoric times, scientists can only speculate about the process (Anderson, 1956). Examples of these speculations are abundant; among the most classic is Harlan's "visualization" of the domestication of cereals (Harlan, 1975),* where he describes a hypothetical scenario beginning with the harvest of wild grass seeds. As people started to plant what they had harvested, he visualizes an inherent selection regime created for interrelated syndromes of characteristics, followed later by a complex series of adaptive responses to human selection. Harlan's scheme is based on inference and deduction from the most detailed knowledge of the morphology and physiology of wild and domestic cereals.

Nonetheless, this process of inference and deduction can lead highly qualified scientists to surprisingly divergent interpretations. Opposing theories on the domestication of maize include the gradual selection of teosinte for ear size and quality (Manglesdorf, 1974) and a "catastrophic sexual transmutation" (Iltis, 1983) whereby maize arose through the single event of foreshortening the lateral branches of the teosinte stalk. The evidence available to the scientists studying maize is essentially the same, yet interpretation varies extravagantly.

The interpretation of crop domestication is continuously changing with new perspectives from diverse fields. Historically, de Candolle (1959; originally published in 1883) and Vavilov (1951; originally published in 1926) established the study with phytogeographic interpretations of botany and plant breeding, respectively. Darwin (1896) found domestication to be fertile ground for evolutionary interpretation, and reciprocally, he found much information on the evolutionary process in the study of domestication. However, Darwin's view that human selection is distinct from natural selection subsequently removed domestication from the mainstream of evolutionary biology. Harlan and DeWet (Harlan, 1975) studied crop evolution from the perspective of plant breeding and agronomy. Anderson (1954, 1968) and Heiser (1973) built crop evolution models based on classical botanical studies and techniques. Archaeologists (e.g., Renfrew, 1969; Lewin, 1988), anthropologists (e.g., Ford, 1978), and geographers (e.g., Sauer, 1952; Harris, 1967) have added their views on history, human actions, and spatial relations. Many have followed in these varied traditions and added individual perspectives until crop domestication as a field of study has become extremely diverse (e.g., Ucko and Dimbleby, 1969). The field has greatly benefitted by investigators with interdisciplinary interests; for example,

* Many of the authors cited have numerous publications on various aspects of domestication. The most synthetic and/or most recent works will be cited with further bibliographies available therein.

Hernández-Xolocotzi (1985) can only be described as a combined botanist, agronomist, anthropologist, and cultural historian. Alcorn (1981) voices the developing view that interactions between native farmers and their plant resources—including plant domestication—are much more complex than simply a response to uses and needs; the interaction is dependent on the social, cultural, and economic milieu. These social factors interact with the autecology of the plant, the agroecological setting, and the evolutionary mutualism between plants and people (e.g., Harlan, 1966; Pickersgill and Heiser, 1976; Rindos, 1984).

From this evolving perspective, it is increasingly recognized that the process of crop domestication is not restricted to the past (Johannessen *et al.*, 1970; Johannessen, 1982). Crop evolution is ongoing and new crops continue to be domesticated. Crops in their traditional agroecosystems evolve continuously (Johns and Keen, 1986; Nabhan, 1985; Altieri and Merrick, 1987; Oldfield and Alcorn, 1987; Salick and Merrick, 1990) and there is a tremendous range in intensity of cultivation and genetic manipulation in traditional agriculture (Posey, 1984). Wild and weedy crop relatives genetically interact with crop populations (Anderson, 1948, 1968; Harlan, 1965, 1976; Harlan and DeWet, 1963; Wilkes, 1977; Pickersgill, 1981; Rabinowitz *et al.*, unpublished manuscript) to the extreme where recently domesticated crops like rubber (*Hevea brasiliensis*) differ little from their wild relatives (e.g., Schultes, 1956). Jointly, this information suggests that crop domestication may be particularly active with underexploited tropical crops (e.g., Johannessen, 1966) which are still traditionally cultivated and hybridize with wild and weedy relatives.

This study on crop domestication combines several of the above perspectives. The focus is on an underexploited tropical crop which is domesticated only rustically. Wild populations grow sympatrically with those cultivated by indigenous people in their traditional agricultural systems. An ecological-evolutionary perspective and rough-hewn population-genetic measurements taken in the field are supported by agroecological (Salick, 1989) and applied agronomic studies (Salick, 1990). The interdisciplinary mix is distinct, but the question remains the same: what are the dynamics of crop domestication and evolution?

COCONA

Solanum sessiliflorum Dunal (Figs. 1 and 2) is among the spiny solanum in the section Lasiocarpa (Solanaceae, nightshade family) monographed by Whalen *et al.* (1981). The 13 species of the section are concen-

FIG. 1. Cocona (*Solanum sessiliflorum* Dunal), a domesticated, nonspiny form with large fruit.

trated in the northern Andes and western Amazon. There are several cultivated species which produce edible fruits (Schultes and Romero-Casteñeda, 1962). Most studied to date is *S. quitoense,* the lulo or naranjilla, of Colombia and Ecuador (e.g., Heiser, 1972). The National Academy of Sciences (NAS, 1975) selected it as one of 36 underexploited tropical plants of promising economic value (see also Vietmeyer, 1986). Agricultural re-

FIG. 2. Cocona fruit of a large-fruited variety grown for eating; small-fruited varieties are grown for juice.

searchers in several countries are now working on improvement and cultivation of the lulo.

Solanum sessiliflorum or cocona is much less studied and less formally cultivated. Previous investigations on cocona include systematics (Whalen *et al.*, 1981; Whalen and Caruso, 1983), economic botany (Schultes, 1958;

Schultes and Romero-Casteñeda, 1962; Patiño, 1962; Heiser, 1968, 1971, 1972; Silva *et al.*, 1989), biology (Fernandez, 1985), and agronomy (Rodríguez and Garayar, 1969; Benza and Rodriguez, 1977; Pahlen, 1977), as well as my own work (Salick, 1986, 1989, 1990).

Regionally, *Solanum sessiliflorum* is known as *cocona* in the Peruvian and Colombian upper Amazon, *topiro* in the Rio Negro drainage, *cubiu* in Brazil, and the *peach-tomato* in English. The English name, though least sonorous, gives the best description of the fruit's flavor if some citrus were added to the already unlikely combination. Cocona is eaten fresh as a fruit, as an ingredient in fruit and vegetable salads, as a flavoring for ices, and as a refreshing tropical fruit juice. The fruit is also cooked as marmalade, desserts, and a vegetable with fish or meat. Its culinary potentials are also underexploited, since the versatile fruit/vegetable favors dishes from pasta to peach-tomato pie (Salick, 1986, Part b). The cocona plant looks somewhat like an eggplant with fruit like tomatoes (Fig. 1) [for botanical descriptions see Schultes (1958) and Whalen *et al.* (1981)].

Cocona is indigenously cultivated by swidden agriculturalists (Schultes, 1958; Boster, 1983; Salick, 1989), with the plants carefully tended in swidden gardens or yard gardens. The seeds, however, are seldom planted, but rather defecated by humans; informal observations suggested that seed germination was greatly improved after seeds passed through our diapered daughter's digestive tract. Cocona thrives on alluvial soils, and yet is commonly cultivated on acid, infertile tropical soils where only a limited number of other crops produce. It is adapted for the wet and humid tropics of the upper Amazon at an altitude of under 1000 m. At the higher altitude, its range is sympatric with lulo, but known hybrids are limited to artificial crosses with limited seed set and viability. In native agriculture, cocona is definitely a minor crop in yard gardens or a minor intercrop in cassava/plantain fields. Nonetheless, it is enthusiastically eaten, especially by children, and may play an important nutritional role. Cocona is high in iron and niacin and contains vitamins A and C (Ministerio de Salud, 1957; Salick, in preparation), all particularly important to women and children in the tropics (e.g., Strongin, 1982). Medicinal benefits of cocona are purported for skin and hair care, for kidney and liver ailments, as a douche, and as a tonic for strengthening the blood.

In the market economy it is generally sold on a limited scale by small farmers, as in Iquitos, Peru (Padoch, 1987), or seasonally, as in Lima, where in the summer there is one small cocona stall in the enormous wholesale fruit market. It is processed locally and sold as juice, ices, and marmalade. One small French–Peruvian canning company (Indalsa) in San Ramon, Peru, cans a very limited amount of juice, which is then shipped to Lima and Italy. Cocona is ripe for development: it yields extremely well (30–60 tons/ha on rich alluvial soils), transports easily, remains fresh for 40 days from harvest, and processes as a delicious juice or preserve.

Cocona also has wild or weedy populations with spiny plants and small fruits (Fig. 3). Since wild or weedy plants are morphologically indistinguishable, it is not possible to differentiate spiny populations with different histories. Spiny forms are found in disturbed habitats especially along river banks and dispersed by cattle in pastures. These wild or weedy populations are selected for survival under heavy grazing pressure, although an occasional spineless cocona may survive in the wild restricted to stream banks or other habitats safe from large herbivores. Seed dispersal in numerous, small, attractive dispersal units (fruit) probably minimizes the cost of fruit production and spreads the risk of seed mortality. In pastures cocona seedlings most often spring from cow dung. In the Peruvian Amazon basin, wild or weedy populations are sympatric with cultivated cocona, cooccurring within a few meters and well within range of the large bees which pollinate cocona. Populations do hybridize readily, although cultivated cocona are distinct in form, habitat, and the selection pressure to which they are exposed.

Cultivated cocona is spineless (Fig. 3) with a great diversity of fruit sizes and shapes (Fig. 4). Cocona domestication relies heavily on the process of culling undesirable plant types; people are persistent in weeding out any spiny cocona that germinate in their gardens or yards. Since seeds are seldom deliberately planted in indigenous gardens, human selection of fruits is lim-

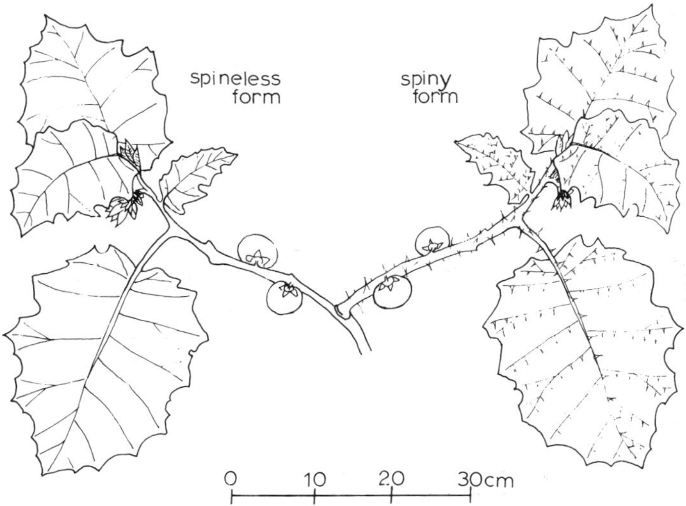

FIG. 3. A comparison of cocona forms with and without spines; fruits are of a small type found on both spiny and nonspiny cocona.

FIG. 4. Radiation of cocona fruit size and shape under domestication; wild cocona have small, round fruit like those on the far left. These cocona were collected over much of the Peruvian Amazon basin (Fig. 5) and were grown in standardized trials at San Ramon, Peru.

ited to eating preferred fruits and then defecating the seeds. Small fruits are preferred for juice, whereas the flesh of large fruits is preferred for flesh. This contrast leads to a diversity of cultivated fruit types.

This evolutionary scenario of gene exchange among wild, weedy, and cultivated populations with ongoing domestication by culling and consuming is distinct from many previous models for major crops. Nonetheless, it falls clearly within neo-Darwinian theory. The problem is to test and detail this evolutionary scenario and come up with alternative hypotheses. What are the ongoing processes of cocona domestication?

STUDY SITES

Cocona is native to the Amazon basin in Peru in the wild, weedy, and domesticated forms. Two study sites were chosen within this region (Fig. 5). One site is near Iscozacin in the Palcazu Valley (Oxapampa, Pasco) at 350 m altitude, where all forms of cocona are common and where it is cultivated indigenously by Amuesha Amerindians. This represents a site of continuing

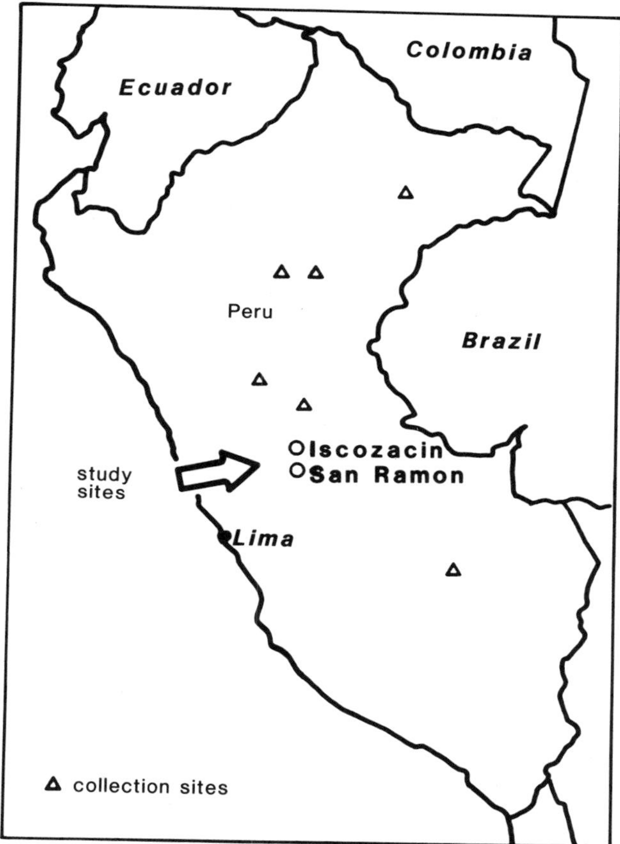

FIG. 5. Map of Peru with cocona collection sites marked with triangles and the study sites at Iscozacin and San Ramon marked with circles.

cocona domestication. The second site is near San Ramon (850 m, Chanchamayo, Junin), where there is limited cultivation of the domesticate by small farmers for sale to the Indalsa canning factory and for the Lima fresh market. At this site the cultigen is physically and genetically isolated from its progenitors. The International Potato Center (CIP) has an experimental station in San Ramon with well-developed facilities for field experimentation. The two sites are distinct and appropriate for two different phases of the investigations.

The Rio Palcazu, a headwater of the Amazon River, has a long, narrow valley opening northward into the Amazon Basin at the confluence of the Pichis and Palcazu where the Pachitea is formed. Amazon air masses squeeze up the Palcazu Valley and rise at the southern end. Rains (Table 1)

TABLE I. Environmental Comparisons of Iscozacin (350 m) and San Ramon (850 m), Peru

Rainfall (mm)

	Jan	Feb	Mar	Apr	May	Jun	Jul	Aug	Sep	Oct	Nov	Dec	Yearly
Iscozacin	1297	738	498	569	424	285	192	251	318	399	631	672	6274 mm
San Ramon	215	277	255	167	84	106	82	102	149	77	130	285	1929 mm

Soils[a]

	Sand (%)	Silt (%)	Clay (%)	Texture	pH	Al	OM	P	K_2O	CEC	Ca	Mg	K	Na
Iscozacin	36	38	26	Clayloam	3.8	1.94	3.79	6.38	198	15	3.36	0.78	0.40	1.40
San Ramon	66	20	14	Sandloam	5.9	0.60	2.4	16	351	12	8.20	1.34	0.20	0.13

[a] Units: For Al, Ca, CEC, Mg, K, and Na, me/100 g; for P, ppm; for OM, %; for K_2O, kg/ha.

are among the highest recorded in the Amazon basin, with a 3-year average of 6274 mm/year and a summer "dry season" in June, July, and August, averaging nearly 250 mm/month. The soils of the area are fairly typical of the upper Amazon basin, being acid, infertile, high in aluminum, and situated on weathered geological terraces (e.g., Sanchez and Benites, 1987). Farming is not easy in the Palcazu Valley (Salick, 1989); slash often does not burn and soils are not productive, but cocona does well in both domesticated and wild forms.

In contrast, the CIP experimental station is situated on some of the richest soils of the notoriously productive Chanchamayo Valley of central Peru (Table I). The sandy-silt-loams are of near neutral pH and well endowed with N–P–K and cations. Additionally, the rainfall is comparatively equitable, with a 4-year average of 1929 mm/year and a 6-month dry season from May through October with an average of 100 mm/month when irrigation is available at the international center as needed. Agriculture is highly mechanized and inputs are abundantly available. Extremely high cocona production was measured under these conditions.

Iscozacin represents the native habitat and indigenous agricultural conditions complete with pollinators, herbivores, and stress. Experiments on habitat comparisons, selection, gene flow, and fertility response were carried out in Iscozacin. At CIP, San Ramon screenhouses were available for seedling germination and uniform field conditions for genetic crossing and varietal comparisons. The relative advantages of these sites were employed in the appropriate experiments.

STANDARD TECHNIQUES

Several methods were used in all experiments and are most conveniently described together. Cocona seeds have a low germination rate if collected directly from the fruit and planted. Experimentally, seeds were soaked in 2% gibberellic acid for 24 hr, which allowed a germination rate of 94% compared to no germination of seeds similarly treated with water (Table II). In subsequent experiments, seeds were always treated in 2% gibberellic acid for 24 hr before being germinated in plastic trays with a standard potting mixture of sand, organic matter, and fertilizer used at the International Potato Center for germination of true potato seed. Seedlings were kept in trays, semishaded, and watered as necessary for 2–4 weeks, when they were transplanted into "Jiffy Pots" and maintained under the same screenhouse conditions for 6–8 weeks. Then, the plants, having received this uniform seedling treatment, were at least 15 cm tall and hardy enough to plant in field experiments in either San Ramon or the Palcazu Valley.

TABLE II. Cocona Seed Germination (%) with Gibberellic Acid Treatments[a]

	Germination (%) at given number of days after treatment		
Giberillic acid treatment	7 days	14 days	21 days
Control (water)	0	0	0
1000 ppm (1%)	11	14	15
1500 ppm (1.5%)	6	9	9
2000 ppm (2%)[b]	94	94	94
3000 ppm (3%)	45	54	54

[a] Powdered gibberellic acid (90% gibberellin) is dissolved in a small amount of alcohol and then added to water to make aqueous solution treatments. Seeds are soaked for 24 hr in these solutions and then removed to petri dishes with moist filter paper.
[b] Treatment used for planting cocona in subsequent experiments.

Uniform experiments in San Ramon which did not attempt to simulate natural conditions received further standard treatments. Seedlings were planted at the beginning of the rainy season. Plant spacing was at 1 m and fertilizer was applied at 50 g/plant around the base of each transplant (500 kg/ha of 160N–160P–200K–60 Mg); this is termed "medium-level inputs" by the International Center and considered minimum for assuring vigorous growth and for making varietal comparisons and experimental crosses. Varieties were planted in plots of 36 plants (5 m × 5 m) with three repetitions. Irrigation and pesticides were applied and weeds removed as deemed necessary by the station manager, who followed recommendations for potatoes, since the pest complex observed was virtually the same.

During the growing season standard plant evaluations included measurements of plant height, length, and number of fully expanded leaves, and the number of reproductive structures (bud, flowers, fruits). A correlation of leaf area with leaf length using 30 leaves spanning the full range of leaf size and shape (shape varies only minimally) showed an extremely close correlation [$r^2 = 0.98$, leaf area (cm^2) = $-866 + 49$ leaf length (cm), $p < 0.001$]. Thus, leaf length could be converted directly to leaf area using this formula and subsequently multiplied by the number of fully expanded leaves per plant to estimate whole-plant leaf area.

Harvest evaluations included weighing and counting fruit on a plant-by-plant basis. There was little variation of fruit size or shape on a single plant, so that ultimately one fruit of each plant was blindly selected to be measured from the harvested fruits. From these selected fruits, length, width, weight, and color were recorded. The number of seeds was counted from 36 fruits spanning the full range of fruit size and shape, including wild fruits.

The number of seeds per fruit did not vary significantly with fruit size, averaging 388 seeds/fruit ($F_{3,4} = 3.94$). Harvesting began after about 6 months of growth in the field, and plants were removed from the field after 8 months because of crop rotations. After 8 months, production had dropped but not terminated completely when the plants were pulled; thus, the figures presented, while not maximum, do represent a reasonable annual harvest within a realistic production schedule. Extensive notes on diseases and insects are available (J. Salick, unpublished observations), but generally pests and pathogens are the same as, though in lower populations than, those on other solanaceous crops in similar tropical areas. Thus, pest management for tomatoes and potatoes is directly applicable to cocona, although less necessary because of cocona's natural pest resistance.

Hand pollination of cocona was carried out as for its relatives. A day before flower anthesis, anthers were removed with forceps. After anthesis of the emasculated flowers, when the stigma was receptive in the early morning, selected pollen was applied using a small pen knife as a spatula. Pollen from selected plants was removed by placing a resonating tuning fork at the base of an inverted anther and catching the released pollen on a small mirror. A few emasculated flowers of each variety were left unpollinated to test for apomixis.

EXPERIMENTS AND RESULTS

Spines

Wild, weedy cocona found outside of cultivation most often has spiny leaves and stems (Fig. 3) and most likely is the progenitor of the domesticate (Whalen *et al.*, 1981). Spininess apparently protects the plants from large herbivores; caterpillars and beetles seemingly are not deterred. People, however, consider the presence of spines more noxious than any other trait and select against it vigorously. To describe the dynamics of domestication as well as the natural selection process, several parameters of spininess were estimated.

Inheritance of spines was investigated by means of genetic crosses. A nonspiny mother was fertilized with pollen from a spiny plant. The F_1 progeny were all spiny; they were then self-pollinated by hand. For the F_2 generation, the ratios of spiny to nonspiny plants in three repetitions are compared to the 3:1 ratio expected for a single gene with dominance. The observed results (Table III) do not differ significantly from the single-dominant-gene

TABLE III. Single-Gene Mendelian Model for the Inheritance of Cocona Spines Tested against Observed F_2 Ratios of Spiny:Nonspiny Offspring[a]

	Observed spiny:nonspiny	Expected spiny:nonspiny	χ^2	df	p
Rep I	81:39	90:30	3.6	1	0.07
Rep II	90:26	87:29	0.4	1	0.5
Rep III	76:28	78:26	0.2	1	0.7
			4.2	3	0.25

[a] A nonspiny mother was fertilized with pollen from a spiny plant. The F_1 progeny were all spiny and self-pollinated. For the F_2 generation, the ratios of spiny to nonspiny plants in three repetitions are compared to a simple Mendelian model for a single-gene dominant of $(1p^2 + 2pq){:}1q^2$ or 3:1. The data do not differ significantly from a single-gene dominant model.

model. Thus, it appears that the loss of spines during domestication could have involved a mutation at a single gene locus.

If cocona spines can be eliminated by a single mutation, the application of basic quantitative genetic models (e.g., Falconer, 1981) can be applied to the study of cocona domestication, at least with regard to this trait. Gene flow and selection are primary components of these models and were estimated experimentally, albeit roughly, in the indigenous conditions of Iscozacin.

Gene flow of a single dominant gene into a population of recessives was tested in a field array of recessive plants with one plant exhibiting the dominant marker. The spread of the dominant marker into the progeny of the recessive parents gives an estimate of dispersion. Two such designs were used here: (1) to measure flow and (2) to find the effect of plant density on gene flow. These estimates of flow were compared to the proportion of spiny progeny found in cultivated cocona. With the Amuesha practice of eliminating all spiny plants in cultivated areas, the percent of spiny progeny derived from seeds of nonspiny cocona fruit is assumed to be an estimate of flow under indigenous agricultural conditions.

The primary estimate of gene flow (m) was accomplished in four plots of 27 nonspiny cocona planted at a distance of 1 m (3 m × 6 m) with one spiny cocona planted in the middle of the long edge (Fig. 6). One fruit per plant was harvested and from each fruit a tray of seedlings (approximately 100) was grown to a stage when spines could be easily detected. The results (Fig. 7) indicate that the log of gene flow decreases linearly with the square root of distance. Additionally, bees, the presumed agent of gene flow via pollen transport and pollinization, were collected entering cocona flowers in these plots, including *Eulaema cingulata* (Fabricius), *Bombus transversalis* (Oli-

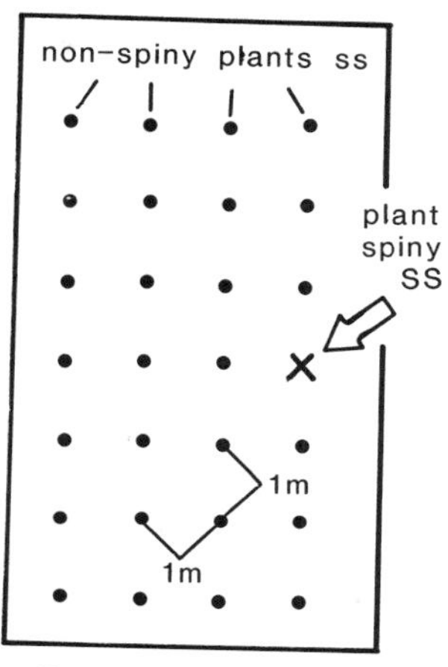

FIG. 6. Experimental design for estimating gene flow among cocona plants. One spiny, homozygote dominant was planted along the edge of a uniform field of non-spiny, homozygote recessives at 1 m spacing. There were four repetitions of this design. Fruit was harvested from the nonspiny plants and seeds grown to a seedling stage, when percent of spiny progeny were assessed.

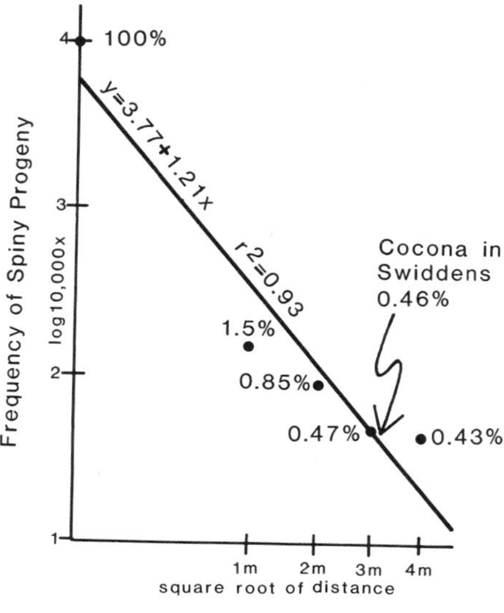

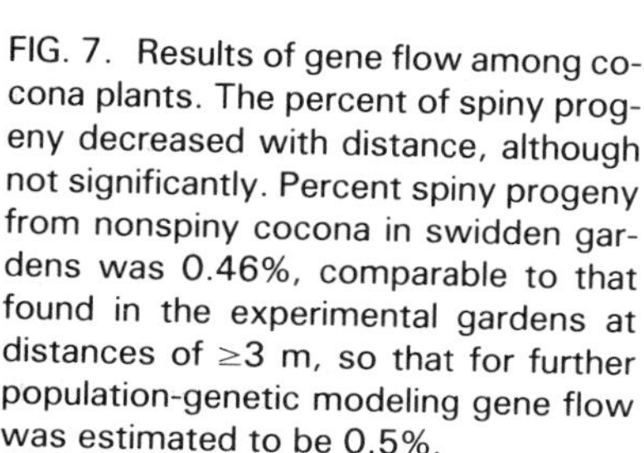

FIG. 7. Results of gene flow among cocona plants. The percent of spiny progeny decreased with distance, although not significantly. Percent spiny progeny from nonspiny cocona in swidden gardens was 0.46%, comparable to that found in the experimental gardens at distances of ≥3 m, so that for further population-genetic modeling gene flow was estimated to be 0.5%.

vier), *Trigona* sp., and *Pseudaugocholropsis graminea* (Fabricius).* It seems doubtful that the latter two bees are cocona pollinators, considering their small sizes.

The effect of cocona density on gene flow was tested in a circular experimental design (Fig. 8) with one spiny cocona in the center surrounded by nonspiny plants. The circle was divided into four quadrants with differing cocona densities and plant spacing at 1, 2, 3, and 4 m. Three fruits were harvested randomly from three plants (or fewer, depending on the treatment) at three distance classes (1–3, 4–6, and 7–9 m). The analysis of variance showed no significant difference in the proportion of spiny cocona found at different densities. This indicates that gene flow is not density dependent at these densities.

In 1000 cocona seedlings grown from fruit harvested in five native gardens, there were 46 (0.46%) spiny progeny (Fig. 7), agreeing well with the experimental estimations of gene flow for distances over 3 m, a realistic

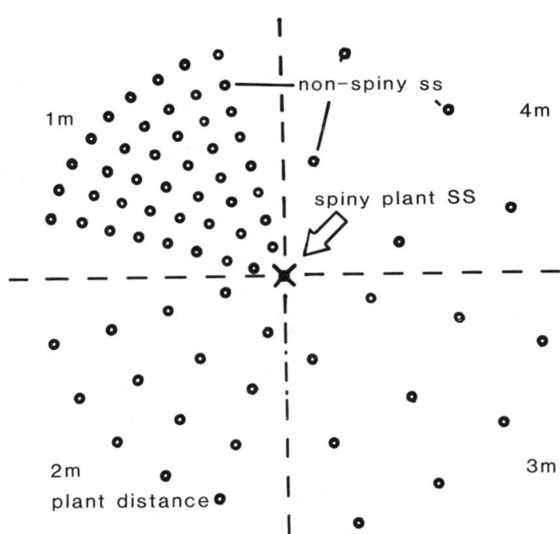

FIG. 8. Experimental design for estimating the effect of plant density on cocona gene flow. One spiny, homozygote dominant was planted in the middle of a circle of nonspiny, homozygote recessives divided into four quadrants with plant spacings of 1, 2, 3, and 4 m. Fruit was harvested from the nonspiny plants and seeds grown to a seedling stage when percent of spiny progeny was assessed. No significant differences were found among densities or distances.

* Professor George Eickwort of Cornell University generously identified the bees.

separation of wild and domesticated cocona. From these three independent assessments, gene flow or rate of migration is assumed hereafter to be 0.5%.

Selection pressure for and against spines and fruit size was estimated with reciprocal transplants of wild and small- and large-fruited cocona in five habitats: a swidden garden, a riverbank, a pasture, a landslide, and secondary regeneration after a swidden garden. The cocona plants in the latter two habitats all eventually died without ever having reproduced, so the data from these unsuitable habitats are not reported. In each habitat three plants of each three types were planted randomly in a plot with 1 m spacing (2 m × 2 m = 9 plants). Monthly evaluations of plant height and leaf area and harvests of fruit were analyzed for differences among plant types within habitats. Although plant height, leaf area, and fruit production are not direct measures of reproductive fitness, more vigorous plants tended to produce more seed as indicated by the correlation within cell column values (and confirmed by further production data in preparation). The results are presented in Table IV. Relative fitness was calculated from the cell means contrasting the character values of the less vigorous fruit types with the most vigorous within a habitat. Thus, in a swidden garden, wild or spiny cocona produced an average of 2097 seeds compared to 2811 seeds for small-fruited domesticates, and so, based on number of seeds, wild cocona has 75% the fitness of the small domesticates in swiddens.

Two adjustments were made in these calculations. Although wild cocona grow and reproduce in swiddens as indicated by the cell means, people relentlessly chop them down. Two approaches helped in estimating an adjustment factor for human selection: First, native farmers were interviewed and swiddens were censused. Each farmer volunteered the opinion that all spiny cocona should be weeded out. In 65 swidden gardens measured for another study (Salick, 1989) there was only one spiny cocona plant present and when brought to the attention of the farmer, she immediately removed it. A conservative estimate of 95% reduction of spiny cocona fitness was used to adjust for this thorough human selection. Second, in nature I never found a large-fruited cocona growing along a riverbank outside of cultivation (although they did well as transplants). For this reason in the study on spines the fitness of wild cocona was compared to that of small-fruited cocona which were found in that habitat.

From the calculated and adjusted fitnesses, selection coefficients (Table V) for spiny cocona in swiddens and on riverbanks and for nonspiny cocona in pastures were calculated based on the four characters measured. The means of these four selection coefficients are used in subsequent models.

Balanced equilibria, or the hypothetical gene frequencies at which gene flow and selection balance one another, can be calculated based on a single-gene model with the above estimates of flow and selection. To test the hy-

TABLE IV. ANOVA Cell Means for Plant Fitness Characters by Habitat and Fruit Types (Wild, Small, Large), ANOVA Results, Relative Fitness for Each Cell, and Adjustments for Human Selection and Lack of Natural Populations of Large-Fruited Cocona in River Habitats[a]

Reciprocal transplant experiment	Swidden garden			Riverbank			Pasture			Significance of F (F value) $df = 12$
	Wild	Small	Large	Wild	Small	Large	Wild	Small	Large	
Characters (means per plant)										
Primary fitness trait										
Seeds (numbers per plant)										
Mean	2097	2811	2730	293	330	668	732	0.0	0.0	****
Fitness	0.75	1.0	0.97	0.43	0.49	1.0	1.0	0.0	0.0	6.13
Adjustment	0.04			0.88	1.0					
Associated fitness traits										
Fruits (numbers per plant)										
Mean	5.73	7.68	7.46	0.8	0.9	1.8	2.0	0.0	0.0	****
Fitness	0.75	1.0	0.97	0.43	0.49	1.0	1.0	0.0	0.0	6.13
Adjustment	0.04			0.88	1.0					
Leaf area (m^2)										
Mean	0.34	0.60	1.09	0.12	0.24	0.55	0.11	0.0	0.0	****
Fitness	0.31	0.55	1.0	0.22	0.44	1.0	1.0	0.0	0.0	7.40
Adjustment	0.02			0.50	1.0					
Height (m)										
Mean	0.65	0.93	0.93	0.37	0.50	0.53	0.34	0.0	0.0	****
Fitness	0.70	1.0	1.0	0.70	0.94	1.0	1.0	0.0	0.0	10.80
Adjustment	0.03			0.74	1.0					

[a] Adjustments for human selection are a 95% reduction of spiny cocona plants estimated from interviews with indigenous swidden agriculturalists. Adjustments in cocona fitness in the riverbank habitat are calculated in relation to small-fruited plants, since there are no naturally occurring large-fruited cocona in these habitats (i.e., in the adjusted data, fitness of small-fruited plants is set at one).

**** $p \leq 0.0001$.

TABLE V. Selection Coefficients for Spiny and Spineless Cocona in Three Habitats Calculated from Fitness Estimates (Table IV) Based on Reciprocal Transplants

	Selection coefficient S		
	Swidden (spiny)	River (spiny)	Pasture (spineless)
Height	0.97	0.26	1.0
Leaf area	0.98	0.50	1.0
Fruits	0.96	0.12	1.0
Seeds	0.96	0.12	1.0
Mean	0.97	0.25	1.0

pothesis that cocona populations are in balanced equilibria, calculated gene frequencies were compared with field observations. In these calculations, gene flow was assumed to counterbalance selection in a manner generally ascribed to mutation, with the difference that gene flow and selection are both substantial for cocona.

Basic quantitative population-genetic models (e.g., Falconer, 1981) devise two cases (Table VI). The first is selection against a recessive gene; this is analogous to selection against nonspiny cocona in pastures. Given the gene flow in Fig. 7 and the selection in Table V, the balanced equilibrium model predicts 5 nonspiny cocona among 1000 spiny cocona. Randomly sampling along transects laid in three pastures in the Palcazu Valley, I found one nonspiny cocona among 300 plants, approximating the predicted frequency.

The second case (Table VI) deals with selection against a dominant gene, as is found where spines are selected against, along river banks and especially in swidden gardens. This equilibrium model predicts 1 spiny cocona among 100 plants in swiddens; random field samples show 1 spiny plant among 300 cocona, within the realm of expectation. Both ungrazed riverbanks and cocona along these riverbanks were comparatively rare and sample sizes had to be reduced to 10 plants per site. Censuses differ from predicted values, with 8 nonspiny cocona found among 30 plants, while the model predicts 28.8 among the same number, suggesting that there is significantly more selection for spines or less gene flow from nonspiny cocona in riverbank habitats than expected from the rough experimental estimates or that a model other than the Hardy–Weinberg model might explain the data.

For pastures and gardens the rough field data on cocona populations do not differ from the predictions based on balanced equilibria of selection and gene flow (Table VI). However, these models are based on the Hardy–Weinberg principle with all of its assumptions. Among these is one crucial to

TABLE VI. Calculation of Expected Population Ratios of Spiny and Nonspiny Cocona Plants Based on Experimental Values of Gene Flow and Selection Pressure[a]

Case 1 (Expected): Selection against a recessive gene (nonspininess)

$$q^2 = m/S$$

Pastures: $q^2 = m/S = 0.005/1.0 = 0.005$, or
5 nonspiny plants among 1000 plants

Case 2 (Expected): Selection against a dominant gene (spininess)

$$q(1 - q) = m/S$$

Riverbanks: $q(1 - q) = 0.005/0.25 = 0.02$, $q = 0.98$, $q^2 = 0.96$, or
4 nonspiny plants among 100 plants
Swiddens: $q(1 - q) = 0.005/0.97 = 0.005$, $q = 0.995$, $q^2 = 0.99$, or
1 spiny plant among 100 plants

Field censuses (Observed)[b]

	Pastures	Riverbanks	Swiddens
Site I	0:100	4:6	99:1
Site II	0:100	3:7	100:0
Site III	1:99	1:9	100:0
Expected	0.5:99.5	9.6:0.4	99:1

Observed populations in pastures and swiddens are *not significantly different* from expected

[a] Observed numbers of spiny and nonspiny cocona plants based on field censuses. Chi-square comparison of expected and observed values indicates that the quantitative population-genetics model does not differ from field observations for pastures and swiddens.
[b] Ratios are nonspiny:spiny.

many *Solanum* species; the Hardy–Weinberg principle assumes random mating when both selfing and self-incompatibility are very common in this genus, instigating the following study on outcrossing.

Outcrossing was estimated by planting three pairs of spiny and nonspiny cocona removed from near neighbors. Seeds from the nonspiny member of the pair were grown to estimate the proportion of spiny offspring or propagules fertilized by the neighbor compared to those self-pollinated. There were 21 spiny cocona among 350 offspring (from 12 fruit), or 6% outcrossing (Table VII), far from evidencing Hardy–Weinberg random mating. Models for inbreeding (e.g., Falconer, 1981) can be used to estimate the effect of this high proportion of self-pollinated offspring.

TABLE VII. Outcrossing Frequency in Cocona and Calculation of Effective Population Size[a]

Outcrossing = 0.06 ± 0.05
21 spiny plants from 350 seeds of fruits on 12 nonspiny plants

$$F = 1.0 - 0.06 = 0.94$$
$$m = 0.005$$
$$F = 1/(4N_e m + 1)$$
$$N_e = (1 - F)/4Fm$$
$$= (1 - 0.94)/4(0.94)(0.005)$$
$$= 3.2 \text{ (95\% confidence limits} \pm 1.5)$$

The effective population size is 3.2, extremely small.

[a] Outcrossing was estimated with paired spiny and nonspiny plants. Selfing or inbreeding (F) is assumed to be the proportion of nonspiny plants. Gene flow (m) is taken from the previous experiments (Fig. 7) and effective population size (N_e) is calculated after Falconer (1981). The extremely small effectively population size allows genetic drift under light selection (i.e., riverbanks).

Effective population size may be interpreted as the number of individuals with which one organism may genetically interact. Selfing reduces effective population size dramatically. Using selfing to estimate inbreeding (F) and using the above estimate of gene flow, the effective population size is calculated to be only 3.2 individuals (Table VII). Inbreeding (F) will little affect the equilibria between gene flow and selection, with the phenotype frequencies affected even less. However, the extremely small population size is critical to genetic drift and fixation, especially where selection is low. This is the case along riverbanks, and may explain the deviations from the expected equilibrium above. Furthermore, drift and fixation with gene flow and limited selection suggest a state of dynamic equilibrium discussed below, rather than a stable balance.

Cocona Fruit

Although spines are of primary concern to people cultivating cocona, they carefully note several other characters as well. My informants enthusiastically discussed cocona fruit size, shape, color, and yield. The first task in studying human manipulation of these characters was to collect the available varieties. With substantial help of colleagues,* a diverse collection of cocona

* I thank Ing. Wanders Chavez and Ing. Homer Tuesta, who brought me cocona both from great distances and of great diversity.

was amassed from much of the Peruvian Amazon (Figs. 4 and 5). Twenty-five of these varieties were planted in replicated blocks at San Ramon experimental station of the International Potato Center (Fig. 9). This parental stock was carefully evaluated and controlled crosses were hand-pollinated to begin the study of cocona fruit genetics.

A complete diallel cross of seven cocona varieties was accomplished (Fig. 10). The parental lines were of small, medium, and large fruit sizes; of flat, round, and long fruit shapes; and of various fruit colors grading from yellow to dark red. The greatest possible diversity of fruit characters was included. The fruits produced after this first cross were genetic products of the maternal parent plant rather than the cross itself; consequently, to evaluate the cross, the fruits were harvested, and seeds extracted and planted. Each parent plant (7 P_1) and plant in the generation following the cross (49 F_1) were evaluated in three repetitions of 36 plants each (5 m $\times$ 5 m) in a randomized block design. Plant characteristics will not be considered in this discussion of domestication because they vary little and because people take little note of them, since fruit is their object. Reported (Table VIII) are mean

FIG. 9. Twenty-five varieties of cocona planted in replicated blocks at San Ramon experimental station of the International Potato Center. This parental cocona stock was carefully evaluated, and controlled crosses were hand-pollinated to begin the study of cocona fruit genetics.

FIG. 10. A complete diallel cross among seven varieties of cocona. All varieties were crossed with all varieties, including self-crosses marked by the diagonal line. Female parents are in the top row and male parents are in the left column. Progeny of any given female and male are found in the box corresponding to the column and row of its female and male parent, respectively. Parents and progeny are illustrated with representative fruit types. The significant result of the diallel is the overwhelming predominance of *maternal inheritance* for the fruit characters measured.

and standard deviation of fruit weight, yield, shape (measured as the ratio of length to width), and color (scored: 0 = yellow, 1 = orange, 2 = red). A subsequent generation (F_2) of hand-pollinated selfed progeny of this diallel cross was planted at the North Carolina experimental station in Yurimaguas, Peru.* Due to inclement weather, soil, labor, and political conditions, only data on fruit weight are available for this latter generation. Nevertheless, these data support the results of the previous generation (Table VIII).

The significant result of the diallel is the overwhelming predominance of a maternal influence† or the fruit characters measured (Fig. X and Tables VIII and IX). A small-fruited mother pollinated by either a small-, medium-,

* Recognition is due to Ing. Beto Pachanasi for carrying out this evaluation.

† I use the general term "maternal influence" to include both maternal inheritance and maternal effects when not distinguished. See below, section on The Models, under Cocona Fruit.

TABLE VIII. Complete Diallel Cross of Cocona[a]

A. P_1 Parents used for diallel cross

Variety	I	II	III	IV	V	VI	VII
Fruit weight, g	29 ± 2	29 ± 1	59 ± 4	39 ± 1	162 ± 12	156 ± 5	141 ± 19
Yield, tons/ha	37 ± 4	33 ± 6	53 ± 8	31 ± 7	44 ± 12	77 ± 4	62 ± 14
Fruit shape	RND (0.9)	RND (0.9)	RND (0.9)	LNG (1.2)	FLT (0.8)	LNG (1.1)	LNG (1.2)
Color	YELLOWOR	ORRED	YELLOWOR	YELLOWOR	ORANGE	YELLOW	ORANGE

B. F_1 Progeny: diallel cross

	Female P_1							
Male P_1	I	II	III	IV	V	VI	VII	Row means
I	27 ± 0	38 ± 1	51 ± 2	39 ± 1	68 ± 2	121 ± 18	186 ± 45	76 g
	25 ± 11	41 ± 11	44 ± 8	44 ± 3	64 ± 5	65 ± 5	60 ± 4	49 tons/ha
	RND	RND	RND	LNG	RND	LNG	LNG	RND (1.0)
	ORANGE	ORRED	ORANGE	YELLOWOR	ORANGE	YELLOW	ORANGE	Orange (0.9)
II	41 ± 1	60 ± 4	51 ± 2	38 ± 1	96 ± 10	142 ± 8	196 ± 14	89 g
	41 ± 10	37 ± 1	44 ± 8	38 ± 3	46 ± 22	71 ± 7	58 ± 3	48 tons/ha
	RND	RND	RND	LNG	RND	RND	LNG	RND (1.0)
	ORANGE	ORANGE	ORANGE	ORANGE	ORRED	YELLOW	ORANGE	Orange (1.0)
III	27 ± 5	57 ± 1	46 ± 2	38 ± 2	201 ± 7	144 ± 5	238 ± 6	107 g
	21 ± 2	40 ± 9	39 ± 9	38 ± 2	50 ± 5	55 ± 18	57 ± 14	43 tons/ha
	RND	RND	RND	LNG	FLT	LNG	LNG	RND (1.0)
	ORANGE	YELLOWOR	YELLOWOR	ORANGE	ORANGE	YELLOWOR	ORANGE	Orange (0.9)

IV	31 ± 3 27 ± 5 RND YELLOWOR	53 ± 3 47 ± 14 RND ORANGE	71 ± 4 48 ± 5 RND ORANGE	40 ± 0 41 ± 1 LNG ORRED	200 ± 2 51 ± 14 FLT ORANGE	14 ± 121 59 ± 17 LNG YELLOW	214 ± 24 59 ± 4 LNG ORRED	107 g 48 tons/ha RND (1.0) Orange (0.9)
V	25 ± 4 22 ± 5 RND YELLOWOR	61 ± 1 38 ± 8 RND YELLOWOR	72 ± 2 32 ± 13 RND ORANGE	55 ± 3 41 ± 11 LNG ORRED	170 ± 5 62 ± 9 FLT ORANGE	150 ± 10 72 ± 10 LNG YELLOW	120 ± 23 43 ± 2 RND ORANGE	93 g 44 tons/ha RND (1.0) Orange (0.8)
VI	31 ± 1 31 ± 2 RND ORANGE	55 ± 4 40 ± 11 RND YELLOWOR	54 ± 6 37 ± 18 RND ORANGE	38 ± 1 50 ± 8 LNG ORANGE	183 ± 14 55 ± 16 FLT ORRED	145 ± 10 65 ± 12 LNG YELLOWOR	151 ± 7 62 ± 14 LNG ORANGE	94 g 49 tons/ha RND (1.0) Orange (0.9)
VII	29 ± 1 22 ± 4 RND ORANGE	56 ± 2 45 ± 16 RND ORANGE	54 ± 6 37 ± 18 RND YELLOWOR	38 ± 1 50 ± 8 LNG ORANGE	173 ± 25 56 ± 4 FLT ORRED	129 ± 9 60 ± 4 LNG YELLOWOR	250 ± 2 49 ± 7 LNG ORRED	104 g 45 tons/ha RND (1.0) Orange (1.0)
Column means	30 27 RND (0.9) or (0.9)	54 41 RND (0.9) or (0.9)	57 40 RND (0.9) or (0.9)	41 43 LNG (1.2) or (1.0)	156 55 FLT (0.8) or (1.1)	139 64 LNG (1.2) yor (0.6)	194g 55 tons/ha LNG (1.2) ord (1.5)	

(continued)

TABLE VIII. Continued

C. F_2 Progeny (selfed) from dialled cross

Male P_1	Female P_1: I	II	III	IV	V	VI	VII	Row means
I	30 RND	35 RND	54 RND	34 RND	47 RND	— —	— —	40 g RND (1.0)
II	38 RND	57 RND	24 FLT	44 RND	— —	87 RND	55 RND	51 g RND (1.0)
III	26 RND	39 RND	46 RND	33 LNG	244 FLT	— —	92 RND	80 g RND (1.0)
IV	— —	25 RND	— —	26 LNG	150 FLT	— —	— —	59 g RND (0.9)
V	31 RND	42 RND	34 RND	66 FLT	188 FLT	114 RND	142 RND	88 g RND (1.0)
VI	23 RND	47 RND	52 RND	33 RND	179 FLT	145 RND	— RND	80 g RND (1.0)
VII	18 RND	18 RND	54 RND	31 LNG	— —	118 RND	196 LNG	73 g LNG (1.1)
Column means	28 RND (1.0)	38 RND (1.0)	44 RND (1.0)	38 RND (1.1)	162 FLT (0.8)	116 RND (1.1)	121g LNG (1.2)	

[a] Data are displayed in three sections: A, for the parent generation (P_1: seven varieties), B, the first generation of the hybrid crosses (F_1: 49 crosses—seven parents by seven parents), and C, the second generation of selfed progeny (F_2: 49 selfed lines from the crosses). Results are summarized for mean (±standard deviation) fruit weight (grams), yield (tons/hectare), shape (length/width), and color. RND, round fruit (length = width); FLT, flat fruit (length < width); LNG, long fruit (length > width); or, orange fruit; rd, red fruit; y, yellow fruit with combinations of these letters (e.g., yor is yellow-orange fruit). Each parent and cross were evaluated in three repetitions of 36 plants each (5 m × 5 m) in a randomized block design. The data presented include cell, row, and column means for the three repetitions.

TABLE IX. ANOVA Results from the Complete Diallel Cross of Cocona (Fig. 10 and Table VIII) and a Second Generation of Selfed Progeny[a]

Character	DF	Male SS	Male MS	*F*	Significance	Female SS	Female MS	*F*	Significance
Diallel cross (F_1)									
Fruit weight	14	16453	2742	3.8	0.001	537637	89606	125	0.0001
Yield	14	693	115	0.8	0.541	19230	3204	23	0.0001
Fruit shape	14	0.056	0.009	2.0	0.065	3.196	0.533	116	0.0001
Color	12	0.100	0.017	1.1	0.372	1.531	0.255	17	0.0001
Selfed progeny of cross (F_2)									
Fruit weight	12	19138	3190	1.9	0.109	91644	15274	9.1	0.0001

[a] There is an overwhelming predominance of maternal inheritance indicated by the highly significant differences among female lines (far right column). Some quantitative inheritance is evidenced by the significance of both male and female parents for fruit weight (first row, significance underlined).

or large-fruited plant produces small offspring closely resembling the maternal parent. Likewise, the large-fruited mother produces large-fruited offspring regardless of the pollen. Some quantitative inheritance is evidenced by the significance of male and female parents in explaining fruit weight. For example, the large, flat female and the small male (column 5 × row 1) produced medium offspring, and the large-fruited cocona which, when selfed (column 7 × column 7), produced outstandingly large offspring (which the local farmers all expropriated immediately for planting in their own gardens). The maternal influence in fruit weight continued through the following controlled self-pollinated generation planted under different environmental conditions at Yurimaguas (Table IX).

Other analyses which confirmed the results of the ANOVAs (Table IX) included nonsignificant mid-parent correlations (F_1: $r^2 = 0.33$, $p > 0.05$; F_2: $r^2 = 0.48$, $p > 0.05$); significant regressions of first and second generations with female parents (P_1) (F_1: $r^2 = 0.74$, $p < 0.001$; F_2: $r^2 = 0.66$, $p < 0.01$); and nonsignificant rank correlations of reciprocal crosses (F_1: $r^2 = 0.18$, $p > 0.05$; F_2: $r^2 = 0.44$, $p > 0.05$). These diverse analyses establish strong maternal influence over characters of cocona fruit.

These unexpected results immediately bring up the question of mechanism: is this apomixis, pseudogamy, cytoplasmic inheritance, or maternal effects? Most of these mechanisms have been reported in *Solanum* species or their near relatives. Apomixis has been described in tomatoes by Rick (1950), but for cocona this is ruled out by the observation that none of the emasculated flowers which were left unpollinated set fruit. Pseudogamy triggered by sterile pollen has been reported for tobacco by Pandy (1975, and previous work cited therein). Pollen viability of both long- and short-style cocona flowers was high, 37% and 38%, respectively (mean of 10 samples each).* Cytoplasmic inheritance has been found in potatoes (e.g., Grun *et al.*, 1962; Grun, 1976; and CIP genetics program). Cytoplasmic inheritance of such a major complex of characters as fruit size and shape has not been reported for *Solanum*, although *Nicotiana rustica* (Solanaceae) shows maternal inheritance for plant stature (Jinks *et al.*, 1972), and a large number of cultivated plants show maternal inheritance for yield (Aksel, 1977). *Brassica campestris* shows maternal effects for a number of the same characters for which *Solanum sessiliflorum* shows maternal influence, including length of fruit and yield (Singh and Murty, 1980).

Electrophoretic analyses to elucidate the mechanisms of the maternal influence in cocona were indecisive. Electrophoretic comparisons of enzymes in parental populations and selected crosses ($N = 25$) uncovered no genetic variation at approximately 40 putative loci.† Further, definitive mechanistic work will require genetic analyses beyond the scope of the present study on domestication.

Selection for fruit size was included in the reciprocal transplant data (Table IV). Large-fruited plants had a comparative advantage on the alluvial riverbank, while in the swidden their numbers were comparable to those of small fruits. On the alluvial soil at San Ramon, cocona production of the parental generation was 55 ± 22 and 52 ± 22 tons/ha for two large-fruited varieties and 29 ± 15 and 37 ± 20 tons/ha for two small-fruited varieties (Salick, 1990).

The effect of soil fertility on the relative production of large- and small-fruited varieties was tested on the nutrient-poor soils of Iscozacin. Large and small cocona were planted at 1 m spacing in a completely random design of five soil fertility levels. Soil fertility levels were (1) no fertilizer, (2) 1.5 tons/ha calcium, (3) 300 kg/ha N–P–K, (4) 300 kg/ha N–P–K with calcium, (5) 600 kg/ha N–P–K with calcium. The results in Fig. 11 are for number of fruit harvested (ANOVA fruit number by fertility, fruit size, and repetition: main effects df = 7, $F = 5.3$, $p < 0.001$, with fertility df = 4, $F = 6.2$, $p < 0.002$). Overall, fruit production increased with increasing inputs of fertilizer (rank correlation of 70%); this overall increase was factored out by analyzing the proportion of small fruits (white sections of the bar graph) compared with

* Appreciation goes to Dr. Noel Palae, physiologist at CIP, for the pollen viability analysis carried out in his laboratory.

† Dr. Norm Weeden and Elizabeth Dixon of the Geneva Experimental Station of Cornell University provided very generous aid in these analyses.

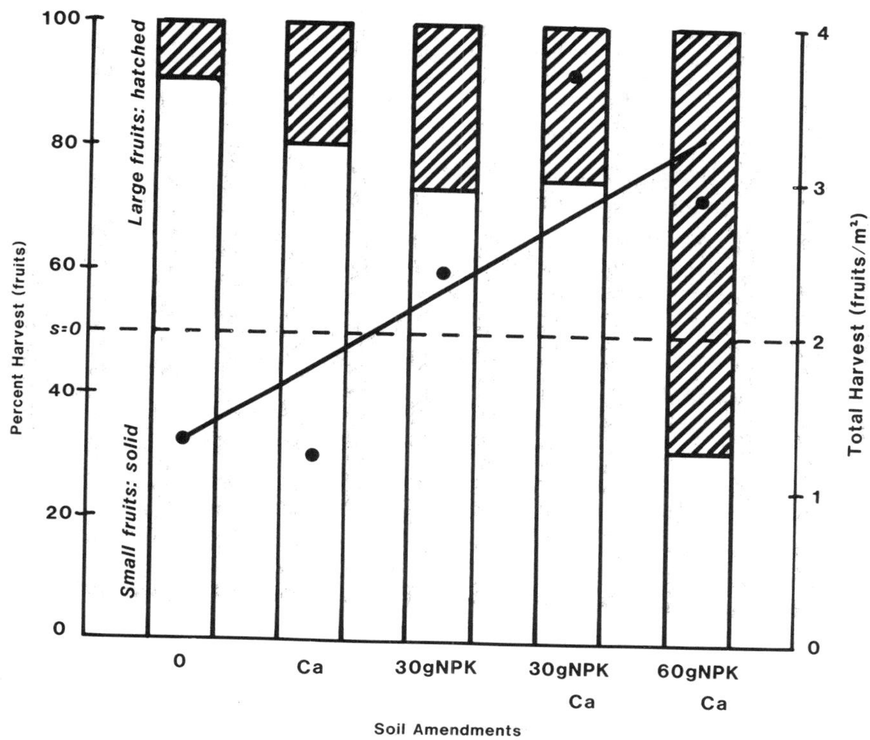

FIG. 11. The effects of soil fertility on cocona harvest and proportion of fruit types. Total fruit production (dots and regression line) increases with soil fertility, as does production of large-fruited varieties (hatched portions of percent-harvest bar graph); small-fruited varieties of cocona (open portions of percent-harvest bar graph) show no response to fertilizer.

the proportion of large fruits (hatched). The proportions of small and large fruits varied with soil fertility level, suggesting that large-fruited plants have a comparative advantage on rich soils. This may indicate why small farmers on poor soils sometimes prefer small-fruited varieties, while large-fruited varieties are grown more often in intensively managed yard gardens or on alluvial soils. These trends are complicated by personal preference and differences in use (small fruits for juice, large ones for pulp).

With the predominance of maternal influence over cocona fruit genetics, the variables needed to construct a model are much reduced, even while theoretical considerations are amplified.

THE MODELS

Spines

A basic model emerges from the field genetic and ecological studies of cocona spines, albeit rough hewn (Fig. 12). There are three distinct populations of cocona (N) in pastures (p), swidden gardens (g), and along riverbanks (r). Each of these populations is further subdivided into subpopulations of the effective population size (N_e = 3.2). Gene flow m among these populations was estimated at 0.5%. Reciprocal transplants defined the selection coefficient S against spines in gardens (S = 0.97) and along riverbanks (S = 0.25), and for spines in pastures (S = 1.0). Cocona selection is heavy in pastures and swidden gardens, dominating gene flow and determining equilibria well skewed toward the selected trait. Riverbanks have less selection pressure against cocona spines.

This is the outline of a domestication model before the consequences of the small effective population size and genetic drift are considered. These added variables suggest a dynamic equilibrium between genetic drift and flow, especially where selection is light (e.g., riverbanks). Genes may flow into a population, but then drift to fixation or elimination where there are both little selection and small samples of genetic diversity passed to subsequent generations. Analyzing the roughly measured population-genetic variables more precisely might reveal the effects of several environmental and organismal characteristics which may contribute to the dynamic between genetic drift and flow (Table X) (Loveless and Hamrick, 1984). However, the

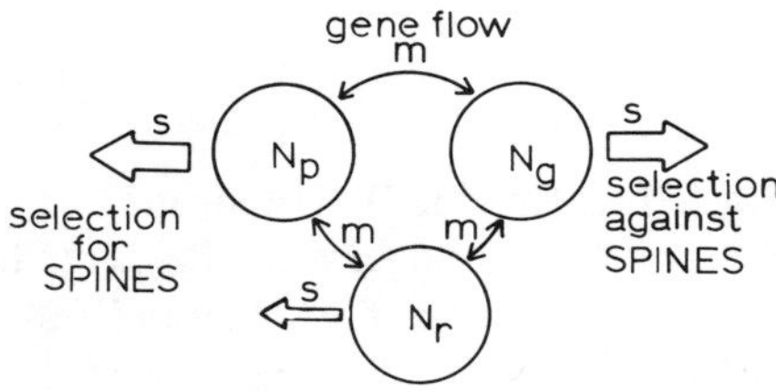

FIG. 12. A simple population-genetic model for cocona spines. There are three populations of cocona (N) in pastures (p), swidden gardens (g), and along riverbanks (r). Genes for spines flow (m) among these populations, while selection (s) acts against spines in gardens and along riverbanks, and for spines in pastures. Selection is heavy in pastures and gardens, represented by large arrows dominating gene flow and allowing equilibrium well skewed toward the selected trait. Riverbanks have little selection, represented by a small arrow, which is balanced by gene flow, although this balance is a dynamic one between gene drift and flow.

TABLE X. Potential Effects of Various Ecological Factors on the Genetic Structure, Crop Diversity, and Evolution of Cocona[a]

Ecological factors	Gene action
Self-pollination	Drift
Annual/biannual life cycle	Drift
Early successional habitat	Drift
Small/fluctuating population size	Drift
Large bee pollinators	Flow
Animal-dispersed seeds	Flow
Continuous flowering	Flow

[a] After Loveless and Hamrick (1984).

human interactions which affect cocona evolution would make some of these variables extremely difficult to assess.

Cocona Fruit

A model for stable maternal inheritance of cocona fruit characteristics (size, shape, color, and yield) is conceptually extremely simple: characters are reproduced consistently. Regardless of the direction of selection, maternal inheritance, if it is stable over generations, should assure that the population response to a constant selection pressure is rapid, approaching fixation over a relatively few generations. There is no gene flow to balance selection since there is no genetic effect of any introduced pollen. Thus, an important result of stable maternal inheritance is the speed and ease with which a population responds to selection. An adapted or preferred phenotype is merely multiplied without alteration in the next generation, very similar to vegetative reproduction. Figure 13 compares the response of a population to (1) stable maternal inheritance, and to nuclear inheritance (2) with gene dominance and (3) without dominance, all under the same selection pressure. Clearly, with maternal inheritance, cocona populations would be able to respond quickly and completely to both human and natural selection of cocona fruit.

Establishing the validity of maternal inheritance is much more difficult than forming the model. In the literature, there is confusion on what constitutes maternal effects and maternal inheritance. Here, I adopt a convention to distinguish between "maternal effects," which are brought about by constituents of the egg cytoplasm or the maternal physiological state, usually persisting for only one generation (Futuyma, 1986), and "maternal inheri-

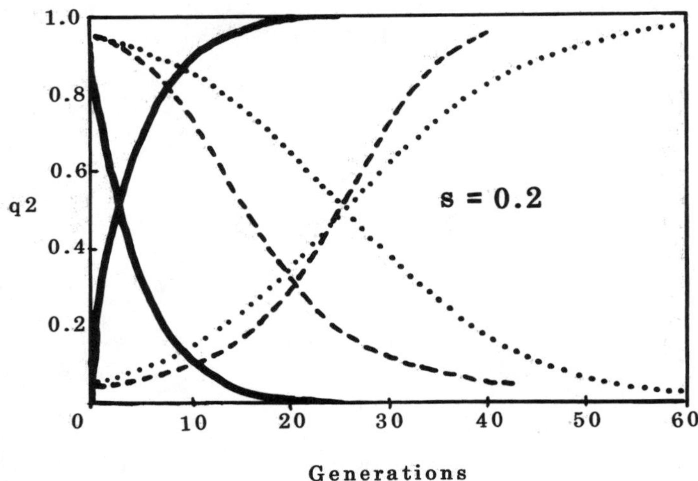

FIG. 13. A comparison of the response of a population to maternal inheritance (solid line) and to nuclear inheritance with gene dominance (dashed line) and without dominance (dotted line), all under the same selection pressure (S = 0.2) (after Falconer, 1981). Clearly, populations with maternal inheritance respond quickly and completely to selection, in this case to human and environmental selection of cocona fruit. Thus, an important outcome of maternal inheritance is the speed and ease with which a population responds to selection. An adapted or preferred genotype is merely multiplied without alteration in the next generation.

tance" (or maternal transmission), which is understood to be nearly synonomous with cytoplasmic inheritance (e.g., Lincoln *et al.,* 1985). I use "maternal influence" when the mechanism is not distinguished.

Roach and Wulf (1987) review maternal effects in plants, which, they stress, can contribute substantially to the phenotype of an individual and have important consequences for the interpretation and design of both ecological and genetic studies. They suggest that the first and best way to validate maternal inheritance is the full diallel cross, which was carried out here for cocona. The second step is to continue selfing and backcrossing through several generations to rule out maternal and environmental effects. One subsequent selfed generation was completed for cocona in a radically different environment through which maternal influence persisted, distinguishing the case of cocona from cases of maternal effects such as snail spirals (Boycott and Diver, 1923). The persistence of maternal traits in cocona fruit over two generations suggests maternal inheritance as the appropriate model—such as that described above. Unfortunately, further research to continue monitoring the stability of the maternal influence has been prohibited by political instability within Peru and restrictions on germplasm transfer of the economically important genus *Solanum* to other countries.

DISCUSSION

Domestication of cocona eliminates spines apparently by the easy process of selecting against one dominant gene. In most environments where cocona grows wild there is strong natural selection for spines as a defense against vertebrate herbivores. Thus, the balanced genetic equilibria evidenced are greatly skewed toward nonspininess under cultivation and toward spininess in most other habitats. This skewedness is found despite significant gene flow among populations because selection pressures are great.

Only in small, patchy environments where there is neither human management of cocona nor herbivore grazing may cocona populations experience genetic drift and fixation even with gene flow because of extremely small population sizes due to inbreeding. Among populations, a statistical balance may be evidenced, but at a very local level drift may fix small populations with or without spines. This situation describes a case of dynamic equilibrium where genes for spines or their absence would circulate among populations, but then the populations would refix at one character state or the other.

These results indicating a complex dynamic of genetic diversity over time may reflect the state of crop genetic diversity more often than not. Such dynamic equilibria should be considered in light of crop germplasm conservation. A germplasm collection at one point in time may not represent the diversity experienced by or available to populations over time. The conservation of static *ex situ* germplasm does not truly represent the genetic composition of a population over time, much less the genetic structure of that population (Salick, 1983; Altieri and Merrick, 1987; Oldfield and Alcorn, 1987; Salick and Merrick, 1990).

Domestication of cocona has also resulted in tremendously diversified fruit sizes and shapes, from small, round fruits to large, long, or flat fruits. Such radiation is common under domestication [e.g., for tomatoes (Rick, 1978)]. It is less common that such crop diversity is maternally inherited as indicated for cocona, and we are left to consider how our understanding of genetic models and evolutionary processes of crop domestication are changed by maternal inheritance. Modern genetics is changing with accumulating evidence that traditional quantitative genetics does not model all populations [reviews on maternal inheritance include Jinks (1964), Grun (1976), and Roach and Wulff (1987)]. Theoretical aspects of extranuclear inheritance are far ranging (e.g., Cosmides and Tooby, 1981; Eberhard, 1980) and the incorporation of non-Mendelian genetics into crop domestication theory is pending.

An outstanding feature of the data and models presented for the evolution of cocona in wild, domesticated, and hybrid populations is their genetic simplicity. A one gene model fits the data on spines, and direct maternal inheritance explains the data on fruit size, shape, color, and yield. Could this simplicity be other than happenstance?

Simmonds (1976, 1979) reviews leading features of 124 major crops. Tabulating from this review, 45% of the crops are inbred and 40% are vegetatively propagated. These represent uniparental inheritance analogous to the maternal inheritance indicated for cocona fruit. Uniparental inheritance has two main advantages: selection of desired characteristics is easy (what you see is what you get), and multiplication of the desired varieties is rapid since all offspring can be used without eliminating undesirable recombinants. Other forms of uniparental inheritance (sexual and asexual) are common among crops in field agriculture and in biotechnology (Table XI). The major impact of biotechnology may be in expanding and refining our ability to propagate crops from a single parent through meristem culture, cloning, or other rapid multiplication techniques.

Indigenous farmers consciously select crops (e.g., Boster, 1985; Franquemont *et al.,* 1988). Although I disagree with Darwin that consciousness makes evolution any less natural, or in any way fundamentally different, there may be certain correlates with conscious selection. It seems reasonable that plants with complicated breeding systems and inheritance patterns might be more difficult to domesticate and select. Natural selection for breed-

TABLE XI. Types of Uniparental Inheritance (Both Sexual and Asexual) in Crops[a]

True seed propagation	Vegetative propagation	Biotechnology
Selfing	Tubers	Meritstem culture
Apomixis	Roots	Embryo culture
Autogamy	Cuttings	Rapid multiplication
Cytoplasmic inheritance	Stakes	Cloning
Mitochondrial DNA	Grafting	
Chloroplast DNA		

[a] Uniparental inheritance has two main advantages: selection of desired characteristics is easy, and multiplication of the desired varieties is rapid.

ing systems has been established (e.g., Rick *et al.*, 1977; Lande and Schemske, 1985; Schemske and Lande, 1985; Waller, 1986; Barrett and Shore, 1987). Human selection may favor breeding systems which facilitate selection, domestication, and cultivation. People may select unintentionally but effectively for simple inheritance patterns. Cocona demonstrates simple inheritance by both a single dominant gene and maternal inheritance; other crops have simple inheritance by apomixis, selfing, vegetative reproduction, and the biotechnology of rapid multiplication. As our appreciation develops for the range and complexity of traits affected by natural selection, our vision of the effects of human selection and crop domestication likewise will expand.

Acknowledgments

My profound thanks go to Dr. Ghillean T. Prance and William Sugrue, who provided the opportunity for me to devote over 3 years in the field to experimental economic botany. Ample funding was provided by the Mellon Foundation and U.S. Agency for International Development/Peru Mission. Institutional support was generously supplied by the Institute of Economic Botany of the New York Botanical Garden, Proyecto Especial Pichis-Palcazu of the President's Office, Peru, and the International Potato Center, Peru. Special appreciation goes to my family and friends, cocona fanatics all.

REFERENCES

Aksel, R., 1977, Quantitative genetically nonequivalent reciprocal crosses in cultivated plants, in: *Genetic Diversity in Plants.* (A. Muhammed, R. Aksel, and R. C. von Borstel, eds.), Plenum Press, New York.

Alcorn, J. B., 1981, Some factors influencing botanical resource perception among the Huastec: Suggestions for future ethnobotanical inquiry, *J. Ethnobiol.* **1**:221–230.

Altieri, M. A., and Merrick, L. C., 1987, *In situ* conservation of crop genetic resources through maintenance of traditional farming systems, *Econ. Bot.* **41**:86–96.

Anderson, E., 1948, Hybridization of the habitat, *Evolution* **2**:1–9.

Anderson, E., 1954, *Plants, Man and Life,* A. Melrose, London.

Anderson, E., 1956, Man as maker of new plants and new plant communities, in: *Man's Role in Changing the Face of the Earth* (W. L. Thomas, ed.), pp. 763–777, University of Chicago Press, Chicago, Illinois.

Anderson, E., 1968, *Introgressive Hybridization,* Hafner, New York.

Barrett, S. C., and Shore, J. S., 1987, Variation and evolution of breeding systems in the *Turnera ulmifolia* L. complex (Turneraceae), *Evolution* **41**:340–354.

Benza, J. C., and Rodriguez, J. B., 1977, *El Cultivo de la Cocona,* Universidad Nacional Agraria, La Molina, Peru.
Boster, J. S., 1983, A comparison of the diversity of Jivaroan gardens with that of the tropical rainforest, *Hum. Ecol.* **11**:47–68.
Boster, J. S., 1985, Selection for perceptual distinctiveness: Evidence from Aguaruna cultivars of *Manihot esculenta, Econ. Bot.* **39**:310–325.
Boycott, A. E., and Diver, C., 1923, On the inheritance of sinistrality in *Limnaea peregra, Proc. R. Soc. Lond. B* **95**:207–213.
Cosmides, L. M., and Tooby, J., 1981, Cytoplasmic inheritance and intragenomic conflict, *J. Theor. Biol.* **89**:83–129.
Darwin, C., 1896, *The Variation of Animals and Plants under Domestication,* Appleton & Co., New York.
De Candolle, A., 1959, *Origin of Cultivated Plants,* Hafner, New York.
Eberhard, W., 1980, Evolutionary consequences of intracellular organelle competition, *Q. Rev. Biol.* **55**:231–256.
Falconer, D. S., 1981, *Introduction to Quantitative Genetics,* Longman, London.
Fernandez, E., 1985, Biologia floral de *Solanum sessiliflorum* e *Solanum subinerme* (Solanaceae), na regiao de Manaus, *Acta Amazon.* **18**:55–65.
Ford, R. I., 1978, *The Nature and Status of Ethnobotany,* Museum of Anthropology, University of Michigan, Ann Arbor, Michigan.
Franquemont, C., Brush, S., Boster, J., Alcorn, J., Salick, J., Plaisted, R., Vander Zaag, P., Thurston, D., and Isbell, B. J., 1988, Inventing Vegetables: A Workshop on Indigenous Vegetable Breeding. Cornell University, Ithaca, New York.
Futuyma, D. J., 1986, *Evolutionary Biology.* Sinauer Associates, Sunderland, Massachusetts.
Grun, P., 1976, *Cytoplasmic Genetics and Evolution,* Columbia University Press, New York.
Grun, P., Aubertin, M., and Radlow, A., 1962, Multiple differentiation of plamons of diploid species of *Solanum, Genetics* **47**:1321–1333.
Harlan, J. R., 1965, The possible role of weed races in the evolution of cultivated plants, *Euphytica* **14**:173–176.
Harlan, J. R., 1966, Evolutionary dynamics of plant domestication, *Jpn. J. Genet.* **44**:337–343.
Harlan, J. R., 1975, *Crops and Man,* American Society of Agronomists, Madison, Wisconsin.
Harlan, J. R., 1976, Genetic resources in wild relatives of crops, *Crop Sci.* **16**:329–333.
Harlan, J. R., and DeWet, J. M. J., 1963, The compilospecies concept, *Evolution* **17**:497–501.
Harris, D. R., 1967, New light on plant domestication and the origins of agriculture: A review, *Geogr. Rev.* **57**:90–107.
Heiser, C. B., Jr., 1968, Some Ecuadorian and Colombian solanums with edible fruits, *Cienc. Nat.* **11**:1–9.
Heiser, C. B., Jr., 1971, Notes on some species of *Solanum* (Sect. Leptostemonum) in Latin America, *Baileya* **18**:59–65.
Heiser, C. B., Jr., 1972, The relationships of the naranjilla, *Solanum quitoense, Biotropica* **4**:77–84.
Heiser, C. B., Jr., 1973, *Seed to Civilization: The Story of Man's Food,* W. H. Freeman, San Francisco.
Hernández-Xolocotzi, E., 1985, Maize and man in the greater Southwest, *Econ. Bot.* **39**:416–430.
Iltis, H. H., 1983, From teosinte to maize: The catastrophic sexual transmutation, *Science* **222**:886–894.
Jinks, J. L., 1964, *Extrachromosomal Inheritance,* Prentice-Hall, Englewood Cliffs, New Jersey.
Jinks, J. L., Perkins, J. M., and Gregory, S. R., 1972, The analysis and interpretation of differences between reciprocal crosses of *Nicotiana rustica* varieties, *Heredity* **28**:363–377.

Johannessen, C. L., 1966, The domestication process in trees reproduced by seed: The pejibaye palm in Costa Rica, *Geogr. Rev.* **56**:363–376.

Johannessen, C. L., 1982, Domestication process of maize continues in Guatemala, *Econ. Bot.* **36**:84–99.

Johannessen, C. L., Wilson, M. R., and Davenport, W. A., 1970, The domestication of maize: Process or event, *Geogr. Rev.* **60**:393–413.

Johns, T., and Keen, S. L., 1986, Ongoing evolution of the potato on the altiplano of western Bolivia, *Econ. Bot.* **40**:409–424.

Lande, R., and Schemske, D. W., 1985, The evolution of self-fertilization and inbreeding depression in plants. I. Genetic models, *Evolution* **39**:24–40.

Lewin, R., 1988, A revolution of ideas in agricultural origins, *Science* **240**:984–986.

Lincoln, R. J., Boxshall, G. A., and Clark, P. F., 1985, *A Dictionary of Ecology, Evolution and Systematics,* Cambridge University Press, New York.

Loveless, M. D., and Hamrick, J. L., 1984, Ecological determinants of genetic structure in plant populations, *Annu. Rev. Ecol. Syst.* **15**:65–95.

Mangelsdorf, P. C., 1974, *Corn: Its Origin, Evolution and Improvement,* Belknap Press, Cambridge, Massachusetts.

Ministerio de Salud, 1957, *Composicion de los alimentos Peruanos,* Ministerio de Salud, Lima, Peru.

Nabhan, G. P., 1985, Native crop diversity in Aridoamerica: Conservation of regional gene pools, *Econ. Bot.* **39**:387–399.

NAS (National Academy of Sciences), 1972, *Genetic Vulnerability of Major Crops,* National Academy of Sciences, Washington, D. C.

NAS (National Academy of Sciences), 1975, *Underexploited Tropical Crops,* National Academy of Sciences, Washington, D. C.

Oldfield, M. L., and Alcorn, J. B., 1987, Conservation of traditional agroecosystems, *Bioscience* **37**:199–208.

Padoch, C., 1987, The economic importance and marketing of forest and fallow products in the Iquitos region, in: *Swidden Fallow Agroforestry in the Peruvian Amazon,* (W. Denevan and C. Padoch, eds.), *Advances in Economic Botany* **5**:74–89.

Pahlen, A. vonder, 1977, Cubiu (*Solanum topiro* (Humb. & Bonpl.)), um fruteira da Amazonia, *Acta Amazon.* **7**:301–307.

Pandy, K. K., 1975, Sexual transfer of specific genes without gametic fusion, *Nature* **256**:310–313.

Patiño, V. M., 1962, Edible fruits of *Solanum* in South American historic and geographic references, *Bot. Mus. Leafl.* **19**:215–234.

Pickersgill, B., 1981, Biosystematics of crop-weed complexes, *Kulturpflanze* **29**:377–388.

Pickersgill, B., and Heiser, C. B., 1976, Cytogenetics and evolutionary change under domestication, *Phil. Trans. R. Soc. Lond. B* **275**:55–69.

Posey, D. A., 1984, A preliminary report on diversified management of tropical forests by the Kayapó indians of the Brazilian Amazon, in: *Ethnobotany in the Neotropics* (G. T. Prance and J. A. Kallunki, eds.), *Advances in Economic Botany* **1**:112–126.

Rabinowitz, D., Linder, C. R., Begazo, O. D., Murguia, S. H., Ortega, D., R., and P. Schmiediche, P., Ecological mechanisms of genetic interchange between cultivars and their wild relatives: Traditional potato farming in the Peruvian Andes. Unpublished manuscript.

Renfrew, J. M., 1969, The archaeological evidence for the domestication of plants: Methods and problems, in: *The Domestication and Exploitation of Plants and Animals* (P. J. Ucko and G. W. Dimbleby, eds.), pp. 149–172, Duckworth, London.

Rick, C. M., 1950, Pollination relations of *Lycopersicon esculentum* in native and foreign regions, *Evolution* **4**:110–122.

Rick, C. M., 1978, The tomato, *Sci. Am.* **239**:76–87.

Rick, C. M., Fobes, J. F., and Holle, M., 1977, Genetic variation in *Lycopersicon pimpinellifolium:* Evidence of evolutionary change in mating systems, *Plant Syst. Evol.* **127**:139–170.

Rindos, D., 1984, *The Origins of Agriculture: An Evolutionary Perspective,* Academic Press, New York.

Roach, D. A., and Wulff, R. D., 1987, Maternal effects in plants, *Annu. Rev. Ecol. Syst.* **18**:209–235.

Rodríguez, F. R., and Garayar, M. H., 1969, *Cultivo de Cocona, Maracuya, y Naranjilla,* Ministerio de Agricultura, Lima, Peru.

Salick, J., 1983, Natural history of crop-related wild species, *Environ. Manage.* **7**:85–90.

Salick, J., 1986, Ethnobotany of the Amuesha. a. Cocona (*Solanum sessiliflorum*). b. A Cocona Cookbook, USAID/Peru, Lima, Peru.

Salick, J., 1989, Ecological basis of Amuesha agriculture, *Adv. Econ. Bot.* **7**:189–212.

Salick, J., 1990, Cocona (*Solanum sessiliflorum*): An overview of production and breeding potentials of the peach-tomato. in: *New Crops for Food and Industry* (G. E. Wickens, N. Haq, and P. Day, eds.), pp. 257–264, Chapman and Hall, London.

Salick, J., In preparation. Nutritional values of an underexploited tropical crop, cocona (*Solanum sessiliflorum*).

Salick, J., and Merrick, L., 1990, Use and maintenance of genetic resources: Crops and their wild relatives, in: *Agroecology* (C. R. Carroll, J. H. Vandermeer, and P. M. Rosset, eds.), pp. 517–548, McGraw-Hill, New York.

Sanchez, P. A., and Benites, J. R., 1987, Low input cropping for acid soils of the humid tropics, *Science* **238**:1521–1527.

Sauer, C. O., 1952, *Agricultural Origins and Dispersals,* MIT Press, Cambridge, Massachusetts.

Schemske, D. W., and Lande, R., 1985, The evolution of self-fertilization and inbreeding depression in plants. II. Empirical observations, *Evolution* **39**:41–52.

Schultes, R. E., 1956, The Amazon indian and evolution of *Hevea* and related genera, *J. Arnold Arbor.* **37**:123–147.

Schultes, R. E., 1958, A little-known cultivated plant from northern South America, *Bot. Mus. Leafl.* **18**:229–244.

Schultes, R. E., and Romero-Casteñeda, R., 1962, Edible fruits of *Solanum* in Colombia, *Bot. Mus. Leafl.* **19**:235–286.

da Silva Filho, D. F., Clement, C. R., and Noda, H., 1989, Variação fenotípica em frutos de cloze introducões de cubiu (*Solanum sessiliflorum* Dunal) avaliadas em Manaus, am Brasil. *Acta Amazonica* **19**:9–18.

Simmonds, N. W. (ed.), 1976, *Evolution of Crop Plants,* Longman, New York.

Simmonds, N. W., 1979, *Principles of Crop Improvement,* Longman, New York.

Singh, J. N., and Murty, B. R., 1980, Combining ability and maternal effects in *Brassica campestris* variety yellow sarson, *Theor. Appl. Genet.* **56**:264–272.

Strongin, J. D., 1982, Machiguenga, medicine, and missionaries: The introduction of western health aids among a native population of Southeastern Peru, Ph.D. thesis, Columbia University, New York.

Ucko, P. J., and Dimbleby, G. W., 1969, *The Domestication and Exploitation of Plants and Animals,* Duckworth, London.

Vavilov, N. I., 1951, The origin, variation, immunity and breeding of cultivated plants, *Chronica Botanica* **13**:1–366.

Vietmeyer, N. D., 1986, Lesser-known plants of potential use in agriculture and forestry, *Science* **232**:1379–1384.

Waller, D. M., 1986, Is there disruptive selection for self-fertilization?, *Am. Nat.* **128**:421–426.

Whalen, M. D., and Caruso, E. E., 1983, Phylogeny in *Solanum* sect. Lasciocarpa (Solanaceae): Congruence of morphological and molecular data, *Syst. Bot.* **8**:369–380.
Whalen, M. D., Costich, D. E., and Heiser, C. B., 1981, Taxonomy of *Solanum* Section Lasiocarpa, *Gentes Herbarum* **12**:4–129.
Wilkes, H. G., 1977, Hybridization of maize and teosinte in Mexico and Guatemala and the improvement of maize, *Econ. Bot.* **31**:254–293.

8

The Support of Hydrostatic Load in Cephalopod Shells

Adaptive and Ontogenetic Explanations of Shell Form and Evolution from Hooke 1695 to the Present

DAVID K. JACOBS

INTRODUCTION

The chambered cephalopod shell has long been a subject of scientific investigation. However, a rudimentary understanding of the function of the shell in supporting external pressure was not achieved until the early 1960s. This functional issue is a relatively simple one: Did the chambered shell contain pressurized gas opposing the external water pressure, or alternatively did the shell structure itself support the load of the overlying column of water? Tests of the hypothesis of internal pressurized gas are also relatively simple to devise. Puncture of the shell while it is immersed in a fluid would appear to be sufficient to demonstrate whether there is a high internal gas pressure. If there were such pressure, a dramatic effusion of gas would occur when the shell was punctured and it should be readily observed. Indeed such a test was suggested in 1832 by the great anatomist Richard Owen (1832). The tests performed in the 1960s (Denton and Gilpin-Brown, 1961*a–c*, 1966; Denton *et al.*, 1961, 1967) were similar in many respects to that suggested by Owen, and they demonstrated unequivocally that the shell does not contain ele-

DAVID K. JACOBS • Department of Invertebrates, American Museum of Natural History, New York, New York 10024.

Evolutionary Biology, Volume 26, edited by Max K. Hecht *et al.* Plenum Press, New York, 1992.

vated gas pressure. Thus the structure of the cephalopod shell does support the forces resulting from the overlying column of water (Fig. 1).

The 130-year hiatus between the suggestion of a test of a hypothesis of shell function in 1832 and its final performance in the early 1960s has interesting implications for the continuity of scientific progress and the transfer of information within scientific traditions. There appear to be several elements contributing to this hiatus. First, the gas pressure model had priority; in 1695, when Robert Hooke suggested that the *Nautilus* shell was a buoyancy compensation device, he also suggested the possible role of gas pressure in expelling water from the shell. Thus, later 19th century workers already had a historically well-rooted explanation to resort to. In addition, study of cephalopods, especially neontological study, was not uniform, but was concentrated toward the end of the 19th century, when cephalopods played a central role in a number of evolutionary theories. In the late 19th century adaptive or functional explanations for morphology were not in vogue. Instead, the evolution of the lineage was thought to parallel an ontogeny, and morphology was thought to be a consequence of the reexpression or recapitulation of the history of the lineage. These views tended to exclude adaptive or functional interpretation of morphology. This conflict between adaptive explanations and evolutionary explanations based on the presumed parallelism of ontogeny and phylogeny is a recurrent theme in this work. It involves both workers who were nominally Darwinian and those who adhered to non-Darwinian evolutionary views, such as the morphogenetic law of Hyatt.

Interestingly, adaptive explanations in the pre-Darwinian Paleyan* tradition are remarkably convergent on neo-Darwinian adaptationist views. The Reverend William Buckland, starting from Owen's skepticism of the gas

* The "Paleyan tradition" here refers to the followers of the Reverend William Paley, who viewed adaptation for precise function in organisms as indicative of "design" and hence of the activities of the creator.

→

FIG. 1. Cephalopods with chambered shells discussed in this work include the modern *Sepia, Spirula,* and *Nautilus* as well as the extinct ammonoids. The chambers are pumped out osmotically by the siphuncle and are filled with gas at less than atmospheric pressure; consequently, the shells must support water pressure when these organisms are submerged. The tubular siphuncle of the externally shelled *Nautilus* and ammonoids and the internally shelled *Spirula* traverses the chambers within the shell. In the internal shell of *Sepia,* the cuttlefish, the siphuncle forms the posterior surface of the chambered shell (this is the familiar cuttlebone of the parakeet cage). All these chambered shells are lighter then sea water when empty and compensate for the muscle mass of the organism, which is denser than sea water. Of these forms, only *Sepia* can adjust its buoyancy rapidly so as to go up and down in the water column. In the other forms, the shells serve to maintain neutral buoyance. [*Spirula* modified from Denton and Gilpin-Brown (1973); *Sepia* and *Nautilus* modified from Denton (1974); ammonoids after Buckland (1836).]

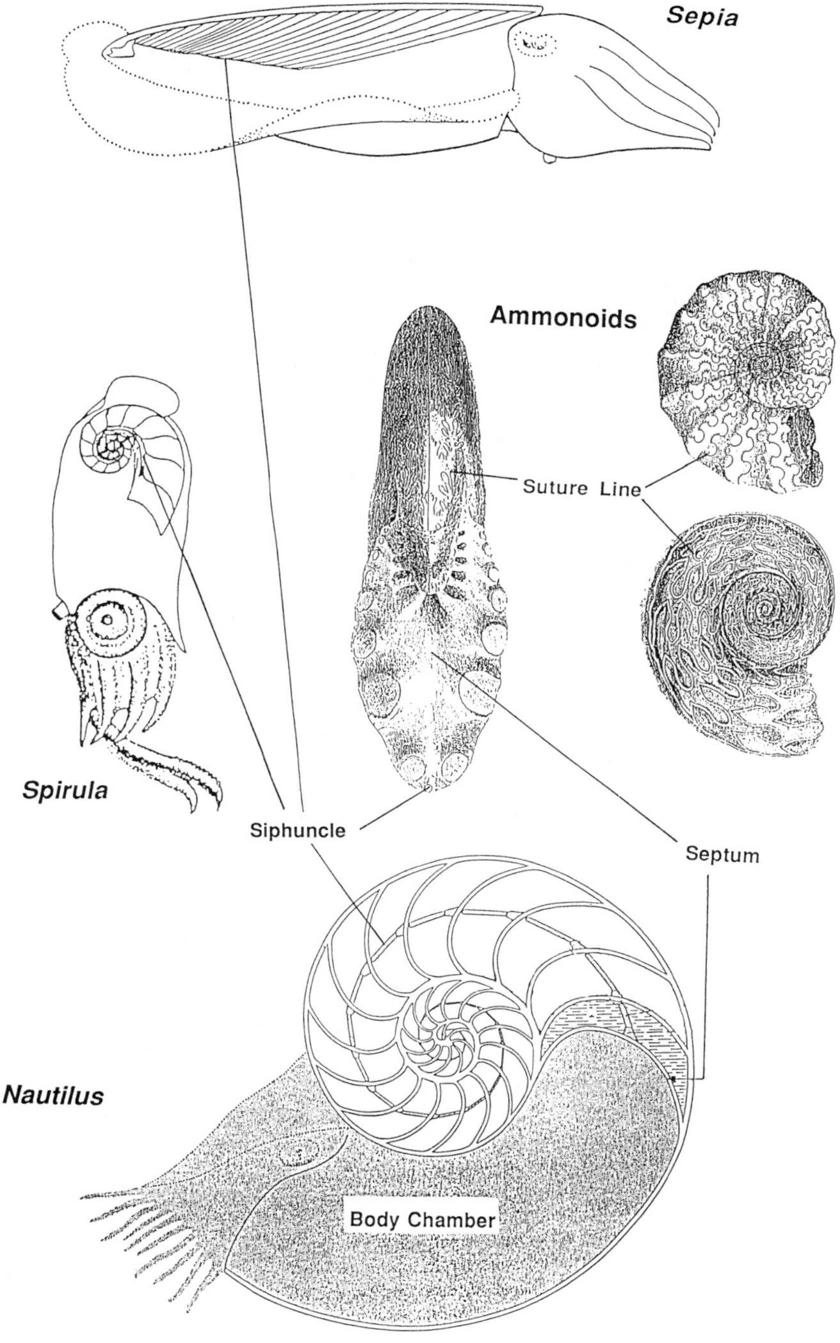
Sepia
Ammonoids
Suture Line
Spirula
Siphuncle
Septum
Nautilus
Body Chamber

pressure model, interpreted the complex structure of many fossil forms in a modern fashion in 1836. Thus, adaptive explanations and functional interpretation were important in the 1830s when the test of the gas pressure hypothesis was first proposed and during the 1960s when it was subject to a definitive test. Non-Darwinian, nonadaptive evolutionary views were conspicuous in the intervening period.

The first part of this work traces the study of the shell through this period from the initial discussions of shell function by R. Hooke through R. Owen to the late 19th century workers such as E. R. Lankester and A. Hyatt who placed cephalopods at the center of their evolutionary theories. A. Willey studied *Nautilus* in the field and was in the best position to examine the gas pressure issue; thus his work is of special interest. This section of the chapter develops the theme that evolutionary interpretations relying on ontogenetic information tended to displace adaptive or functional interpretation of organismal design.

Perhaps as a consequence of their more intimate knowledge of the diverse morphological features of fossil cephalopods, including the complexly evolving ammonoid septal-suture, a number of paleontologists (Buckland, 1836; Pfaff, 1911; Spath, 1919; Westermann, 1956) considered the possibility that the cephalopod shell functioned to support hydrostatic load. However, in each case the concept of structural support was not successfully transmitted to the rest of the paleontological community. The work of Buckland, Spath, and Westermann are each examined in some detail. Buckland (1836) invoked adaptationist explanations, and held what appear to be very modern views, both in terms of hypothesis testing (sometimes via experiment) and in terms of his use of adaptation in explanation of form and function. His views were eclipsed by nonadaptive, post-Darwinian views of cephalopod shell form such as those of Hyatt.

Spath's (1919, 1933) views are especially interesting; during his career he conducted a major offensive against nonadaptive evolutionary explanations, especially against Hyatt's views, which were still influential in paleontological circles in the first third of the 20th century. Spath argued for adaptive explanation, including the structural support function of cephalopod shells. These arguments were part of a larger debate that preceded the New Synthesis. In this dialogue, paleontologists, who favored recapitulative notions of evolution, faced off against modern developmental biologists such as W. Garstang and G. de Beer. Spath sided with the modern workers and provided the coup de grace to the recapitulationists. The argument for cephalopod shell function, however, was overlooked relative to this larger debate.

In one final episode just prior to the test of the gas pressure hypothesis and the rediscovery of cephalopod shell function, Westermann (1956) argued for a functional interpretation of the septum and suture of ammonites. Westermann was confronted by Arkell (1957*a*), who feared that this approach might unsettle traditional taxonomy, and again the function of the cephalopod shell became embroiled in, and obscured by, a debate rather than judged on its merits.

Once the great difference between the internal and external pressures of the chambered shells was robustly demonstrated in modern cephalopods (Denton and Gilpin-Brown, 1961*a–c*, 1966; Denton *et al.*, 1961, 1967), the issue of structural support of hydrostatic load was actively investigated in both fossil and modern shelled cephalopods. These recent studies emphasizing the function of the cephalopod shell in supporting hydrostatic load constitute the last portion of this history of the concept of structural support of hydrostatic load in the shells of cephalopods.

THE PROBLEM PRIOR TO 1960

Hooke

Hooke was the first to suggest that the chambers of *Nautilus* were emptied of water (Derham, 1726). He made this initial suggestion in a presentation to the Royal Society of London in 1695. However, it was not until the mid-20th century that Denton and Gilpin-Brown (1961*a–c*, 1966; Denton *et al.*, 1961, 1967) demonstrated that the shells of modern cephalopods bore hydrostatic load while submerged as a consequence of the empty condition of their chambers. This is especially surprising, given the nature of the observations needed to confirm that the shell supports hydrostatic load; observation of the pressure of the gas contained in the chambers is all that is required to demonstrate that hydrostatic load would be supported in the shell, rather than by internal gas pressure, when the cephalopod was submerged. The experiment is simple enough; the cephalopod shell is punctured while it is immersed in a liquid and the entrance of the liquid into, or the exit of gases from, the shell is observed in order to determine the internal pressure relative to atmospheric pressure. This methodology was advocated by Owen (1832, 1878):

> It would be advisable in the event of another fortunate capture of the *Nautilus*, to lay open the chambers under water, when the presence of gas in any of them would be ascertained; and it might be received and analyzed; the contents of the central tube, if gaseous, would at once be detected. . . . The same might be urged in the capture of a living *Spirula*. [Owen (1878), p. 974.]

Nautilus were subsequently captured on the *Challenger* expedition, and freshly caught specimens were studied in great detail by Willey at the end of the 19th century (Davis, 1987). *Sepia* and *Spirula* were also widely studied in the late 19th and early 20th centuries. So why did it take 130 years to determine empirically the gas pressure within the shell?

Two major factors inhibited researchers from carrying out the simple experiments required to demonstrate that gas pressure in the shell was low, and that hydrostatic loads must be borne by the shells of cephalopods. First, there was a well-entrenched model, involving compressed gas, that did not require loads to be borne by the cephalopod shell. This model was supported by the analogy to the swim bladders of fish. Second, during the period of most active research on modern cephalopods in the late 19th century, adaptive explanations were not actively sought; instead, morphological features were viewed in the context of the perceived relationship between ontogeny and phylogeny.

The Compressed Gas Model

A number of mechanisms can be envisioned that could empty the cephalopod shell and impart a positive buoyancy to the shell. Analogy to man-made floatation tanks, or to the swim bladders of fish, suggests that the way to empty a vessel under water is to increase the partial pressure of the gas inside the vessel until it equals or exceeds the exterior hydrostatic pressure. At this point the pressurized gas expels or displaces the water. The first to propose this mechanism for the emptying of a cephalopod shell appears to have been Robert Hooke, who addressed the Royal Society on December 16, 1695 (Derham, 1726) as follows:

> . . . for the Solution of such a Phaenomenon as this, of the floating and sinking of the *Nautilus,* which I had discoursed of the last Meeting but one. It seem'd, indeed, very strange, how that Creature could so, at his Will fill, and empty, the Cavities of his Shell, with Water; it is easy to conceive, how he could fill his Shell, with Water, and so sink himself to the Bottom; but then how (when there, at such a Distance, from the Air) he could evacuate the Water, and fill the Cavities with Air, that was difficult to comprehend, especially being under so great a Pressure of Water: But if Nature had furnish'd him with a Faculty of producing an artificial Air, then the Riddle would be quickly unfolded. I found, therefore, that by Art it was feasible to produce such an artificial Air, and that it was endued [sic] with a very great Power of Expansion, so that it would not only make itself room to expand notwithstanding the incumbent pressure of the Air on all Sides; but, if sealed up in strong Glasses, it would break out the Sides there of, which might have as much Power of Expansion as might counterpoise, nay, nay out-power both the Pressure of the Air, and also the Water too, though 100 times greater than

> that of the Air. It will be, I confess, a difficult Matter for me to prove, that the *Nautili* have such a owner, for that I could never yet get a Sight of that Fish that inhabits those Shells. [Derham (1726), pp. 309–310.]

At the next meeting of the Royal Society, Hooke made the analogy to the swim bladders of fish (Derham, 1726). Thus, Hooke can be viewed as setting the standard interpretation to which many other workers adhered.

Owen

The major 19th century skeptic with respect to the pressurized gas mechanism of shell emptying was Owen; he felt that the vascularization of the siphuncle was insufficient to perform the gas generating function: "The size of the artery seems barely adequate to support the vitality of the membrane, much less to effect a secretion, for which in a fish an ample gland appears to be indispensable" (Owen, 1832).

Buckland (1836) cites Owen, but other workers did not share Owen's concern. Moreau (1876) did experimental work on the function of the swim bladder of fish and this work, while in progress, inspired Bert (1867; see Denton and Gilpin-Brown 1966) to interpret the shell of *Sepia* as a pressurized gas vessel. Bert went to the trouble of determining that the gas contained in the cuttlebone was nitrogen, but did not examine the internal gas pressure. Thus, the entrenched idea of internal gas pressure expelling water from the shells of cephalopods received frequent support from the analogy to, and observations of, the swim bladders of fish.

The Dominance of Ontogenetic Studies

Willey

Arther Willey was in an especially good position to examine the internal pressure of gas in the chambers of *Nautilus;* during his field studies he regularly trapped specimens in deep water and kept them alive in his attempts to raise embryos. He was also familiar with Owen's classic work, *Memoir on the Pearly Nautilus* (Owen, 1832), which described the anatomy of *Nautilus* in detail and contained the admonition to examine the gas pressure of the shell. Oddly, Willey (1902) devotes only a few sentences in his lengthy monograph, *Contributions to the Natural History of the Pearly Nautilus,* to something approaching the course of study advocated by Owen (1832, 1878). Willey observes that "If the shell, with the live *Nautilus* in it, be perforated over the chambers under water, the air bubbles gently out as the water enters" (Willey, 1902, p. 747). This observation is not construed as a test of a hypothesis and its interpretation does not address Owen's concern:

> The chambers are not individually air-tight since they are perforated by the siphuncle, but collectively they are rendered an air tight and water tight hydrostatic apparatus, owing to the fact that the animal itself completely closes up the entrance to the chambers in virtue of its adherence to the shell by the muscles and anulus (girdle of Owen). Any loss by diffusion might be made good by the siphuncle, but apart from this I see no reason to imagine that the air which fills the chambers undergoes any appreciable fluctuation of pressure. It is, I am convinced, an error to suppose that variations of pressure of the air in the chambers enable *Nautilus* to rise or sink as the case may be. The air simply renders the shell buoyant once for all. [Willey (1902), p. 747.]

This interpretation raises another stumbling block; muscular force and soft tissues are required to keep the passage to the siphuncle blocked against the force of external water (or internal gas) pressure. Willey trapped *Nautilus* at depths greater than 500 feet (Ward, 1988), where water pressure exceeds 230 lb/in.[2] This is in excess of the forces sustainable by soft tissue. Willey's brief statements on gas pressure led Denton and Gilpin-Brown (1966) to conclude that Willey had a "poor understanding of the pressures exerted by water at depth and the difficulties that the organism might have in emptying or maintaining an empty vessel against such pressure."

As for a mechanism of gas generation, Willey (1902) cites the article written by E. R. Lankester, for the 1878 edition of the *Encyclopaedia Britannica,* which compares gas generation in cephalopods to that of parallel cases in the swim bladders of fish, as well as a number of examples of gas generation in invertebrates. Neither Willey's empirical investigation nor his interpretation of his results appear to have been extensive enough, or detailed enough, to resolve with authority any of the issues surrounding the gas pressure argument.

Ward (1988) contends that Willey's ineffectual treatment of the gas pressure question was the source of "his most grievous errors." Ward also suggests that the editorial influence of Lankester may have been prevented Willey from adequately interpreting his observation of gas pressure, and the related issue of the presence of liquid in the chambers. Ward (1988) accuses Lankester of suppressing this information due to the priority he placed on his authorship of the preseptal gas hypothesis.

Ward's interpretation seems unlikely. Swinnerton and Trueman (1918) indicate that J. F. Blake had priority in terms of the preseptal gas hypothesis and there is no mention of Lankester in this regard. Additionally, Lankester was widely known for a large body of other work. In his F.R.S. obituary

(Goodrich, 1930) there is no mention of cephalopod work at all. Given that the preseptal gas hypothesis did not originate with Lankester, and that he was not widely known for it, there would appear to be no reason to assume that he editorially suppressed Willey's observations as a consequence of his support for it.

I would suggest that both Denton and Gilpin-Brown's (1966) and Ward's (1988) harsh assessments of Willey are based on the presumption that Willey should have been interested in and should have adequately addressed the issue of gas pressure or its absence, because it has *subsequently* proved to be of great interest. Willey had a different research objective involving the ontogeny of *Nautilus.* Furthermore, this objective was a component of a research program in molluscan embryology pursued by Lankester, Willey's superior; this research program was typical of late 19th century biological thought. In the late 19th century, ontogeny was thought to provide direct evidence of, and sometimes even a causal explanation for, the pattern of evolution.

The most striking thing about Willey's discussion of gas pressure is its brevity. This issue, which was of great interest to Denton and Gilpin-Brown (1966) and Ward (1988), was obviously not at the top of Willey's agenda. What was at the top of his agenda is apparent: "he undertook as a condition of his election to the studentship to proceed to New Britain in order to attempt the investigation of the embryology of the Pearly Nautilus . . . Willey was assisted in his equipment by a grant from the Government Grant Committee of the Royal Society of London" (Willey, 1895, p. 405).*

This costly research was undertaken because the specific objective of securing an ontogenetic series of *Nautilus* was of great interest to Willey, Lankester, and late 19th century science in general. Willey's activities reflect the proviso attached to the funding he received. In an attempt to achieve the desired goal, Willey scurried from island to island for 3 years, suffering from malaria and often risking his life in small boats (Ward, 1988). He succeeded in catching *Nautilus,* and was able to keep some alive in regions where the surface water was cool enough. He was even occasionally tantalized by the production of eggs by his captive *Nautilus,* but these eggs were never fertile. He did not obtain the ontogenetic series that his field work was intended to provide.

* This was written by Lankester and appended to a published version of a letter he had received from Willey (1895), hence the reference to Willey in the third person.

Lankester

The investigation of the ontogeny of *Nautilus* fits nicely into Lankester's research program on molluscan embryology.* Early in his career Lankester (1876) made a detailed comparative study of the development of a number of mollusks which was presented to the Royal Society. Lamellibranch, Nudibranch, and Pulmonate gastropods, a "Heteropod," as well as the cuttlefish and squid, *Sepia* and *Loligo,* were included in the analysis. In this work it is evident that Lankester believed that the ontogeny of cephalopods had a special explanatory role in unraveling the evolutionary relationships within the mollusca:

> If, as seems justifiable, the Cephalopoda are to be regarded as more nearly representing the molluscan type than do the other classes, or, in other words, more closely resemble the ancestral forms than they do, we might look in the course of the development of the less typical Mollusca for some indication of a representation of the internal pen of the higher Cephalopoda. We might expect to find some indication of the connection between this and the calcareous shell of other forms; in fact the original shell of all Mollusca should be an internal one, or bear indications of a possible development into that condition. [Lankester (1876), p. 35.]

With this notion of molluscan development in mind, it is apparent that the development of externally shelled *Nautilus,* whether it showed embryological evidence of derivation from an internally shelled form or not, would be critical to Lankester's evolutionary interpretation of the mollusks. The role of cephalopods in approximating the molluscan archetype from which other taxa were derived relates to the emphasis Lankester placed on degeneration from a type as an important evolutionary mode. In "Degeneration: A Chapter in Darwinism," Lankester (1890) characterizes the contemporary vision of evolution as progressive: "It has been held that there have been some six or seven great lines of descent . . . and that along each of these lines there has been always and continuously a progress—a change in the direction of elaboration" (Lankester, 1890, p. 22). In contrast to this view, Lankester believed that three modes of evolution were evident: maintenance of the *status quo,* which he referred to as "Balance," indicated by the long fossil record of the modern "Lingula and Terebratula, the King-crabs, and the Pearly Nautilus"; increase in complexity of structure, termed "Elaboration";

* This is only one of a wide range of activities in which Lankester engaged; his career started in the paleontology of Devonian fish, he expressed himself regarding the role of science in society, and worked extensively on the development of vertebrates as well as mollusks. The most consistent theme of his diverse career was the investigation of development. Similarly, Willey and Bashford Dean (a contemporary of Willey's who also briefly examined *Nautilus*) both worked in the development of vertebrates as well as invertebrates. Thus, all these workers were concerned with development in general, more than with the study of any particular group.

and decrease in complexity of structure, termed "Degeneration." Lankester argued that "the hypothesis of Degeneration will . . . be found to render most valuable service in pointing out the true relationships of animals which are a puzzle and a mystery when we use only and exclusively the hypothesis of Balance or the hypothesis of Elaboration" (Lankester, 1890, p. 25).

Lankester considered degeneration as well as elaboration to have proceeded by terminal addition, and to be discernible by the recapitulation of phylogeny by ontogeny. He supports this view by reference to the apparently degenerate terminal stages of barnacles, tunicates, and parasites. The higher taxonomic affinities of these taxa were known only through investigation of their early ontogenetic stages; the adult forms were too "degenerate" to be properly classified. This degenerative conception of evolution was thought to be generally applicable within each of the types; "Amongst the Mollusca—the group of headless bivalves, the oysters, mussels and clams . . . are . . . clearly enough explained as degenerated from a higher type of head-bearing active creatures like the Cuttlefish" (Lankester, 1890, p. 45). Thus, degeneration was given a major role in Lankester's evolutionary scheme.

Hyatt

Parallels can be drawn between Lankester's ideas and those of other influential late 19th century evolutionary thinkers such as Haeckel and Hyatt. Hyatt's ideas strongly influenced cephalopod workers in the late 19th and early 20th centuries (Spath, 1919, 1933). Hyatt's conception of evolution is similar to Lankester's in that they both invoke (1) recurrent evolution from an ancestral stock, termed a "Radical" by Hyatt, (2) an ascendant and a descendant or degenerative phase to evolution (Hyatt's degeneration parallels ontogeny and is explained as "phylogerontic" senescence of the lineage), and (3) recapitulation as the means of unraveling evolutionary relationship. Unlike Lankester, who was a Darwinian, Hyatt's concept of evolution was neo-Lamarckian. As a consequence of the Lamarckian evolutionary mechanism, an adaptive metaphor is employed: "Food and opportunity would have acted, in such localities, as stimulants to new efforts for the attainment of more perfect adaptation and for changes of structure useful to that end. . . . That this process should end in the production of structures suited to the environment is inevitable" (Hyatt, 1894, p. 367).

However, in Hyatt's vision, this adaptive effect was confined primarily to the formation of a new type (higher taxon) from a preexisting stock termed a radical and to the terminal addition of adult characters during acceleration; evolution was otherwise subject to the "law of morphogenesis," in which "the phenomena of individual life are parallel with those of its own phylum . . . not only can one indicate the past history of groups from the study of the young . . . it is also possible to prophecy what is to happen in the future history of the type from study of the corresponding paraplastic phenomena in the development of the individual" (Hyatt, 1894, p. 391). Thus the history of a group is thought to have ontogenetic properties that include preordination, just as the ontogeny of the individual has predictable or actuarial qualities.

According to Hyatt, many of the morphological changes associated with the old age or senescence of the lineage had negative adaptive value that could ultimately lead to the demise of the group: "The apertures and forms of the retrogressive shells all show that they were exceptional, that they had neither well-developed arms for crawling nor powerful hyponomes for swimming. This retrogression was in itself unfavorable to prolonged existence" (Hyatt, 1894, p. 380). Thus, the power of morphogenesis could lead to changes in the mode of life of the organism in the face of adaptive forces. It is apparent that the evolutionary ideas of Hyatt contain much to distract the worker from the interpretation of form as a product of adaptation and consequently from pursuing studies on the functional nature of morphology.

Similarly, Lankester, although a Darwinian, did not concern himself with adaptation. In his comparative study of molluscan development he does not refer to adaptation, and expresses surprise that in a direct developing freshwater clam "there is nothing that corresponds to the velum of the '*veliger* form' of Gastropod development . . . And it is even still more curious to note that not even at the earliest stage . . . is there any thing which indicates or corresponds in the remotest degree morphologically to a head" (Lankester, 1876, p. 8; italics in original). This is not surprising from an adaptationist standpoint; the veliger stage in marine larvae functions to gather food while the larvae are in the plankton; the velum, or feeding organ of the veliger, would be of much more limited adaptive value in the larval stage of a mollusk that develops within the egg. The surprise expressed by Lankester is a consequence of his understanding of the type or ancestral mollusk. He thought the type mollusk expressed a veliger stage in development; consequently, a veliger form should be retained and expressed in the recapitulation of phylogeny, evident in the ontogeny of the descendant mollusks. A similar predicted or hypothetical veliger stage played an important role in Hyatt's (1889) understanding of the phylogenetic relationship between nautiloids and ammonoids.

Despite the differences between the Lamarckian and Darwinian mechanisms of evolution to which they subscribed, Hyatt and Lankester both felt that virtually all of ontogeny was sequestered from adaptive influence and resultant evolutionary change. Thus, they both minimized the role of adaptation in evolution.

Certainly Willey was influenced by Lankester's ideas prior to his departure for the South Pacific in 1894. In fact, Willey had already published with Lankester on the development of *Amphioxus* in 1890 (Goodrich, 1930). In one of his first publications on *Nautilus,* "Zoological observations in the South Pacific" (Willey, 1897), written while he was still in the South Pacific, Willey employed Hyatt's terminology to define an early ontogenetic feature, the "nepionic" constriction (Landman and Waage, 1982), a constriction of the *Nautilus* shell associated with hatching (Table I). In this work, Willey refers to several of Hyatt's publications, and refers to Hyatt's publication, "Embryology of the Cephalopods," as "well known." Thus we can confirm that the developmental ideas of Lankester and Hyatt were much on Willey's mind in his lonely hours of research in the Pacific, very likely pushing him to observe and interpret anatomical observations in a developmental light, to the exclusion of adaptive or functional morphological interpretations.

Willey even judges the veracity of the natives on the basis of their ability to accurately report development, rather than other aspects of natural history, "My guide had seen the lizard making this nest some three days previously, and on opening the eggs I found the young embryos to be at the corresponding stage of development, with the heart beating. This was another welcome proof of the trustworthiness of the natives" (Willey, 1895, p. 407). More recently evolutionists interested in speciation, such as Mayr (1963), have dropped by the native's hut to determine if he or she names the

TABLE I. Hyatt's Terminology[a]

	Ontogeny		Phylogeny	
	Embryonic		Phylembryonic	
Anaplasis	Nepionic	Phylanaplasis	Phylonepionic	Epacme
	Neanic		Phyloneanic	
Metaplasis	Ephebic	Phylometaplasis	Phylephebic.	Acme
Paraplasis	Gerontic	Phyloparaplasis	Phylogerontic.	Paracme

[a] In Hyatt's terminology (after Hyatt, 1894) there were five ontogenetic stages: an embryonic or larval stage, the "young or immature" nepionic stage, the "adolescent" neanic stage, the "mature or adult" ephebic stage, and the "senile or old" gerontic stage. All stages could be further subdivided by the use of the prefixes ana, meta, and para if further subdivision into distinct ontogenetic stages could be made. In all cases comparable terms were employed in phylogeny by the addition of the prefix "phyl" to the ontogenetic term. Also in the table are Haeckel's terms, where the "acme" refers to the maximum taxonomic richness or peak diversity of the group; "epacme" is the rise to this peak, and "paracme" is the fall from it. In addition to these sets of terms for the ontogenetic cycle in phylogeny, there is the general term "anagenesis" for the rise toward the apex in the morphogenetic cycle and the term "catagenesis" for the subsequent decline.

local bird species with the accuracy of western science. The native must wonder about the changing talisman of these naturalist-supplicants.

Adaptation versus Recapitulation

The exclusion of an adaptationist perspective by the focus on recapitulation is not only evident in the work of Lankester (1876, 1890) and Hyatt (1889), both of whom felt that the developmental series of *Nautilus* held keys to the larger issues of cephalopod, and even molluscan, phylogeny. This perspective is also evident in the thinking of Willey himself and was characteristic of most cephalopod workers and many other naturalists in the late 19th and early 20th centuries. Anatomists and naturalists avidly sought, not just single specimens, but a sequence of specimens showing the various stages of development. Adaptive issues were peripheral to the working out of the grand sweep of evolution likely to be depicted in ontogeny.

At the turn of the century cephalopod workers applied Hyatt's methodology, using ontogenetic schemes to unravel phylogeny. J. P. Smith performed a number of detailed ontogenetic examinations of ammonites, including "The Development of *Lytoceras* and *Phylloceras*" (Smith, 1898), "The Development of *Glyphioceras* and the Phylogeny of the Glyphioceratidae" (Smith, 1897), and "The Development and Phylogeny of *Placenticeras*" (Smith, 1900*a*). In his examination of *Lytoceras,* Smith refers to this Triassic form as a "Silurian nautiloid" when it was forming its first septum with a straight suture; when it formed its second septum containing a lobe it "became a Devonian ammonoid." So recapitulation was conceived of as the literal revisiting of the ancestral condition in the accelerated sequence of ancestors that could be discerned in the ontogenetic sequence.

Verrill (1896) was one of the few who took recapitulation theory to its logical end; he concluded, on the basis of molluscan ontogeny, that the type ancestor of Mollusca was a veliger larva. In another often cited work of the time, Bather (1897) defended the integrity of cephalopod ontogeny for recapitulatory purposes against the "Intussusception hypothesis" of Dr. E. Riefstahl. The intussusception hypothesis supposed that the shell was alive and grew without the agency of the mantle; chambers of the cephalopod shell formed not at the body chamber, but further back in the shell, by the growth of the shell wall between septa. This of course would introduce elements in the middle of the ontogenetic sequence preserved in the shell, and would have been fatal to the reading of the recapitulated phylogeny from ontogeny. There was great relief when Bather (1897) disposed of this competing hypothesis and the rule of law (biogenetic or morphogenetic) was reestablished in the land.

Hyatt's influence is apparent not just in work on cephalopods; it is evident in work on all major invertebrate groups, especially among paleontologists, from 1890 to after the turn of the century. A great many workers employed Hyatt's nomenclature (Table I and Fig. 2) in detailed studies of the ontogeny of fossil and modern forms and then used the ontogenetic sequence to help reconstruct the lineage of groups as recorded in the fossil record. Work following this formula includes studies by Jackson (1890) at Harvard on "the Pelecypoda, the Aviculidae and their allies," Beecher at Yale on brachiopods (Beecher, 1889, 1891*a*, 1892, 1897), cephalopods (Beecher, 1890), corals (Beecher, 1891*b*), and trilobites (Beecher, 1893, 1895), Cummings at Yale on brachiopods (Cummings, 1903) and at Indiana University on bryzoans (Cummings, 1904, 1905), Raymond at Harvard on brachiopods (Raymond, 1904) and trilobites (Raymond, 1914), and Grabau at Columbia on gastropods (Grabau, 1902, 1903, 1907) and crinoids (Gra-

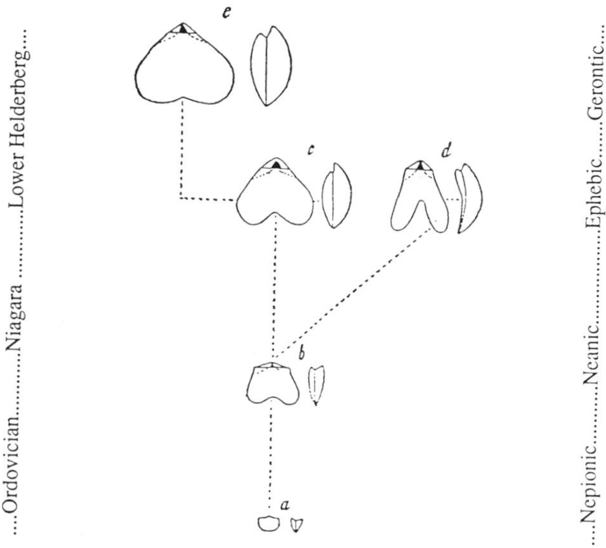

FIG. 2. Beecher's (1901) phylogenetic tree of the *Bilobites* brachiopod lineage (now referred to as *Dicoelosia*) illustrates the use of Hyatt's nomenclature. This tree depicts the succession of forms through the stratigraphic section with the ontogenetic terms of Hyatt applied to the lineage. Note that the acme of the lineage, in Haeckelian terms, would be where forms *c* and *d* both occur, resulting in the maximum diversity. This is comparable to the ephebic or adult stage of the lineage in Hyatt's nomenclature. *a,* Nepionic stage. Ordovician type, such as *Platystrophia bifurcata. b,* Nealogic period at which divergence begins. *c, Bilobites bilobus.* Niagara Horizon. *d, Bilobites verneuilianus. e, Bilobites varicus.* Lower Helderberg Horizon.

bau, 1906). Hyatt's basic scheme, in which there was first progressive and then senescent or gerontic evolution, accompanied by tachigenesis or acceleration of the terminally added features to earlier stages in development, is apparent in many of these works. Grabau went so far as to apply Hyatt's cycle of coiling to gastropods. Hyatt indicated that cephalopods went through a sequence of first straight forms and then coiled forms that subsequently straightened out again in a senescent reduplication of the youthful condition in the evolutionary sequence. Similarly, Grabau indicates that the uncoiling of vermetid gastropods is a senescent evolution within the gastropods.

That many of these workers investigated more than one phylogenetic group suggests that they viewed themselves as examining evolution, or phylogeny, via the analysis of ontogeny rather than as students of a particular taxonomic group. This is further borne out by the publication of theoretical or methodological work on recapitulation theory and its application: Jackson (1899), "Localized Stages in the Development of Plants and Animals," Smith (1900*b*), "The Biogenetic Law from the Standpoint of Paleontology," and Cummings (1909), "Paleontology and the Recapitulation Theory" (1909).

From this body of evidence it should be apparent that the examination of the ontogenetic series as an aid to the interpretation of phylogeny was a dominant mode of studying invertebrates at the turn of the century. This appears to have been especially true in those groups which had a record of ontogeny preserved in the fossil record. In addition, investigators employed both neontological and paleontological data in their analyses; the fields of paleontology and biology were not as distinct as they are today. Despite the questioning of the biogenetic law by Hurst (1893) and Sedgewick (1909), the preeminence of the law was not seriously threatened even among neontologists until Garstang (1922) rose to the occasion to advocate his ideas regarding larval adaptation, which he referred to as "paedomorphosis."

As a consequence of the broad influence of the biogenetic and morphogenetic laws, adaptation was only infrequently resorted to in evolutionary discussions and interpretations of morphology in the late 19th and early 20th centuries. This period, when adaptive interpretations were not in vogue, corresponded with the most active investigation of shelled cephalopod anatomy; subsequently, a period of inactivity set in (Davis, 1987). After that of Willey (1902), the next widely cited paper concerning *Nautilus* was published in 1962 by Bidder (1962). Denton *et al.* (1967) report that recovery and anatomical study of *Spirula* were major objectives of a number of oceanographic voyages around the turn of the century. There were similar peaks in the intensity of study of the other modern shelled cephalopods in the late

19th century and a subsequent period of lack of scientific interest and few publications. Subsequent to the turn of the century, comparative natural historical studies of all kinds diminished with the rise of laboratory genetics and the model organism approach to development and physiology. This emphasis reduced interest in cephalopod morphology to a single nerve, the giant axon of the squid.

WILLIAM BUCKLAND AND ADAPTATION

Buckland (1836) showed great insight into the structural issue, but appears to have been largely ignored by those few 20th century paleontologists who expressed similar ideas. Buckland came to a modern, and in my view accurate, understanding of the adaptive function of the cephalopod shell, operating from a religious agenda. Many subsequent workers were unable to attain this understanding due to the nonadaptive nature of their evolutionary ideas.

Buckland argued that cephalopod shells were filled with air at atmospheric pressure, that shells of cephalopods were subject to great pressure, and that shell features such as the external shape of the shell, ornament, and the septal-suture of ammonoids were adaptive features designed in detail to resist pressure. He was also the first to invoke arch function in discussion of the shell and the first to apply the analogy of the architecture of gothic cathedrals to the ammonite shell. In these regards Buckland was essentially modern in his understanding of the problem and its solution (see also Jacobs, 1990).

Buckland cites Owen's (1832) observation that there is no gas-generating gland in *Nautilus* analogous to "that which is supposed to secrete air in the air-bladder of fishes" and Buckland used this argument for justification of "the theory which supposes *the chambers of the shell to be permanently filled with air alone, and the siphuncle to be the organ which regulates the rising or sinking of the animal*" (Buckland, 1836, p. 332; italics in original).* Unlike Willey, Buckland knew that "the pressure of the sea, at no great depth, will force a cork into a bottle filled with air, or crush a hollow cylinder

* Buckland envisioned a physical pump which would force fluid into the siphuncle causing it to expand into the chambers, thus adjusting bouyancy. Current understanding suggests that the siphuncle serves as an osmotic pump to empty the chambers. In this regard alone Buckland's understanding of shell function departs from the modern assessment.

or sphere of thin copper; and as the air chambers of Ammonites were subject to similar pressure, whilst at the bottom of the sea, they required some peculiar provision to preserve them from destruction" (Buckland, 1836, p. 345).

Buckland used this understanding in the formation of a hypothesis of necessity or requirement, the first stage in an adaptive argument (subsequently, adaptation is demonstrated by showing that these needs or requirements are met). Buckland describes the constructional problem faced by shelled cephalopods in these terms, "it was necessary that they should be thin, or they would have been too heavy to rise to the surface: it was not less necessary that they should be strong, to resist pressure at the bottom of the sea; and accordingly we find them fitted for this double function, by the disposition of their materials, in a manner calculated to combine light-ness and buoyancy with strength" (Buckland, 1836, p. 338).

Buckland then explains how the shape of the shell, the sutures, and ribbing function to meet this combination of structural needs: "The entire shell, is one continuous arch, coiled spirally around itself in such a manner, that the base of the outer whorls rests upon the crown of the inner whorls, and thus . . . is calculated to resist pressure . . ." (Buckland, 1836, p. 339).

In reference to the shells of ammonites, Buckland goes on to describe how the complexities of the septal-suture (he refers to the septum as the transverse plate) are functionally related to the shape of the external shell:

> Thus on the back or keel . . . where the shell is narrow, and the strength of its arch greatest, the intervals between the septa are also greatest, and their sinuosites comparatively distant; but as soon as the flattened sides of the same shell . . . assume a form that offers less resistance to external pressure, the foliations at the edges of the transverse plates approximate more closely; as on the flatter forms of a Gothic roof, the ribs are more numerous, and the tracery more complex, than in the stronger and more simple forms of the pointed arch.
>
> The same principle of multiplying and extending the ramifications of the edges of the transverse plates, is applied to other species of Ammonites, in which the sides are flat, and require a similar increase of support; whilst in those species to which the more circular form of the sides gives greater strength . . . the sinuosities of the septa are proportionately few. [Buckland (1836), pp. 348–349.]

Buckland considers ribbing an additional adaptation for strengthening the shell:

> From the disposition of these ribs over the surface of the external shell, there arise mechanical advantages for increasing its strength, founded on a principle that is practically applied in works of human art. The principle I allude to, is that by which the strength and stiffness of a thin metallic plate are much increased by corrugating, or applying *flutings* to its surface. A common pencil-case, if made of corrugated or fluted metal, is stronger than if the same quantity of metal were disposed in a simple tube. Culinary moulds of tin and copper are in the same way

> strengthened, by folds or flutings around their margin, or on their convex surfaces. The recent application of thin plates of *corrugated* iron to the purpose of making self-supporting roofs, in which the corrugations of the iron supply the place, and combine the power of beams and rafters, is founded on the same principle that strengthens the vaulted shells of Ammonites. [Buckland (1836), pp. 339–340; italics in original.]

Buckland's understanding of the structural significance of shell architecture was equal to or superior to that of subsequent workers, even to those who considered the shell to function in resisting pressure. Spath (1919) appears to be the only 20th century ammonite worker who refers to Buckland. Although Spath discusses the correlation of shell shape with sutural complexity and Buckland's argument regarding pressure, he does not relate them in the unequivocal and forceful manner of Buckland's adaptive argument. In addition, Spath (1919, 1933) reviews other possible functions of the complex septal surface, including attachment to the shell and additional area for the secretion of gas into the shell! Spath is therefore not arguing for a specific point regarding either gas pressure or structural support of the shell; he is mustering his forces in support of adaptive explanations in general. Thus, Spath is at best agnostic on the role of structural support of hydrostatic load. His real objective in the arguments expressed in his earlier more theoretical work (Spath, 1919) and in his later paper (Spath, 1933) was to oppose the use of ontogeny as a causal explanation for the series of forms present in the phylogeny, a concept derived from Hyatt's morphogenetic law (as I will discuss shortly).

Other than Spath, Pfaff (1911) and Westermann (1956) were the only workers to argue in favor of shell support of hydrostatic load prior to 1960. Spath (1919) and Westermann (1971) both cite Pfaff (1911) in this regard; Pfaff was concerned with the action of hydrostatic pressure acting through the body chamber and on the last septum. Pfaff also calculated the strength of the concave nautiloid septum (Westermann, 1971) and concluded that the convex ammonoid septum would be substantially stronger.

Westermann (1956) expressed similar views to those of Spath (1919) and Buckland. He was immediately assaulted for his views by Arkell (1957*a*), who advocated the air-pressure mechanism. Westermann did not return to this topic until Westermann (1971), well after the work of Denton and Gilpin-Brown (1961*a–c*, 1966; Denton *et al.*, 1967) had demonstrated that shelled cephalopods bore hydrostatic load on their shell structure. In this and subsequent work Westermann (1971, 1975) demonstrated the relationship between whorl shape and the elaboration of sutural elements. Westermann's later work was more quantitative than Buckland's observations. However, Westermann did not show the insight into the arch function of the exterior shell of the ammonite that is evident in Buckland's work. This argument of Buckland's was not reiterated by recent workers until Jacobs (1990).

We can conclude that Buckland had an understanding of the function of ammonite and other cephalopod shell features that was equivalent to or better than that of many other workers who considered these issues. Even *after* the work of Denton and Gilpin-Brown demonstrated unequivocally that the shell supports hydrostatic load, few workers were able to surpass Buckland's understanding of the structural function of the ammonite shell.

Parallels between Buckland's Views and neo-Darwinian Adaptation

Buckland's awareness of the forces exerted at depth and of Owen's (1832) anatomical argument regarding the lack of vascularization sufficient for gas generation was a critical prerequisite for his adaptive understanding of the cephalopod shell. Buckland's keen insight and his capacity for imaginative analogy to human contrivances (such as fluted pencil cases and culinary molds) also played a role in his capacity to unravel the adaptive value of the cephalopod shell. As a result of these factors Buckland was able to properly interpret the lack of gas pressure within the cephalopod shell and was subsequently able to interpret the functional properties of the shell. However, these points speak to intrinsic features of the argument. They do not answer the equally interesting question of why Buckland was investigating adaptive questions, questions comparable to those of current interest in paleontology. Nor do they explain why paleontologists in the later 19th and early 20th centuries were not pursuing such a regimen.

Buckland's investigations ranged over a vast array of subject matter, but nearly always turned on issues that have also been of recent interest in paleontology. The following are examples of shared interests between Buckland and modern paleontologists that are evident in his chapter on mollusks alone:

1. The ecological concepts of trophic levels and guilds are apparent in Buckland's work; he suggests that predatory gastropods replaced ammonites in the economy of nature. Referring to ammonites, Buckland wrote, "Their sudden and nearly total disappearance at the commencement of the Tertiary era, would have caused a blank in the 'police of nature,' allowing the herbivorous tribes to increase to an excess, that would ultimately have been destructive of marine vegetation, as well as of themselves, had they not been replaced by a different order of carnivorous creatures, destined to perform in another manner, the office which the inhabitants of Ammonites . . . then ceased to discharge" (Buckland, 1836, p. 300). The writings of late 19th and

early 20th century paleontologists do not contain paleoecological interpretations of organismal interaction comparable to those expressed by Buckland.

2. Buckland also observed a rapidly increasing incidence of shell-boring from the middle Jurassic to the Tertiary:

> Most collectors have seen upon the sea shore numbers of dead shells on which small circular holes have been bored by the predaceous tribes, for the purpose of feeding upon the bodies of the animals contained within them; similar holes occur in many fossil shells of the Tertiary strata wherein the shells of carnivorous Trachelipods [gastropods] also abound; but perforations of this kind are extremely rare in the fossil shells of any older formation. In the Green-sand and Oolite, they have been noticed only in those few cases where they are accompanied by the shells of equally rare carnivorous Mollusks; and in the Lias, and strata below it, there are neither perforations, nor any shells having the notched mouth peculiar to perforating species. [Buckland (1836), p. 299.]

Observations comparable to these are the basis of a recent modern thesis of biotic escalation championed by Vermeij (1987).

3. In addition, Buckland observes that the replacement of ammonites by gastropods represents the replacement of a higher (more complex) form by a lower one, leading to the conclusion that progress is not a continuous process—a point that Gould (1989) has championed of late.

These examples should suffice to indicate that the problems that were of interest to Buckland are still of interest to invertebrate paleontologists today and frequently meet with similar resolution.

Hypothesis Testing: Buckland Contrasted with Late 19th Century Workers

Those who employed an adaptive paradigm, such as Buckland, and more recently the neo-Darwinian school, differ in many ways from the late 19th and early 20th century paleontologists and natural historians who operated with the morphogenetic or the biogenetic law in mind. In the purest conceptualization of Haeckel's biogenetic law the ultimate *mechanical* cause of ontogeny is the recapitulation of phylogeny. This mechanical approach admitted of no causal intermediary either within or without the organism, and this argument was expressed as a law, rather than as a testable hypothesis. This contrasts with an argument involving adaptation. In adaptive arguments, ultimate causation, be it the hand of the creator or the action of natural selection, does not make precise predictions. The investigation proceeds at an intermediate level where individual cases or classes of adaptation are investigated. At the intermediate level, hypotheses of adaptation are formed by the identification of a requirement for existence imposed on the organism (such as the need for the cephalopod shell to resist hydrostatic load). Such a hypothesis is confirmed by identification of particular features (of the shell) that function to fulfill the identified need.

There is no intermediate stage of hypothesis generation and testing in the application of Hyatt's or Haeckel's laws. In fact, these laws eliminated not only the testing of adaptive hypotheses in evolution, but also the testing of causal interactions in development itself. This is the reason that de Beer (1958) takes such exception to Haeckel's biogenetic law: "Clearly, if phylogeny was the mechanical cause of ontogeny as Haeckel proclaimed, there was little inducement to search for other causes . . ." (de Beer, 1958, p. 172). De Beer, being a developmentalist, is interested in examining the causal relationships within development. He realized that he could not pursue this kind of research if the scientific community believed that recapitulationist laws had already explained all of development in a causal sense.

The developmental agenda of de Beer, His, Spemann and other 20th century developmentalists became an experimental program (Oppenheimer, 1967). This is distinct from the more passive observation of the developmental system practiced by Willey and Lankester among others. This recapitulationist developmental program consisted in the precise observation of developmental series as an indication of phylogeny. With this lack of experimental experience or motivation, it is perhaps not surprising that Willey was so unsuccessful at setting up the experiment to observe gas pressure in the appropriate manner. In contrast, earlier workers, including both Buckland and Owen, conducted or suggested experiments that directly tested hypotheses. Aspects of experimentation and hypothesis testing are evident in Buckland's investigation of the buoyancy adjustment issue:

> The following experiments shew that the weight of fluid requisite to be added to the shell of a Nautilus, in order to make it sink, is about half an ounce.
>
> I took two perfect shells of a Nautilus Pompilius, each weighing about six ounces and a half in air, and measuring about seven inches across their largest diameter; and having stopped the Siphuncle with wax, I found that each shell, when placed in fresh water, required the weight of a few grains more than an ounce to make it sink. As the shell, when attached to the living animal was probably a quarter of an ounce heavier than these dry dead shells, and the specific gravity of the body of the animal may have exceeded that of water to the amount of another quarter of an ounce, there remains about half an ounce, for the weight of fluid which being introduced into the siphuncle, would cause the shell to sink; and this quantity seems well proportioned to the capacity both of the pericardium, and of the distended siphuncle. [Buckland (1836), p. 330.]

Buckland's methodology contrasts with that of Willey (1902) (quoted on pp. 293–294). Willey does not describe his puncture of the shell of *Nautilus* as an experiment nor does he indicate what hypothesis he is testing through his investigation. His puncture of the *Nautilus* shell is simply one more observation. It is not imbued with the notion of experimentation in which Owen (1832) originally suggested the activity, or with the notion of experimentation inherent in Buckland's examination of the buoyancy of *Nautilus* shells. The lack of an *a priori* question may have resulted in the lack of resolution produced by Willey's investigation. When later observers attempted to employ Willey's observations to resolve the gas pressure issue, the observation proved insufficiently precise for this task.

Buckland (1836) uses observations from the fossil record to *test* the idea of regular progress. The observation that the lower group, the gastropods, ecologically replaces the higher ammonite group "affords an example of *Retrogression* which seems fatal to that doctrine of *regular Progression*" (Buckland, 1836, p. 312; italics in original). Thus, Buckland has defined progress and tested that concept of progress against the record, finding that it fails the test.

This contrasts with Lankester's (1876) observation of the absence of a veliger stage, or a head, in the larval development of a direct developing bivalve (see quotation on p. 298). These observations are in conflict with predictions derived from the biogenetic and morphogenetic laws, which dictate that ontogeny recapitulates phylogeny. According to Lankester's views, the clams are descended from creatures with a head, and consequently they should recapitulate this stage in their development. That they do not, indicates that either Lankester's phylogenetic hypothesis is incorrect, or the law of recapitulation has exceptions. Unlike Buckland, Lankester has not set his observations up as a test or a proof of a hypothesis; consequently, Lankester is not obliged to accept either of these possible deductions. He merely states that he finds the results "curious."

The laws of recapitulation, biogenesis, and morphogenesis do not appear to have been subjected to test in the same sense or degree as hypotheses of adaptation. In fact, as numerous observations of exceptions to the the law of recapitulation mounted, addenda to the law, such as acceleration or tachygenesis, were invoked which explained the absence of developmental stages. Life history became so compressed that the whole history of the race could not be recorded in the palingenetic unfolding of development; gaps resulted in the observable phylogenetic sequence recorded by development. Development including these absences was referred to as lipopalingenesis. Thus, departures from predictions of the "law" do not lead to its rejection. Granted, in an adaptationist program, the premise of adaptation is also not directly tested, but testing of individual adaptations is conducted at least in terms of consistency of observations with prediction.*

* Oddly, the biogenetic law, since it makes direct predictions without an intermediate stage of hypothesis generation, may be directly subjected to test through observation. Due to the intermediate stage of hypothesis generation, adaptation may be intrinsically less testable (Gould and Lewontin, 1979).

Paleyan Adaptation

In addition to the similarity between Buckland's methodology and the modern scientific approach, the use of adaptation in explanation has a modern gestalt; it appears similar to that of much modern Darwinian or neo-Darwinian argument. However, Buckland's adaptive arguments precede the publication of Darwin's evolutionary arguments. This argument "of design" constitutes a program of investigation that is derived directly from the Reverend William Paley (Bowler, 1976), and is the theme of the *Bridgewater Treatises,* of which Buckland's work is but one of nine volumes. Furthermore, the adaptations that Buckland discusses occur through geologic time. Each successive fauna of the earth is adapted to its particular surroundings; the trophic structure is balanced in each case with predators to eat the herbivores. This impression of change with time and frequent references to "affinities" between the taxa present in the successive stages give Buckland's discussion an evolutionary flavor. However, Buckland's ideas were not evolutionary in the modern sense; he ascribed the cause of the changes he observed in the stratigraphic succession as the consequence of the repeated interference of creative power.

Buckland's notion of frequent interference by the creator and his dedication to the investigation of adaptation was typical of the natural theology of his day. This agenda was not the intellectual ball and chain currently associated with dogmatic creationists; rather, it was a very general program, which though theistic, generated a multitude of hypotheses that could be tested by empirical investigation.

SPATH VERSUS MORPHOGENESIS—VITALISM

By the first two decades of the 20th century Hyatt's law of morphogenesis had imbibed of the elixir of vitalism. This invigorating brew rendered the morphogenetic law more flexible and virulent. The facile use of this combination of ideas to explain detailed morphological features produced a reaction from Spath (1919), who took the opposite path and championed adaptive explanations. This dialectic between the adaptive notions advocated by Spath and the morphogenetic law in its more virulent form advocated by Lang (1919a, b), among others, comes under the purview of this discussion because the operation of gas pressure during ontogeny is one of the vital forces incorporated into morphogenesis. As discussed previously, Spath, in considering adaptive alternatives to the morphogenetic law, touched on the possible function of the septal-suture in resisting hydrostatic load.

One aspect of the law of morphogenesis as originally propounded by Hyatt was that a lineage or group evolved along a predetermined trajectory that paralleled ontogeny; first there was ascendant or "anagenetic" evolution to an acme and then subsequent "catagenetic" evolution that produced gerontic features characteristic of old age, harking back in a senescent manner to the morphology of forms earlier in the phylogenetic series. Thus, in Hyatt's original concept an underlying ontogenetic force controlled evolution in a causal sense, resulting in parallelism between the ontogenetic and phylogenetic series. This contrasts with Haeckel's biogenetic law, where ontogeny is a mechanical recapitulation of phylogeny (de Beer, 1958). Oddly, these two laws are not mutually incompatible and Hyatt also believed firmly in recapitulation as well as the direct control of phylogeny by ontogeny. Thus, although the biogenetic law and the morphogenetic law are often muttered in the same breath, there is the distinct added element of a predetermined ontogenetic force governing the direction of evolution in the morphogenetic version. Vitalism invoked similar internal forces or evolutionary potentialities. Many of these vital forces were conceived of as ontogenetic and were compatible with the morphogenetic law. This combination of vitalism and morphogenesis is related to a variety of evolutionary concepts current in the early part of the century, including some aspects of orthogenesis.*

In orthogenesis, evolution was thought to follow along a preset trend as a consequence of underlying potentialities. These underlying potentialities were also purported to be causally related to the frequent observation of homeomorphy, the repeated evolution in different stocks of similar evolutionary trends producing very comparable forms. The self-avowed vitalist Lang (1919*a*) advocated this view of repeated evolution, as opposed to the Darwinist or "mechanistic" view, as he called it in "Old Age and Extinction in Fossils":

> Above all living matter has potentiality and the actualization of this potentiality is Evolution. Now phenomena proclaim the *manner* of Evolution, while its *method* is expressed in our theories. For instance, the phenomena of Homeomorphy and Periodicity suggest that Evolution is limited in its directions and fluctuating in its progress, while "expression points" or "bursts" of evolutionary activity in a lineage irresistibly introduce the idea of a barrier or inhibition removed and allowing full play to variation. . . . As potentialities are actualised, so they become exhausted in a given direction, causing the appearance of periodicity in evolution;

* Not all examples referred to as orthogenesis are vitalistic, involving the realization of intrinsic evolutionary potential; others involve what would be referred to today as evolutionary constraint. In this regard, Lang (1919*a*, p. 104) cites Crampton: "the limits within which a lineage can vary become less as evolution proceeds, because structures already formed cannot evolve beyond the scope of their fundamental architecture. In this case potentiality is choked rather than exhausted."

> but until exhaustion begins, actualization increases in intensity, so that exaggeration of structures are often produced which may themselves cause the extinction of the lineage exhibiting them. [Lang (1919*a,* pp. 103–104; italics in original.]

In Lang's (1919*a, b*) vision, homeomorphs were a consequence of the similarity of intrinsic evolutionary potential. This view of homeomorphy is similar to, and derivative of, the cycle of morphogenesis of Hyatt (1894), in which lineages went on their own cycles. The grand cycle of Hyatt was the cycle of coiling; first, cephalopods such as orthocones and bactritids were straight, then during the anagenetic period of their evolution they coiled into nautiliforms. Coiled forms represented the acme of the cephalopod lineage. Subsequent uncoiling was the senescent phase of evolution, where the earlier stages of life were successively revisited.

These cycles were evident to the observers of the time in both the ontogenetic series within the organism and the phylogenetic series. The phylogenetic series, or lineage, consisted of a succession of forms following the cycle. Very similar ideas were applied to other taxonomic groups. For example, Grabau (1907) described the vermetid gastropods as the uncoiled senescent forms of the turretellid lineage.

Clearly, Eimer, who was the first to expound on orthogenesis, perceived a linkage between his ideas and those of Hyatt. In *On Orthogenesis and the Impotence of Natural Selection in Species-formation,* Eimer (1898) contrasts the ideas of "selection fanatics," as he termed Darwinians and especially the followers of Weismann, with "the beautiful investigations of Hyatt on Ammonites (Arietidae) . . . researches which give striking examples of orthogenesis and of the non-utility of numerous characters relating to sculpturing" (Eimer, 1898, p. 12).

Thus, we can see that many workers of the period viewed evolution as a consequence of intrinsic vital properties or forces leading evolution down particular paths. Eimer and Lang are probably representative in viewing their vitalist ideas as directly conflicting with mechanist evolution via natural selection.

The combination of the morphogenetic law and vitalism was particularly potent. Vital forces could be identified with physiological processes active through ontogeny, providing an intermediate level at which predictions could be made that were compatible with the evidence. This allowed the morphogenetic law in conjunction with vitalism to develop an explanatory power equivalent to that of adaptation, often capable of explaining similar sets of observations.

It is in light of this ongoing conflict—between adaptation (selectionism, mechanism) and morphogenesis/vitalism—and the association of vital forces with subsidiary physiological hypotheses, that Spath (1919) takes up the sword in his adaptationist crusade. His initial assault is directed at Swinnerton and Trueman's (1918) paper, "The Morphology and Development of the Ammonite Septum." These authors had constructed an elaborate machine to be used in conjunction with the grinding of the ammonite specimen. In so doing, a contoured plot of the septal surface of the ammonite could be produced. This was intended to provide an alternative for examination of the ontogenetic series of septal-sutures. That is, by examining the shape of a single septal surface from the medial region to the periphery it was hoped that one could reconstruct the sequence of sutures present in ontogeny. However, it was not this fixation with the ontogenetic series that troubled Spath. Earlier, Spath (1914) had performed a detailed ontogenetic study of the ammonite *Tragophylloceras*. What troubled Spath was the causality ascribed to some of the ontogenetic changes in sutural pattern observed in this work.

Swinnerton and Trueman (1918) suggested that changes in sutural shape result from the waning force of gas pressure: "In minor details the later septum seems at first more intricately wrinkled, a feature which is still better exhibited in other specimens some of which show these and other 'aging' characters through a wide range of septa. This complexity is strongly suggestive of the wrinkling of a collapsing or flaccid bladder as opposed to the simpler and more turgid outline of the folioles on earlier sutures, and suggests a diminution in the vigor of gas-secretion in the declining period of life" (Swinnerton and Trueman, 1918, p. 39).

According to Swinnerton and Trueman, gas secretion is not involved just in the evacuation of the shell as Hooke (Derham, 1726) envisioned it. They thought that gas pressure functioned during ontogeny in septal formation. Secretion of gas behind the body of the animal is presumed to push the body forward within the shell to form a new chamber prior to septum formation.* Thus, the gas pressure hypothesis is used to explain the observed outward convexity of the ammonite septum, and details of the ammonite suture where the septum meets the shell are presumed to depend on gas pressure. Gas pressure in turn is presumed to be a function of the waning life force of the animal. In Swinnerton and Trueman's (1918) paper the detailed frilling of the septa is explained by the reduction of gas pressure late in ontogeny; this period of ontogeny corresponds to the later "life history" of the lineage. Thus, through the intermediate action of gas pressure the detailed changes in

* The view that pressure builds up behind the body, pushing it forward to accommodate chamber formation, may be accurate. However, an aqueous solution, not gas, is likely to be involved. The fluid in the chamber of *Nautilus* is an aqueous solution produced by the transport of ions into the forming chamber (Ward, 1979). If the ionic concentration were to exceed that of sea water, an osmotic pressure greater than sea water would be produced. This would account for the outward bulging of the septum observed in ammonites.

morphology of the suture are rendered explicable by morphogenesis (the ontogenetic trend of the lineage, in this case senility).

There are other examples of the use of senescence to interpret the series of septa in the cephalopod shell. The closer spacing of the last few septa of ammonites and nautiloids has often been taken as indicative of maturity. In addition, this sutural approximation had been interpreted as a consequence of the waning of the life force (Hyatt, 1889; Bather, 1897; Willey, 1902; see Spath 1919). Similarly, increased sutural complexity in ammonites is viewed by Lang (1919b) as an "extreme actualization of potential" in vitalist terms, but also as serving the physiological function of getting rid of "superfluous calcium carbonate." It is this admixture of cooked-up physiological explanation that provides the flexibility of explanatory power, a power of hypothesis generation, to the morphogeneticist/vitalist; this flexibility permits the elaboration of hypotheses that, at least after the fact, conform to observation. It is this explanatory potential at an intermediate level of hypothesis formation that is comparable to, and directly competitive with, adaptive explanations of similar observations.

By the beginning of the 20th century a lot more was riding on the gas pressure hypothesis than a mechanism of shell emptying. The action of gas pressure during cephalopod ontogeny had become associated with and interwoven into the dominant forms of evolutionary explanation of the day.

Spath's Attack on Morphogenesis

In his "Notes on Ammonites," Spath (1919) responds to the morphogeneticist/vitalist school; he is specifically inspired by the gas pressure argument of Swinnerton and Trueman (1918). It is in this context that Spath begins his crusade. This crusade became a major theme in his life's work. His proadaptation arguments proceed on all available fronts, he attacks the inconsistencies of the individual morphogenetic arguments, and he demonstrates that there are adaptive alternatives to individual morphogenetic arguments. He also argues that adaptive ideas should be preferred *a priori* and that the vitalist underpinnings of the morphogenetic law are fundamentally unscientific.

Regarding inconsistency, Spath (1919) takes exception to the use of sutural complexity as an indication of senescent evolution in Swinnerton and Trueman's (1918) work. These authors also indicate that greater simplicity of the suture, rather than complexity, should be taken as indicative of senescence. Thus, both increased and decreased complexity of the suture were explained by the same stage in the ontogenetic control of phylogeny.

Spath in his low-key style of criticism suggests that, "The association of complexity with decline may seem contradictory . . ." (Spath, 1919, p. 28).

Spath presents evidence suggestive of the adaptive role of the suture line, that the suture line is highly correlated with whorl shape and ornament, and points out that adaptive or mechanical evolution in conjunction with familial or hereditary factors may be sufficient to explain sutural evolution: "in the mechanical adjustment to a wider side, either by spreading-out of the lateral lobes and saddles or by the addition of auxiliary or adventitious elements, similar suture-lines may result in different stocks, yet the modified shells can generally be referred to their ancestral stocks by means of some retained family characters" (Spath, 1919, p. 29). Therefore, homeomorphy does not result from similar intrinsic potentialities of separate lineages, but from similar, "mechanical," adaptive responses in the organism. In cases where the precise adaptation is unknown, "the writer would be inclined to favor the theory of adaptation, therefore, even if the special mode of adaptation be not quite clear in some cases, rather than speak of decline of vitality or phylogenetic degeneration" (Spath, 1919, p. 33).

Bursts of evolution, "periodicity of elaboration" of the suture line, do not result from the release of pent up internal potentialities as they do in an orthogenetic argument. In Spath's conception these are the consequence, of repeated adaptive radiation, which in turn respond to the ". . . changes of structure and diversity of life' and are probably 'directly related to the physical conditions of habitat', so that the 'stability of organic forms is in direct ratio to the stability of the conditions of existence" (Spath, 1919, p. 33).

Spath also opposes the vitalist precepts embodied in the senescent component of cyclic-morphogenetic evolution: "the representation of Ammonite phylogeny as a series of cycles 'which is in direct contradiction to a causal explanation of their development', as it conveys the impression of an inborn racial necessity of a predestined character" (Spath, 1919, pp. 32–33). Thus, Spath counters orthogenetic arguments in addressing the same points that Lang (1919*a,b*) brought up in his work in the same year.*

Recapitulation, Development, and Paleontology

One might think that Spath's forceful attack of 1919 would have had a broad effect on the profession and that the role of morphogenetic and vitalis-

* Spath does not cite Lang, despite his familiarity with his ideas. Spath depended on the material at the British Museum (Natural History) where Lang was a curator. He worked closely with him and may not have wished to confront him directly.

tic explanation would have diminished. That this did not occur is especially surprising given the frequent repetition of arguments against the biogenetic law by neontologists such as Hurst (1893), Sedgewick (1909), Garstang (1922), and de Beer (1958). These workers all argued essentially the same points. Similarity of forms early in ontogeny did not indicate recapitulation of adult forms. Even if it was generally true that larval forms were more similar than the adults of divergent taxa, *à la* Von Baer, this was not the same as recapitulation of adult form. It was a consequence of a shared ancestor with a similar form early in its ontogeny, not its adult stage. All these authors also pointed out the frequency of cases where even Von Baer's law was violated; they all advanced examples of Caenogenesis, radical differences in early, usually larval, ontogeny of forms with very similar adult stages.

Smith (1900*b*) and Bather (1893) argued that recapitulation theory was applicable often enough to have been successfully applied by Hyatt's students such as Jackson and Beecher (Smith also studied with Hyatt). In addition, they argued that the succession of forms in the fossil record was in accordance with their recapitulatory interpretations. This point was raised in response to claims based on the development of modern forms. Paleontologists (Smith, 1900*b;* Bather, 1893) would argue that the recapitulation of forms only pertained to succession in a lineage, and therefore the comparison of the developmental stages of modern forms did not directly address the question of recapitulation. In the minds of these paleontologists a historical, i.e., fossil, sequence of several ontogenetic series was required for a real test of recapitulation, and they believed that recapitulation had passed this exacting test.

Flexibility of Recapitulation

An additional factor that contributed to the longevity of recapitulation theory and derivative forms of Hyatt's and Haeckel's ideas was the flexibility and success of their application by cephalopod workers subsequent to Hyatt. Workers such as Buckman and J. P. Smith were much less dogmatic in their application of the theory than their often ardent support of it might suggest. These workers, unlike Hyatt, were relatively careful in interpreting the fossils in their stratigraphic succession.

Hyatt was often not aware of the stratigraphic relationship of the fossils he examined; this is particularly true of his most detailed phylogenetic analysis, *Genesis of the Arietidae* (Hyatt, 1889), in which he admits that "the researchers were conducted almost wholly in Museums, because it was found impracticable to study stratigraphical superposition in the field. This part of the work has already been accurately done by local geologists, and my notes were largely made upon their collections. More extended studies might have made the work more accurate than it is, but this was not possible for me" (Hyatt, 1889, p. vii).

The Jurassic age of the material in question required Hyatt to study European material, there being no exposed marine Jurassic rocks in eastern North America. In Europe, the Jurassic sequence containing the ammonites in question is widely exposed, and had been the subject of classic studies by Quenstedt, Oppel, and D'Orbigny; subsequently Quenstedt, Neumayr, and Uhlig had generated phylogenetic schemes based on this material (Hyatt, 1889). In addition, the Jurassic contained the greatest diversity of ammonites, representing the acme (in Haeckel's terminology) of the group. It was therefore natural that Hyatt should turn to material of this age to showcase his evolutionary ideas. Indeed, the limited knowledge of the section allowed a certain freedom of interpretation, permitting an appearance of consistency between the morphogenetic law and the stratigraphic succession of ammonites.

Whatever Hyatt's shortcomings in regard to the order of appearance of forms in the section, the work of some of his followers, such as Buckman and J. P. Smith, though not necessarily perfect (Spath, 1919), was conducted on their home turf and with specimens whose order in the section they had some familiarity with. These workers elaborated on the expected pattern of acceleration in development in order to make the sequence in the sections consistent with the morphogenetic law. In other words, observations inconsistent with the law resulted in elaboration of the law rather than rejection of it.

Buckman

Buckman (1919–1921), in his enthusiasm for Hyatt's ideas, devised detailed morphogenetic cycles for keels and ornament of the ammonite shell published in *Type Ammonites* (see Fig. 3). Buckman (1922–1923) even constructed his geologic time scale using the coiling cycle of the ammonoid lineage; the Jurassic period became the "Ammonoid" and was followed by the "Baculoid" time period equivalent to the Cretaceous when uncoiled forms such as *Baculites* were prominent.

Buckman, in his *Type Ammonites,* introduced a number of new terms to accommodate departures from the acceleration of development; acceleration was the expected result of terminal addition and recapitulation in an ontogeny of fixed length:

> The ontogenetic record is preserved excellently among Ammonites, but its recapitulatory fidelity is often marred. Irregularity of record may, it is here suggested, be grouped as follows:—

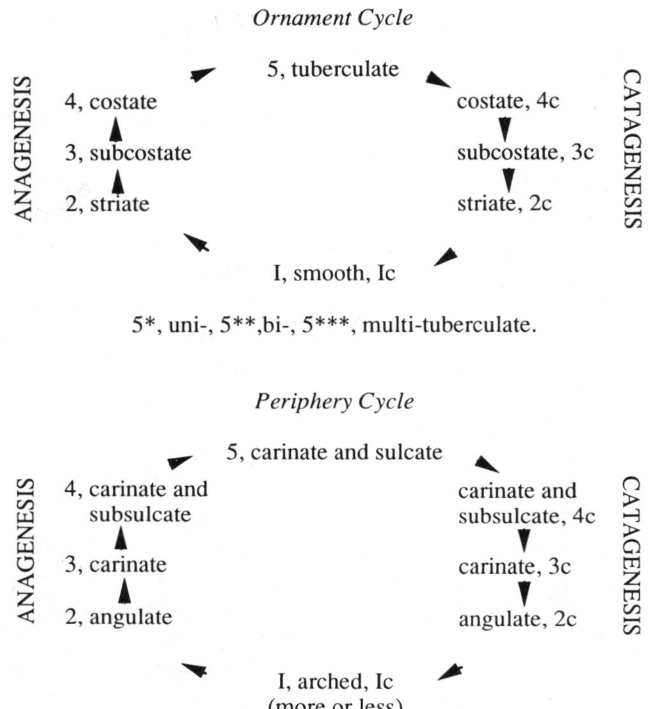

FIG. 3. Examples of morphogenetic evolutionary cycles devised by Buckman for ammonites. Note that there is an anagenetic or progressive phase and a catagenetic descendant or senile phase of evolution for each set of characters. This follows the form of Hyatt's (1889) coiling cycle. Other workers devised similar cycles for other taxa. Also note that the cycles operate independently from one another and consequently can be used to report what would now be termed heterochronic change in a codified manner.

Palingenesis,

saltative	skipping of stages,
cunctive	delaying development,
precedentive	unequal acceleration. [Buckman (1909–1912), p. vii]

In subsequent volumes of his *Type Ammonites,* Buckman (1919–1921, 1922–1923) introduced new, redundant, terms for the same phenomena, including "bradygenesis" for the slowing down of stages, "tachygenesis" for the acceleration of adult forms into earlier ontogeny, and "lipopalingenesis" for the skipping of stages in development.* This multiplicity of deviations

* Buckman employed the mutation theory of De Vries to explain the skips in recapitulation.

from the strict acceleration as well as the disconnection for the various cycles, coiling, ornament, keels, etc., gave the morphogenetic approach as practiced by Buckman the same flexibility of interpretation as heterochrony provides to certain ammonitologists working today.*

Spath does not spend much effort debunking Buckman, perhaps because Buckman was performing the job adequately himself. Buckman's understanding of the origin of gas generation is instructive:

> The early Orthocones utilizing the gas-effusion, which resulted from the temporary indigestion under the nervous apprehension of danger, found that a cone thus made more buoyant was a help in rapid retreat from foes. They brought the feature to perfection—a systematic gas-generator: and containing chambers. But retreat from foes was only incidental: bottom-crawling in search of food was the normal occupation. As the plant was still of service to lighten the load of the superincumbent conch. [Buckman (1919–1921), p. 18.]

Despite these fanciful interpretations, Buckman, unlike Hyatt, was on his home turf and devoted much effort to biostratigraphic interpretation. Indeed, his divisions of the Jurassic section of Dorset employed a stratigraphic unit termed the "hemera," which was finer than the zonal stratigraphy. It proved to be useful only in the local sections.

J. P. Smith

Smith, like Buckman, was an advocate of the utility of recapitulation. Despite his interpretation of recapitulation as an actual revisitation of the ancestral condition during the development of the individual, Smith was relatively pragmatic in his usage of recapitulation in phylogenetic interpretation. In "The Biogenetic Law from the Standpoint of Paleontology," Smith (1900*b*) indicates that there are two types of development, fetal (direct developing) and larval, and "characters that are useful in a free state, but not in a foetal state, are liable to be lost. Thus in the foetal type the tendency is toward loss of the record through omission of stages or obscuring them, for many organs that would be highly developed in mature forms, or in free larvae, will be either suppressed or undifferentiated" (Smith, 1900*b,* p. 414). Smith also viewed caenogenesis as playing a large role in early development, perhaps even introducing entire larval stages.

* Several French workers are currently engaged in research programs that involve the heterochronic interpretation of the evolution of ammonite lineages (Dommergues *et al.,* 1989; Meister, 1988). These workers employ a methodology that involves the notation of types of ornament, whorl shape, etc., through ontogeny and comparison of ontogenetic changes in these features with evolutionary changes in the lineage. Modern heterochronic terminology is employed. However, there is no overt reference to a predetermined cycle.

Although these concepts may appear crude or self-evident, they are far more realistic than the earlier arguments of Lankester and Verrill. These authorities did not recognize that features of free-swimming veliger larva were unlikely to appear in forms that did not leave the egg during larval development (see previous discussion, pp. 298 and 300). In addition, the morphogenetic law predicted the presence of a veliger stage early in the ontogeny of cephalopods; consequently, a veliger stage was central to Hyatt's (1889) arguments regarding the ontogeny of ammonites and nautiloids. Hyatt advocated this position despite the fact that the early ontogeny of *Nautilus* was unknown,* and a veliger stage had not been observed in any cephalopod. Similarly, Verrill (1896) argued, on the basis of acceleration, that the veliger larva was the ancestral type from which all mollusks were descended.

Even in the adult, Smith felt that the characters were dissociated and accelerated at different rates and were thus heterochronic in terms of the currently accepted nomenclature. Thus, the methodologies of Smith and Buckman at least admit of, and in some sense incorporate, the arguments of later workers. For example, 20 years later Garstang (1922) emphasized the introduction of novel features in the larval stages of development in his criticism of recapitulation theory. Similarly, the work of Buckman and Smith also permitted the evolutionary rearrangement of different features in development. This point was emphasized by Needham (1933) in his later criticisms of recapitulation theory, "On the Dissociability of the Fundamental Processes in Ontogenesis."

Although Spath did not agree with the methods of Smith and Bather, these workers were much less vulnerable to criticism because they did not indulge in the vitalistic excesses of Lang and admitted to exceptions to recapitulation. In addition, Smith (1900*b*) and Bather (1893) made a case for the superiority of the study of actual lineages through time. Embryologists merely compared contemporaneous modern forms which lack the ancestor/descendant relationship that Smith and Bather considered to be required in any assessment of the recapitulation theory. Thus, a stalemate persisted, the neontologists maintaining that the recapitulation glass was half empty, the paleontologists insisting that it was half full.†

* Cephalopods are now all thought to have direct development and to lack a veliger stage.

† Neontologists were often more concerned about other developmental biologists than with paleontologists. Garstang even wrote humorous lyrics making fun of Macbride (published posthumously—Garstang, 1966), who had committed the sin of writing an embryological textbook based on a recapitulatory premise.

Spath's Second Assault

Perhaps because of the limited effect of his earlier efforts, Spath was obliged to take another stab at the followers of Hyatt in his 1933 paper, "Evolution of the Cephalopoda," in which he attempted

> . . . to show that some of the opinions expressed by Hyatt fifty years ago and still adopted by certain workers were based on mere conjecture. Thus it is commonly asserted that (1) the recent *Nautilus* can be traced back in an evolutionary series through spiral gyrocones and curved cyrtocones to the straight *Orthoceras*. This series is held (2) to be characterized by the absence of a calcareous protoconch. Then it is stated (3) that an *Orthoceras* . . . gave rise to a straight ammonoid shell ('Bactrites') with ventral siphuncle. Next (4) this is said to produce the loosely coiled 'Mimoceras' and the involute goniatities which in turn gave rise to the ammonites. [Spath (1933), p. 420.]

Spath then proceeds to demonstrate that all these notions derived from Hyatt violate morphological evidence and grossly violate the stratigraphic succession of fossils in the rock record. Thus, Spath is no longer harping on the adaptive point of view, and in this work there is no mention of gas pressure. Spath is now attempting to demonstrate that Hyatt's results, while still widely espoused, are factually incorrect.

Even after all the efforts of Spath and the neontologists, the true believers in recapitulation theory and biogenetic law were still holding the fort. This is evident in the ardent support of George (1933) in "Palingenesis and Palaeontology" and in subsequent publications of Swinnerton, "Development and Evolution" (Swinnerton, 1938) and *Outlines of Palaeontology* (Swinnerton, 1947), a textbook that still showcases Hyatt's theory and nomenclature.

Trueman Overturned

Spath was able to convince at least one of his nemises of the invalidity of Hyatt's methodology. Spath demonstrated that the previous interpretation of the evolution of the Lipoceratidae, explicated by Trueman (1919) using the Hyattian assumption of progressive involution of coiling, had resulted in an interpretation that was exactly opposite to the stratigraphic succession of forms. Spath literally overturned Trueman's results in a stratigraphic sense; the specimens Trueman thought were ancestral proved to be at the top of the section and hence younger than the forms that Trueman thought were descended from them. The overturning of Trueman's result must have been especially sweet vindication for Spath; he used specimens collected, with the necessary stratigraphic precision, by W. D. Lang himself. Lang was duly impressed and wrote a flattering preface to Spath's (1938) monograph on the subject. Thus, Lang was converted, and paid penance for his orthogenetic excesses.

This paper of Spath's, as well as comments to similar effect in earlier work, was important in that it demonstrated that recapitulation theory was, in many cases, not supported by the succession of forms in the fossil record. The evidence that the succession of forms was often misapplied due to circular reasoning on the part of paleontologists was put to good use by de Beer (1958), who cites Spath's work widely in this context.* Thus, Spath as a paleontologist was able to undermine the notion that recapitulation theory was consistent with the record of lineages in the fossil record. That recapitulation theory could and had been successfully tested by the consistency of the stratigraphic and ontogenetic successions had been the trump card of paleontological defenses against the likes of Sedgewick (1909) and Garstang (1922). Thus, the fact that de Beer succeeded in advancing the issue beyond the statements of these earlier workers is largely dependent on the work of Spath. In this sense Spath eventually contributed to the resolution of this issue, to the extent that it has been resolved.

WESTERMANN VERSUS ARKELL

Westermann (1956), in "Phylogenie der Stephanocerataceae und Perisphinctaceae des Dogger," resurrected the structural ideas of Pfaff (1911) and applied them to the taxonomy of Middle Jurassic ammonites. This aroused the ire of Arkell, the preeminent ammonitologist of the mid 20th century and the supreme authority on the Jurassic. Arkell was interested in the utility of a stable classification of ammonites for geological purposes. He feared that Westermann's use of functional interpretation in phylogenetic analysis would disrupt this stability.

Pfaff (1911) argued that hydrostatic load had been borne by the shell; his primary concern was the action of hydrostatic force through the body chamber laterally against the last septum (Westermann 1971). Westermann (1956) realized that the septum played a functional role in supporting the shell at the suture line against exterior hydrostatic load. He also perceived that there was a relationship between the sutural lobes and the shape of the shell and cited Spath (1919) in this regard. This understanding led Westermann to consider the whole septum in this functional context. Septa of

* Other workers, such as Pavlov (1901) and Schindewolf, are also cited by de Beer in this regard. Pavlov appears to have had no effect in his own time. His resurrection, at least in the English-speaking world, may be due to his citation by Spath (1919).

ammonites are folded (see Fig. 1); in this pattern of folding Westermann perceived a functional role relating to the transmission of forces generated by the support of the shell wall at the suture. He considered this functional relationship to be of phylogenetic import. Consequently, he employed the evidence supplied by these folds in conjunction with the traditional characters, the suture line and external features of the shell, in his analysis of the Middle Jurassic ammonite lineages. This novel approach yielded novel taxonomic relationships that were not uniformly in agreement with those advocated by Arkell.

In response, Arkell (1957*a*), in "Sutures and Septa in Jurassic Ammonite Systematics," objected both to the resultant phylogenetic ideas generated by Westermann and to the premise of his structural hypothesis; namely, that the shell supported external hydrostatic load. In order to undermine Westermann's position, Arkell (1957*a*) directed his argument regarding hydrostatic load toward Pfaff (1911). Westermann (1956) had used Pfaff's (1911) work in constructing his premise of structural support of hydrostatic load. Arkell characterizes Pfaff as having three lines of evidence against the gas pressure hypothesis. Arkell attacks these objections with three counterarguments supporting the role of gas pressure in the shell, and opposing the notion of structural support of hydrostatic load. However, in each case Arkell's argument is inaccurate or overstated. It is also apparent that Arkell was aware of the inaccuracy of some the positions he took. I will examine each of the positions that Arkell attributed to Pfaff, and Arkell's response to each position.

Arkell cites Pfaff's first argument:

> 1. In *Nautilus pompilius* no organ has yet been discovered which could act as a kind of compression pump to build up such pressures in the camerae, in excess of atmospheric pressure. [Arkell (1957*a*), p. 243.]

Arkell's response to this point is:

> 1. Atmospheric pressure has no relevance below the surface of the sea. At depth the external pressure is determined by the weight of the overlying column of water, and if the gas in the camera started at atmospheric pressure, the internal pressure could be brought up to equal the external by the animal's merely generating gas, which would then be sucked into the camerae until pressures were equal (see 2). [Arkell (1957*a*), p. 243.]

Unfortunately for Arkell's argument, water pressure at depth is not directly comparable to the pressure of gas at depth. Arkell is correct in asserting that atmospheric pressure is an increasingly trivial component of the total pressure as one goes deep into the sea. However, total pressure is not the issue here; the issue is the partial pressure of gases dissolved in a liquid. In this case, the ocean is no different from other liquids exposed to gas. The operative chemical generalization which applies is that the partial pressure of a gas dissolved in a liquid is equal to the partial pressure of the gas over the liquid. Otherwise dissolved gas pressure at depth will be out of equilibrium with the surface pressure. Contrary to Arkell's position, the partial pressure of gases in the ocean tends to approach the atmospheric values even at depth.* Thus, if a cephalopod were to produce gas in its tissues, the gas would constantly diffuse out of the tissues into the surrounding environment. It would not be passively "sucked into" the chambers. Gas pressure in the chambers in excess of atmospheric pressure would require constant transport of gas in the blood, which in turn would require a well-developed vascular system. This was Owen's (1832) point; when he failed to find the requisite vascularization—typical of the swim bladders of fish—in *Nautilus,* he correctly assumed that gas generation was not feasible. Arkell's misunderstanding of this point appears to be an honest mistake; he previously expressed the same point earlier (Arkell, 1949).

This appears not to be the case in regard to Pfaff's second point:

> 2. Examination of the siphuncle shows no openings, without which such a circulation of gas could not take place. [Arkell (1957*a*), p. 243.]

Arkell responds as follows:

> 2. Although no perforations in the siphuncle are visible, it is recorded that when a camera of a living *Nautilus* is punctured under water, gas bubbles out (Willey, 1902, p. 747). The fact that in fossil ammonites the siphuncle sheath differs from all the rest of the shell by being phosphatic suggests that it had a special function, which may have been the transmission or even generation and reabsorption of gas in response to changing pressures inside and out. [Arkell (1957*a*), p. 243.]

This response is a bit disingenuous of Arkell. First, this paraphrase of Willey omits the word "gently" before "bubbles out" and misrepresents Willey's opinion. Willey believed that the shell was merely filled with gas once and did not undergo subsequent pressure changes. It is clear that Arkell was aware of this point, since in a review paper (Arkell, 1949) he had previously represented Willey as saying as much. Second, Arkell's argument regarding

* The main exception to this rule is the production and consumption at depth of the metabolic gases such as CO_2 and O_2. However, such situations are out of equilibrium with the surface and are unstable. Instability of dissolved gas at depth in excess of atmospheric partial pressure was forcefully demonstrated by the explosive degassing of a lake in Cameroon (Kling *et al.,* 1987). CO_2 was produced at depth by organic decay; it built up in solution in excess of the atmospheric partial pressure of this gas. When some of the water rose toward the surface, the partial pressure of gas in solution exceeded the total pressure and gas bubbles formed, causing the water to rise faster still, resulting in an explosive degassing of the lake and CO_2 poisoning of the local population. This serves to illustrate the point that the dissolved gas pressure of the ocean cannot long remain much in excess of atmospheric pressure.

the phosphatic composition of the siphuncle of ammonites certainly does not pertain to *Nautilus,* where the siphuncle is organic with carbonate septal necks. Consequently, even if one were to accept Arkell's premise that all ammonoids had phosphatic siphuncles* and his special pleading regarding the potential gas-generating ability of phosphate, this argument could not pertain to the most available and familiar form, modern *Nautilus,* which Arkell mentions in the same breath.

Pfaff's third point as translated by Arkell is:

> 3. Since the ammonites are ideally adapted in their whole structure and down to the smallest details to resist water pressure from outside, and since they have thin shells, they would burst when they rose rapidly towards the surface of the sea if the camerae contained gas at high pressure. [Arkell (1957*a*), p. 243.]

Arkell's response:

> 3. If one wished to design a means of preventing the ammonite whorls from bursting when the animals swam rapidly upwards, one could scarcely devise anything more perfect than the frilled and interlocking edges of the septa, which afford maximum attachment of the sets to the whorls. Besides this they are built into the shell wall in such a way . . . as to provide a superbly strong structure. Moreover, the fact that in all ammonites the septa are convex forwards, and that the forwardly directed frills (saddles) are more rounded than the backwardly directed (lobes), suggests that during their formation the membrane secreting the septa was under the influence of pressure from within rather than from without. [Arkell (1957*a*), p. 243.]

There are two reasons why the septa of ammonites cannot be interpreted as being equally or better suited to resist pressure from within rather than pressure from without. First, the carbonate shell material (nacre), being brittle, is two to three times stronger in compression than in tension (Currey and Taylor, 1974; Currey, 1976). Internal pressure is naturally going to result in tensile stress on the septum and outer shell wall. External pressure results in compressive stresses, and the specific arrangement of sutural complexity minimizes tensile stresses due to bending in the shell wall (Jacobs, 1990). However, it is doubtful that Arkell was aware of the material properties of nacre and its consequences for shell design. However, the statement that the septum is "interlocking" with or "built into" the shell wall is inaccurate. The shell wall could be better described as "resting on" the septum at the suture. Such a situation is compatible with external load resulting in compression, but not with internal gas pressure resulting in tension. Any tensile strength between septum and shell wall is due to adhesion rather than the structural suggested by the terms "interlocking" and "built into."

* Work subsequent to 1957 indicates that ammonites do not have phosphatic siphuncles. Their siphuncles are made of the conchoilin and carbonate materials that modern cephalopod siphuncles had long been known to contain [see Bandel and von Boletzky (1979) for review].

Arkell's last point regarding the convexity of the ammonoid suture is well taken, in that this convexity is suggestive of higher pressure behind the organism shaping the septum-secreting surface of the body. However, there is a tendency to overestimate this curvature because the typical septal section studied passes down the midline of the shell (Spath, 1919) where the dorsal and ventral lobes of the suture occur. In addition, Arkell's argument does not account for the concavity of the septum in nautiloids. We now know that the fluid behind the first septum of *Nautilus* is an aqueous fluid that diffuses into the space behind the animal prior to septal formation because of the transport of ions by the organism into this space (e.g., Ward, 1979). Any positive pressure responsible for the convexity of the septum and the ballooning of the saddles of ammonoid septa toward the aperture is likely due to aqueous fluid pressure under the osmotic control of the organism.

So why did Arkell feel compelled to mount this vehement argument against the hydrostatic support function of the shell as presented by Westermann? The answer would seem to lie in Arkell's conservative perception of the nature and purpose of classification. It was the novel application of the structural argument that appears to have upset Arkell. In addition, in 1957 Arkell was terminally ill and had just completed his *magna opera. The Jurassic Geology of the World* (Arkell, 1956), and the ammonite volume of the *Treatise on Invertebrate Paleontology* (Arkell, 1957*b*). Presumably, Arkell viewed Westermann's novel use of structural information in phylogeny reconstruction as a threat to his legacy.

Competing Uses of Taxonomy

Arkell's vision of taxonomy is apparent in "Paleontology and the Taxonomic Problem" (Arkell and Moy-Thomas, 1940). Arkell argues that the utility of paleontological classification in biostratigraphy is in conflict with the alternative goal, the use of classification to express phylogenetic relationships:

> Palaeontological classification, like all classification, must primarily be useful. It should provide a practical means by which fossils can be identified and compared. Since geologists date the strata by the contained fossils, it is essential to their science that they should be able readily to identify the fossils . . . classification has come to aim, not only at providing an easy means of recognizing fossils, but also at giving a summary of existing knowledge of phylogeny. It is because these dual objects frequently tend to produce conflicting results that the problems of taxonomy have arisen. [Arkell and Moy-Thomas (1940), p. 395.]

This conflict manifests itself in the following manner:

> The first of these difficulties faces, in particular, the student of ammonites at an early stage of his investigation. Schindewolf (1928) has shown in detail that a number of lineages of Ammonoidea can be traced up through the Devonian rocks, all undergoing more or less parallel evolution. They may be classified in two ways: either 'vertically', making the genera correspond with the lineages, along which the successive forms are marked by parallel species, or 'horizontally', making the genera correspond with successive grades irrespective of the lineages. [Arkell and Moy-Thomas (1940), p. 397.]

There is no question that Arkell preferred to see horizontal classification employed, where stratigraphic information is retained at the expense of phylogenetic information. For those pedants who were concerned with something as trivial as detailed phylogenetic relationship, they could, if they felt compelled to, incorporate this information in brackets at the subgeneric level, "Purists who wish to express every subtlety of phylogenetic theory in their classification may surmount the difficulty by using the names *Gryphaea* and *Exogyra* as genomorphs, writing such species as *Gryphaea dilatata* in the form advocated by Dr. Lang for certain corals, *Ostrea* {*Gryphaea*} *dilatata*" (Arkell and Moy-Thomas, 1940, p. 405).

The utility of taxonomy for stratigraphic work is also dependent on the stability of the nomenclature: "Endless unnecessary complication is caused by not pitching the scale of the classification in such a way that the familiar genera of the literature of the last century or more remain as genera so that they can still be used for faunal lists and records" (Arkell and Moy-Thomas, 1940, p. 406). Because of the practical application of taxonomy, Arkell held that conservatism was in order. Thus, Arkell's complaints about Buckman's work tend to center not on the incorporation of morphogenetic cycles in Buckman's phylogenetic approach, but on the fact that Buckman was an extreme splitter, "Every new genus and species should fit as nearly as possible into a uniform scale of values. . . . Examples of flagrant disregard of this rule are Buckman's innumerable genera made by plotting up contemporary species of the single good Liassic ammonite genus *Dactylioceras*" (Arkell and Moy-Thomas, 1940, p. 406). It is an example of Arkell's conservatism and adherence to precedent that so many of Buckman's genera appear in the *Treatise* volume written by Arkell (1957*b*) despite Arkell's conviction that many of them were only subgenera or species.

Arkell's vehemence is undoubtedly a product of personal experience. It is probable that no one before or since has gleaned as much geologic information by comparison of faunal lists as Arkell did. Arkell's (1956) *Jurassic Geology of the World* is a monumental work of global correlation treating every major exposure of Jurassic rock. In addition, Arkell believed that comparable information spoke to such important geologic topics as the nature of geosynclines, biogeography, and continental drift (Arkell, 1949). He (Arkell and Moy-Thomas, 1940) thought that recasting the taxonomy of ammonites in phylogenetic terms was unwarranted, "The only alternative to an arbitrary horizontal classification would be a complete recasting of the nomenclature . . ." (Arkell and Moy-Thomas, 1940, p. 398).

It was the potentially revolutionary nature of Westermann's taxonomic approach, rather than any underlying importance that Arkell placed on the function of the septum, that led Arkell to disagree so vehemently with Westermann. Arkell (1957*a*) also attacked Westermann's taxonomy itself on a point by point basis. Thus, it seems likely that Arkell was motivated by the threat to taxonomic stability inherent in Westermann's (1956) novel method, coming just as he had put the last touches to his legacy. He wanted this legacy to be treated with the same respect for precedent with which he had treated others. He did not want his good works of synonymy, his ammonite systematics of great utility for a number of geologic applications, immediately overturned by someone with an excessive fascination with phylogenetic veracity and a new methodology to boot.

Ironically, Westermann (1958) conceded the structural point, which had not been Arkell's main concern in any case. He now admitted that "Pfaff's hypothesis that the pressure compensation could not be equalized while the animal descended may partly be disproved (Arkell, 1957*b,* p. 243). However, the pressure compensation would not have been perfect: consequently, normal pressures from both sides as well as peripheral pressure and traction on the septum acted on descent and ascent" (Westermann, 1958, p. 443). However, Westermann did not give up his phylogenetic arguments, but defended them vigorously. Thus, the major effect of Arkell's (1957*a*) attack on Westermann (1956) was to suppress one more time the peripheral issue of structural support of hydrostatic load. Westermann was able to argue his phylogenetic case without this point in his response to Arkell's criticisms.

Even as Westermann was equivocating in print on the structural issue, Denton and Gilpin-Brown were about to embark on a research project on the gas pressure in the cuttlefish, *Sepia.* This work (Denton and Gilpin-Brown, 1961*a–c*) was the first of several studies demonstrating that modern shelled cephalopods do not contain pressurized gas, and are subject to hydrostatic load.

THE POST-1960 EMPHASIS ON FUNCTIONAL STUDIES

The investigations of Denton and his collaborator Gilpin-Brown in the 1960s demonstrated that the shells of cephalopods are not supported by gas pressure, but support hydrostatic load in the structure of the shell itself. The work of Denton and Gilpin-Brown led to a resurgence of interest in the adaptive function of cephalopod shells, especially in regard to the interpretation of the life habits of fossil forms. Prior to 1960, Westermann (1956) was already conducting an investigation based on the premise that cephalopod shells supported hydrostatic load. Following the work of Denton and Gilpin-Brown, Westermann returned to the analysis of the structural function of cephalopod shells and was joined in this effort by several other workers.

These developments in the study of the function of the cephalopod shell occurred at a time when a more theoretical approach became acceptable in paleontology. The work of Raup (1966, 1967) on shell coiling gave theoretical studies of morphology in cephalopod paleontology a legitimacy that they had not enjoyed when Arkell was the reigning ammonitologist. Previously, theoretical comments on fossil cephalopods were limited to the introductions of monographic works such as Spath (1938), or the few journals such as the *Biological Review* which dabbled in paleontologic ideas of a broader or more theoretical nature. In the 1960s and 1970s new journals such as *Lethaia* and *Paleobiology* were founded. These journals devoted much attention to theoretical and organismal issues in paleontology, rather than just the phylogenetic and geological interpretations which had been of overwhelming importance to Arkell (Arkell, 1956, 1957*a, b;* Arkell and Moy-Thomas, 1940).

In a more general sense, there seems to have been a resurgence in the 1960s in interest in the organismal and naturalistic aspects of both biology and paleontology. Investigators no longer focused on the application of a detailed ontogenetic record to the interpretation of phylogeny. Operating within the neo-Darwinian adaptationist paradigm, these workers took a more holistic approach, concentrating on the autecology of *Nautilus.* In cephalopod studies, Bidder (1962) led the way with her study of swimming in *Nautilus.*

In the 1970s, the funding agencies (primarily the National Science Foundation) were convinced of the importance of sending expeditions to the South Pacific to investigate *Nautilus.* These expeditions, conducted by Saunders and Ward among others, elevated *Nautilus* from an obscure creature whose life was little understood to perhaps the best known deep-water invertebrate. Even Willey's original objective of procuring living embryos has now been accomplished. These field efforts, along with collateral efforts on anatomy, physiology, and phylogenetic relationships based on allozyme studies, conducted in the laboratory and the aquarium have culminated in the recent publication of *Nautilus,* a comprehensive volume summarizing the recent work on the genus (Saunders and Landman, 1987).

This more recent body of work can be compared to Willey's turn of the century efforts funded by the Royal Society. One of the justifications of Willey's study of *Nautilus* was that it would provide ontogenetic information thought to be important for understanding the phylogenetic relationships among living and fossil cephalopods. More recent expeditions were largely instigated by paleontologists and were justified, in part, because of the presumed importance of the results for understanding fossil cephalopods. One objective was to understand the mode of life of fossil cephalopods by analogy to ecological and functional studies conducted on *Nautilus*. This approach is indicative of the breadth of issues relating to organismal biology that had become of interest to paleontologists.

Denton and Gilpin-Brown

Denton and Gilpin-Brown began their researches on shelled cephalopods using *Sepia officinalis*, which was available near the Plymouth laboratory.* In 1961 Denton and Gilpin-Brown published a series of papers on the buoyancy of *Sepia* (Denton and Gilpin-Brown, 1961*a–c;* Denton *et al.*, 1961). Subsequently, they published similar work on the buoyancy of the other living shelled cephalopods, *Nautilus* (Denton and Gilpin-Brown, 1966) and *Spirula* (Denton *et al.*, 1967).

The investigation of *Sepia* (Denton and Gilpin-Brown, 1961*a*) involved weighing the animals in sea water shortly after capture, in order to determine their density. The density of *Sepia* varied, changing during the course of captivity; some individuals became denser and some lighter. Dissection revealed that changes in density were a consequence of changes in the density of the chambered shell, the cuttlebone, rather than the other tissues of the *Sepia* (see Fig. 1). Further investigation revealed that this variation was not attributable to changes in gas pressure in the cuttlebone comparable to the gas bladders of fish as had previously been supposed; this was evident from two observations. First, when *Sepia* was trawled up from depth, gas did not emanate from the cuttlebone as a consequence of the sudden ascent and decompression; nor was there evidence of gas in the tissue surrounding the cuttlebone. Second, a number of methods were employed to determine the internal gas pressure, including the puncture of the shell under water. These observations always revealed that water entered the shell rather than gas emanating from it.

Additional analysis revealed that the little gas contained in the cuttlebone was primarily nitrogen and that the pressure of the gas in the shell was near 0.8 atm; this value is equivalent to the partial pressure of the nitrogen in

* The Plymouth Laboratory was founded by E. Ray Lankester.

air, suggesting that the gas in the chambers was in diffusive equilibrium with the blood. This diffusive model was confirmed by Denton and Gilpin-Brown (1961*b*) when they examined the pressures in the succession of chambers formed during the growth of the cuttlebone. The most recently formed chambers had a much lower internal pressure than those formed earlier, suggesting that there had not been sufficient time for the gas to diffuse from the siphuncle into the chamber. It was not until the ninth most recently formed chamber that the gas pressure value approached 0.8 atm. Diffusion experiments suggested to Denton and Gilpin-Brown that it would take over 1 month for the "nitrogen to go halfway to equilibrium" (Denton and Gilpin-Brown, 1961*b*, p. 380).

Denton and Gilpin-Brown (1961*c*) made the observation that the density of *Sepia officinalis* varied diurnally in response to light. *S. officinalis* transported water into the chambers, attaining a high density and remaining on the bottom during the day; at night they pumped water out of their chambers, presumably in order to feed more actively in the water column. Similar density differences could be induced in the laboratory by exposure to light.

Possibly the most important observations published on *Sepia* in 1961 was Denton *et al.*'s (1961) investigation, "The Osmotic Mechanism of the Cuttlebone." This paper reported that the fluid contained in the chambers was much less saline than sea water. This indicated that the method of emptying the shell was osmotic and depended on the removal of salts from the cameral fluid by some ion transport mechanism.

Similar studies on *Nautilus* in (Denton and Gilpin-Brown, 1966) and on *Spirula* (Denton *et al.*, 1967) were logistically much more difficult. Study of *Nautilus* involved a journey to the South Pacific by Denton and Gilpin-Brown. Study of *Spirula* involved research on an oceanographic vessel. Despite the different settings, the results are quite similar; *Nautilus* and *Spirula* shells are pumped out osmotically; the shells subsequently fill by diffusion and are in equilibrium with the gases in the blood; the shells contain nitrogen at approximately 0.8 atm of pressure.

Denton and Gilpin-Brown's research program had demonstrated that all modern shelled cephalopods empty their shells with an osmotic pump and that the internal gas pressure in these shells is only a fraction of atmospheric pressure. They summarized their work and extended it by analogy to fossil shelled cephalopods in "Floatation Mechanisms in Modern and Fossil Cephalopods" (Denton and Gilpin-Brown, 1973). Similar results were presented by Denton (1974) in the Croonian Lecture given to the Royal Society in 1973.

The low internal gas pressure in the cephalopod shell demonstrated by Denton and Gilpin-Brown indicated that the hydrostatic pressure of submerged shelled cephalopods was borne by the shell itself, not by internal gas pressure. This result vindicated the few workers on fossil and modern cephalopods who had held this view. Puncture of the shell while it was immersed in liquid was one of the investigative techniques employed by Denton and Gilpin-Brown (1961*a,* 1966; Denton *et al.,* 1967). With this and similar techniques they demonstrated the low cameral gas pressure of shelled cephalopods. Despite the similarity of their approach to that suggested by Owen (1832, 1878), Denton and Gilpin-Brown do not cite Owen in this context.

Other workers whose ideas were vindicated by Denton and Gilpin-Brown's work include Buckland (1836), who emphasized the adaptive role of septa, sutures, and ornament in supporting the cephalopod shell against hydrostatic load, Pfaff (1911), who also thought cephalopod shells were designed to resist hydrostatic load, and Westermann (1956), who developed Pfaff's ideas and applied them in a phylogenetic scheme. Westermann was the only one alive to appreciate the turn of events and returned to the study of the cephalopod shell strength in his later work.

The Limits of Osmotic Pumping

It would seem that the work of Denton and Gilpin-Brown provided all the answers to the question of how cephalopod shells are evacuated. However, there was one question they were aware of (Denton and Gilpin-Brown, 1966; Denton *et al.,* 1967), but were unable to answer. That is, if there is an osmotic pumping mechanism which removes water from the shell, it should be constrained by the osmotic difference (the difference in concentration of ions) between fresh and salt water. The ability of ions to attract, or pump, water against a pressure gradient is dependent on the concentration of ions in solution. The concentration of salt in sea water generates an osmotic pressure relative to fresh water only sufficient to pump against the weight of a column of water 240 m deep. At greater depths a void space cannot be directly created by removing the salt across a membrane through which water can follow by diffusion. Presumably this should be the limiting depth below which the chambers of cephalopods could not be emptied using a simple osmotic pump; at greater depths fluid would tend to diffuse into the chambers rather than out; consequently, the chambers would cease to function in buoyancy compensation. However, observations of *Nautilus* and, especially, *Spirula* indicate that they spend the greater part of their lives below 240 m; *Spirula* lives at 3–5 times this depth (Bruun 1943, 1950; cited by Denton, *et al.,* 1967).

The fact that the osmotic difference between fresh and sea water is only equivalent to 240 m of water and that shelled cephalopods lived below this depth was a dilemma that troubled some of the more insightful workers who considered the osmotic mechanism of fluid removal from the chambers of shelled cephalopods. Evidently Bruun (1943), who worked on *Spirula,* considered the possibility of osmotic pumping: "Undoubtedly, the most simple explanation [of chamber emptying] could be given if it could be proved, that the pressure inside the shell is always about 1 atm. . . . In this case we need only assume that a chamber . . . is first filled with fluid, which is absorbed and replaced by air, by a fairly simple osmotic mechanism" (Bruun, 1943, p. 11). This proved to be a prescient description of the actual emptying process; however, Bruun was aware that the depths at which *Spirula* lived were in excess of the depth at which a simple osmotic pump would work. Presented with this discrepancy between the observed habitat depth and the osmotic force, Bruun reluctantly accepted the alternative, the gas pressure mechanism of shell emptying, despite the evidence that *Spirula* shells did not burst when rapidly raised to the surface.

Because Denton and Gilpin-Brown started their work on shelled cephalopods with *Sepia,* which lived at shallow depth, they were not immediately confronted with the issue of the insufficiency of the osmotic concentration of sea water. However, in their later publication on *Nautilus* (Denton and Gilpin-Brown, 1966) they were confronted by the dilemma that *Nautilus* evidently used an osmotic pump and was also frequently captured below 240 m. In response to this dilemma Denton and Gilpin-Brown (1966) invented the concept of "decoupling," in which the pumping material of the siphuncular tube would be out of contact with the fluid in the chamber after a certain amount of fluid had been removed from the chamber. This was presumed to buffer the ionic system in some way and allow the organism to make excursions below 240 m:

> The slowness of diffusion of salts within the narrow chambers of the cuttlebone greatly limited the rate of equilibration between this liquid and that deeper within the chamber. The cuttlefish can, therefore, make appreciable excursions in depth, provided these are not too long lasting, without changing the composition of the liquid which lies within the chambers but which is away from the siphuncular wall. A similar situation obtains in *Nautilus;* as soon as a chamber has been about half emptied of its liquid the only 'coupling' between the liquid and the siphuncular epithelium is by the spreading of liquid by the thin wettable pellicle lining the septum and the 'wick' formed by the calcareous and horny siphuncular tubes. Very likely in *Nautilus* as well as *Sepia* change of depth will involve change in the equilibrium concentration of the salts on the low-pressure side of the siphuncular wall. [Denton and Gilpin-Brown (1966), p. 742.]

Even greater problems were confronted when Denton *et al.* (1967) examined *Spirula;* Bruun (1943) had suggested that *Spirula* had been caught at depths as great as 1750 m. Denton *et al.* (1967) preferred arguments suggesting that *Spirula* was limited to regions higher in the water column, where water temperatures were in excess of 10°C. They inferred "that *Spirula* does not live at so great a depth as was at first thought and that its limit is not much below 500 m" (Denton *et al.*, 1967, p. 189). They also concluded "that when *Spirula* is in its normal [head-down, see Fig. 1] swimming position, this liquid [in the chambers] will be almost completely de-coupled from the permeable region" of the siphuncle (Denton *et al.*, 1967, p. 188). By minimizing the depth and invoking decoupling, Denton *et al.* (1967) hoped to diminish the discrepancy between the observations and the simple osmotic model: "We do not therefore have to postulate a mechanism which can pump salts and liquids against a hydrostatic pressure as high as 175 atm. Like *Nautilus*, however, *Spirula* must often go below 240 m, which is the lower limit at which the simple osmotic mechanism, which we found adequate to explain our results on *Sepia*, could work" (Denton *et al.*, 1967, p. 189).

There are problems with Denton and Gilpin-Brown's hypothesis of decoupling during depth excursion. However relevant decoupling might be above 240 m for *Sepia officinalis*, such a mechanism cannot overcome the 240-m-depth barrier where the water pressure gradient across the siphuncle of a partially emptied or entirely emptied chamber will exceed 24 atm. In this situation, the entire siphuncular area permeable to water will have fresh water forced through it regardless of whether it is in contact with gas or liquid on the cameral side.

No doubt Denton and Gilpin-Brown were aware of the tenuous nature of their decoupling hypothesis; in Denton and Gilpin-Brown (1973) they mention an alternative which would enable the cephalopod to exceed the 240-m depth limitation imposed by the salinity of sea water. This alternative is the "standing gradient osmotic flow mechanism" first described by Diamond and Bossert (1967). These authors came to the conclusion that minute intracellular canaliculi are present in a variety of secretory cells, and that similar intracellular channels occur in tissues involved in osmotic regulation. An example that appealed to Diamond and Bossert (1967) was the nasal gland of marine birds, such as the albatross, which removes salt from the blood. Using these minute channels, organisms maintain high osmotic gradients, high enough to essentially desalinate marine water in ocean-going birds. This system could also function as a high-pressure osmotic pump in cephalopods, evading the limitation imposed by the salinity of sea water.

The system works by concentrating salt in the small intracellular channels via an ion transport mechanism; water would diffuse into these small intracellular channels through the relatively large surface area of the slender tubes; because of the small internal diameter of the channel, diffusion through the walls would greatly exceed the diffusion up the channels from the opening. Consequently, water could be forced out of the end of the intracellular channel at pressure equivalent to the osmotic concentration maintained in the channel; the water would then exit into the siphuncle. This mechanism could pump against almost any pressure, the limitation being the concentration of ions that could be maintained in the tube, not the ionic strength of sea water.

Despite the mechanism of localized osmotic gradient pumping suggested by Denton and Gilpin-Brown (1973), the notion of decoupling (the separation of fluid in the chambers from the siphuncle) did not disappear. Numerous workers interpreted and reconstructed the morphological features of cephalopod shells with decoupling in mind; the shells were figured with retained decoupled fluid in the chambers; examples include *Spirula* and fossil nautilids (Ward, 1980) and ammonites (Bandel and von Boletzky, 1979).

Greenwald *et al.* (1982) demonstrated that the siphuncle of *Nautilus* has all the characteristic features of a standing gradient osmotic pump: electron micrographs revealed a featherlike arrangement of myriad fine canaliculi leading to larger channels entering the siphuncle; numerous mitochondria are present in the siphuncular epithelial cells; and the "sodium-pump" enzyme Na–K ATPase is present in high concentration. This set of observations confirms the suggestion of Denton and Gilpin-Brown (1973). After Greenwald *et al.*'s (1982) observations, and X-ray studies of freshly caught *Nautilus* that revealed little cameral liquid present in the older chambers (Ward *et al.*, 1981), interest in the decoupling of cameral liquid has declined.

Structural Investigations

Despite the uncertainty of the greatest depth at which an osmotic pump could function, Denton and Gilpin-Brown's research program in the 1960s demonstrated that shelled cephalopods did not contain pressurized gas, but supported hydrostatic load in the structure of their shells. This understanding led to investigations of the structural function and strength limitations of cephalopod shells. The objectives of these studies were often directed at determining the relationship between habitat depth and the failure strength of cephalopod shells. There were two approaches taken in this effort. One group of studies was grossly empirical, involving examination of the crushing limits of the entire shell or whole organism. The second approach was more refined; it involved the calculation of the depth at failure of various components of the shell based on their material properties and geometries. These calculated results could also be tested on modern shelled cephalopods and had the added advantage of potential analogous use in the study of fossil forms.

The Empirical Approach

Empirical examination of the strength of cephalopod shells began with Bruun's experiments on *Spirula* (Brunn, 1943; see Denton *et al.,* 1967). Bruun placed *Spirula* shells that had been cast up on the beach in a pressure vessel and found that they imploded at pressures of 40–65 atm; Bruun presumed that the living organism could withstand pressures equivalent to 500–750 m depth. Denton *et al.* (1967) placed fresh shells in a pressure chamber and found that they withstood much greater pressure. In fact, they withstood pressures equivalent to 1500 m, twice the maximum depth estimated by Bruun. This discrepancy was probably a consequence of post mortem deterioration of shell strength.

In *Nautilus* dead shells were also studied first (Denton and Gilpin-Brown, 1966; Raup and Takahashi, 1966; Westermann, 1973; Saunders and Wehman, 1977). These studies produced a wide range of implosion depths, from less than 300 to over 700 m. Subsequent testing of live animals revealed that crushing depths were all slightly in excess of 750 m in depth (Ward *et al.,* 1980; Kanie *et al.,* 1980). So again, the fresh or living organisms were substantially stronger and more consistent in their response than the dead shell material.

The osmotic pumping mechanism was first identified in *Sepia* (Denton and Gilpin-Brown, 1961*a–c*; Denton *et al.,* 1961), indicating that the cuttlebone supports hydrostatic load. Despite the early identification of this structural problem, *Sepia* was the last of the living shelled cephalopods in which crushing strength was examined. Denton and Gilpin-Brown (1973) briefly mention a crushing strength for dead *Sepia* (presumably *S. officinalis*) of 200 m.

In a study conducted in the Mediterranean, Ward and von Boletzky (1984) were able to compare crushing strengths of *S. officinalis, S. orbignyana,* and *S. elegans.* They demonstrated that among the species of *Sepia,* crushing strength varies and corresponds to gross shell morphology. In *Sepia,* failure is often gradual rather than catastrophic, with implosion of part of the cuttlebone, often of the juvenile chambers, not necessarily leading to total failure of the shell or death of the animal. Therefore, it is harder to ascribe a single depth as limiting. However, failure involving severe tissue damage occurs in *S. officinalis* at 200–400 m, suggesting that 200 m may be limiting. Implosion in *S. orbignyana* did not occur at all until a depth of 550 m was attained and damage was minor, involving juvenile chambers only, to a depth of 750 m.

S. officinalis is limited to a shallow habitat as indicated by implosion and by limitation of frequent capture to depths less than 150 m; this form has a flattened cuttlebone. *S. orbignyana* and *S. elegans,* which crush at greater depth and are caught down to 430 m, have longer, more cylindrical shells and more closely spaced septa. Ward and von Boletzky (1984) observe that the smaller radius of curvature of the surface of the *S. orbignyana* and *S. elegans* cuttlebones are likely to render the shells stronger, as would the close spacing of the septa.

On the other hand, the flattened cuttlebone of *S. officinalis* could be advantageous in shallow water, because it provides a larger siphuncular surface area relative to the shell volume, perhaps facilitating the diurnal filling and emptying of the shell that has been observed in this species (Denton and Gilpin-Brown, 1961*c*). This tradeoff between shell strength and pumping rate, as mediated by siphuncular area, may be an important constraint on shelled cephalopods (Ward, 1982). Ward observed that only sepiids and the Paleozoic endoceratids have a large enough area of siphuncle relative to cameral volume to permit buoyancy change on a daily basis. This is primarily a consequence of the limitations on the size of the siphuncle resulting from strength requirements. Tubes of smaller diameter are stronger but necessarily have a smaller surface area for pumping. This pumping limitation may constrain cameral emptying rate sufficiently to limit overall growth rate in the deepest-water forms. The evolution of the cuttlebone in sepiids, where the siphuncular surface forms a portion of the enclosing wall of the chambered shell supported by the numerous septa, appears to represent a significant advance in the functional anatomy of shelled cephalopods. The siphuncular sheet is supported at close intervals by the septa (see Fig. 1). Thus, the siphuncle forms a tentlike tensile element. The organic material of the siphuncle is strong in tension, but has little of the strength in compression necessary to support compressive or bending stresses. This placement of the siphuncle in an exterior position allows it to be of larger area, permitting more rapid pumping, yet Ward and von Boletzky (1984) cite evidence that some sepiids are found to 1000 m depth. So the use of the siphuncle as an external portion of the internal cuttlebone appears to be a robust structural solution.

The work of Ward (1982) puts to rest one of the longest-standing ideas in the study of cephalopods, that shells of ectococliate cephalopods such as *Nautilus* and ammonites served in buoyancy adjustment, allowing the organism to ascend and descend in the water column. The idea of buoyancy adjustment was first expressed by Hooke in 1695 (Derham, 1726) and had been a standard concept invoked by virtually all students of modern and ancient cephalopods discussed in the previous section of this study. Among modern cephalopods only sepiids fulfill this expectation and actually adjust their buoyancy on a daily basis.

Stress Calculations

Stresses in thin-walled cylindrical and spherical pressure vessels are amenable to calculation. The shell wall stress S in a thin-walled cylinder is equal to the pressure difference P across the cylinder wall multiplied by the radius R of the cylinder and divided by the shell wall thickness t: $S = PR/t$. Stress in a sphere or hemisphere subjected to pressure is just half that of a cylinder: $S = PR/2t$. In addition, if one knows the failure strength S_{max} of the material composing the pressure vessel, one can calculate the pressure (or depth) at which failure would be likely to occur. For example, the pressure that a thin-walled cylinder can withstand is $P = tS_{max}/R$. The simplicity of these membrane stress equations led to their application to components of the cephalopod shell. Both tubular siphuncles and hemispherical septa have been approached in this manner.

Denton and Gilpin-Brown (1966) calculated the expected strength of the *Nautilus* siphuncle, employing the strength of silk as S_{max}. The result was a predicted maximum depth at failure of 500 m. A test of siphuncular material preserved in the lab yielded an S_{max} that was substantially lower and resulted in expected failure depth of 350 m. This value proved to be half the depth that live *Nautilus* are able to survive without implosion. This again demonstrates that specimens deteriorate, and that soon after death they cannot be relied upon in strength estimates. Collins and Minton (1967) examined fresh material. Their results suggest that the *Nautilus* siphuncle could withstand over 480 m depth.

Westermann (1971) used the curvature and thickness of the siphuncle to develop the "siphuncular strength index." This allowed him to compare a variety of fossil cephalopods to *Nautilus* in terms of their siphuncular strength and their habitat depth. Forms such as lytoceratids and phylloceratids, thought to live in deep water, proved to have greater siphuncular strength index values than did other ammonites, many of which had substantially weaker siphuncles than *Nautilus*. Further experimentation on the rupture strength of *Nautilus* siphuncles (Chamberlain and Moore, 1982) and the implosion tests on live *Nautilus* (Ward *et al.*, 1980; Kanie *et al.*, 1980) provided better constraints on the failure strength of *Nautilus* siphuncles. This information, in conjunction with the examination of more and better-preserved ammonite siphuncles, allowed Westermann (1982) to better calibrate his siphuncular strength index, but did not substantially change his earlier results (Westermann, 1971).

The last septum of *Nautilus*, and of other shelled cephalopods, is subject to hydrostatic force acting through the body and body chamber. In *Nautilus*, *Spirula*, and many fossil cephalopods the septum, or portions of the septum, form concave surfaces to which the equation for spherical pressure vessels has been applied. Evidently, Pfaff (1911; see Westermann, 1971) was the first to apply stress calculations to the *Nautilus* septum. He compared the concave septum of *Nautilus* to the convex septum of ammonites and concluded that the ammonite septum was six times stronger.

Westermann (1973, 1977) developed a "septal strength index" similar to his siphuncular strength index using data on thicknesses and radii of curvature of the septa in fossil cephalopods. This index was calibrated with implosion data from *Nautilus* and *Spirula.* Westermann analyzed members of a large number of fossil cephalopod orders in this fashion.

The septal strength index has been called into question by Chamberlain and Chamberlain (1985). On the dubious basis of explosion rather than implosion of *Nautilus* shells (they pumped fluid into the shell through the siphuncular opening rather than exposing it to increased external pressure), these authors claim that failure occurs not in the septum, but in the shell wall. Nacre is two to three times stronger in compression than in tension (Currey and Taylor, 1974). By applying pressure to the interior rather than the exterior of the shell, the shell wall normally in compression would be subject to tension. In addition, the septa which are normally under tension are in compression and push out on the shell wall. Consequently, it is not too surprising nor particularly relevant to the living *Nautilus* that the shell wall rather than the septum fails when the interior of the shell is subject to pressure.

There have been a number of responses and clarifications regarding the ontogenetic variation of the septal and siphuncular strength indices (Saunders and Wehman, 1977; Chamberlain and Chamberlain, 1985, 1986; Westermann, 1985*a;* Hewitt and Westermann, 1987*b,* 1988). There is also some question as to how similar the failure strength of material and the structural arrangements of septum, siphuncle, and connecting ring of modern forms are to those of the fossils (Westermann 1982). These comments and exchanges have resulted in only minor modification of the original septal strength index.

The safety factor issue—how close the organism approaches the point of mechanical failure of the system in its daily life—has also raised questions regarding the accuracy of Westermann's indices. In terms of interpreting the depth of the water in a facies of interest in which a chambered cephalopod shell is preserved, the additional relationship of the organism to the substrate, whether it had a benthic mode of life or lived in the water column, must be known. Despite these complications, the siphuncular strength index and the septal strength index are approximate guides to depth, and provide intriguing insight into the relative habitat depth of the various fossil cephalopods examined.

One class of studies does not suffer from the uncertainty of the safety factor issue. These are taphonomic studies. Westermann (1985*b*) examined a Silurian fauna where a number of species had failed in post mortem descent through the water column, whereas another species with a higher septal strength index had reached the bottom without implosion. An analysis such as this would appear to provide upper and lower limits on the actual water depth.

When cephalopods implode, the chambers are occasionally filled with the same sparry calcite or other crystal growth that usually accumulates in the unimploded chambers. This observation indicates that the shell imploded above the bottom during post mortem descent. Otherwise, implosion of the shell would suck in benthic mud. Insights such as this guided Hewitt (1988) in his elegant study of the taphonomy of the Nautilids of the London Clay.

Function of the Complex Septal Suture of Ammonoids

The external shell of cephalopods and its internal septal supports combine to form a complex geometry. This complexity makes it much more difficult to calculate the structural properties of the phragmacone and its septal supports than it is to calculate the strength limits of simple tubular features such as the siphuncle or hemispherical regions of nautiloid septa. This is particularly true of ammonites, where the shell wall is supported by a complexly folded septal-suture. Over the years numerous authors have ascribed a number of functions to the suture of ammonoids. These have included attachment of the animal to the shell, which was apparently first suggested by Von Buch (Buckland, 1836), and the decoupling of fluid in the chambers, among numerous other explanations (Kennedy and Cobban, 1976). Since the work of Denton and Gilpin-Brown, a structural role in shell support has been most frequently alluded to as the primary function of complex ammonoid septal-sutures.

In the previous discussion of Buckland (1836) it was demonstrated that as a consequence of his adaptive, Paleyan research approach and his familiarity with Owen's (1832) doubts about the role of gas pressure, he was able to make an adequate structural interpretation of shell function. Buckland realized that (1) the shell surface of externally shelled cephalopods formed a vaulted surface with variable curvature, (2) the flatter regions of the vault, such as occur on the flanks of the shells, would be weaker under external pressure than the narrow-vaulted venter of the shell, (3) these weaker, more broadly vaulted surfaces were supported by more closely spaced elements of the suture pattern because of the greater complexity of the suture in these regions of the shell, and (4) ribbing had a strengthening effect analogous to corrugation employed in human construction. Unfortunately, Buckland's efforts have not been considered by recent workers.

Although Buckland has not been widely read by 20th century ammonitologists, some of his ideas have been rediscovered. As was discussed in the previous section, the correlation between the external shape of the whorl and the number of lobes was taken up by Spath (1919), who reviewed the evidence of correlation available at the time. More recently there has been renewed interest in the relationship between the number of lobes and the compression and evolution of the whorl which produces the broadly curved weak flanks of the shell. Prior to the general recognition of the structural support of hydrostatic load indicated by Denton and Gilpin-Brown (1961*a–c*, 1966; Denton *et al.*, 1967), Westermann (1956, 1958) recognized that the suture covaried with the external shape of the shell, as did the folding of the septum itself and that these relationships related to support of the shell wall. Westermann returned to this thesis in his "Model for origin, function and fabrication of fluted cephalopod septa" (Westermann, 1975). In this work Westermann further develops his argument that the folds and flutes of the septa serve to support the shell walls at the septal-suture. Subsequently, attempts have been made to quantify the relationship between the number of lobes and the whorl section; these include efforts by Checa (1986) and Hewitt and Westermann (1987*a*). This quantitative approach provides evidence of a relationship between sutural pattern and the external shape of the shell. Following a suggestion of Westermann's (1971), Ward (1980) quantitatively demonstrated that the suturally complex ammonoids attained a wider range of shell shape than nautilids.

Hewitt and Westermann (1986, 1987*a, b*), using failure strengths of nacre from *Nautilus* (Currey and Taylor, 1974; Currey, 1976), applied a number of engineering equations to calculate stress in a variety of the elements of the suture and the shell wall. The problems they attempted to analyze included application of membrane stress equations to the flutes of the septum, buckling equations to the medial portion of the septum, and a bending stress equation to the shell wall. To treat the complex of the septum, suture, and shell wall with these simple engineering equations, simple geometric elements were isolated and engineering equations applied to them. However, it is the nature of complex structures that the loading of one portion of a structure puts stress on and induces strain in attached elements of the structure. This phenomenon is known as indeterminacy and requires that all the stresses and strains of the interacting elements must be calculated simultaneously. Hewitt and Westermann freely admit to this limitation of their analysis. In addition, it can be argued that Hewitt and Westermann applied equations which overly simplified the circumstances of even the limited elements they approached. For example, they calculate bending in a curved portion of the shell wall using the equation for a flat plate.

Jacobs (1990) modeled the ammonoid shell as a series of interacting vaults spanning between the complex sutural elements. This analysis suggests that the suture functions to divide the shell surface into vault elements that have comparable reaction forces. As a consequence of the comparability of these reaction forces, tensional forces are minimized and a strong light shell can be built out of the brittle nacre.

SUMMARY

In 1695 Hooke was the first to state that the chambered shell of cephalopods served as a buoyancy compensation device (Derham, 1726). By use of analogy to the air bladder of fish, he inferred that the shell was emptied by and contained pressurized gas. Despite calls for the test of this gas pressure concept by Owen in 1832 and 1878, the gas pressure thesis remained the generally accepted concept of the function of the cephalopod shell. Surprisingly, this concept was not subjected to adequate testing until the 1960s. Despite the lack of test and the general acceptance of the gas pressure model, a few workers inferred that the shell did not contain pressurized gas and served to support water pressure.

Early on Buckland (1836), employing an adaptationist agenda derived from Paleyan natural theology, came to a very accurate understanding of the role of the shell, suture, and septa in supporting water pressure. This perspective was not retained by later paleontologists. As a consequence of the influence of the biogenetic and morphogenetic laws which advocated recapitulation rather than adaptive explanation, paleontologists as well as neontologists of the late 19th and early 20th centuries did not emphasize adaptive or functional explanations in their evolutionary constructs. Darwinians such as Lankester as well as the neo-Lamarckian followers of Hyatt minimized the role of adaptation in evolution and used cephalopods as a primary example in their evolutionary ideas.

Interest in cephalopods waned among biologists for the first half of the 20th century. Cephalopods were no longer among the central evolutionary models. In contrast, paleontologists maintained their interest in cephalopods throughout this period. In the early part of the century Pfaff (1911) argued in favor of structural support of the shell. However, vitalistic notions of evolution ultimately derived from Hyatt's ideas of morphogenesis, but incorporating subsidiary levels of physiological explanation, were in vogue in the paleontological community after the turn of the century. These fundamentally morphogenetic ideas competed with adaptive explanations. Swinnerton and Trueman (1918) incorporated gas pressure as a physiological factor in the vitalist causal nexus they used to explain the succession of morphological features in evolution. Spath (1919) opposed Swinnerton and Trueman, advocated adaptive explanations in general, and argued in favor of structural support. However, the argument for the structural function of the shell was only one of numerous lines of evidence he presented in support of adaptation. In his later work Spath (1933, 1938) dropped the adaptive argument entirely and concentrated on pointing out the stratigraphic inaccuracies of those who employed the morphogenetic law in phylogenetic analysis. Spath's persistence helps to highlight the debate between the paleontological community and developmental biologists over the recapitulation issue. This debate in the early 20th century preceded the New Synthesis and may help explain the relatively small role played by developmental biologists in evolutionary theory later in the century.

There was one more attempt, by Westermann (1956), to resurrect the structural hypothesis prior to its ultimate proof by Denton and Gilpin-Brown in the 1960s. However, Westermann's initial effort was also suppressed because it was linked to phylogenetic arguments that apparently offended Arkell's conservative views on phylogeny. So, despite several episodes where the hydrostatic load issue was broached, the view that the cephalopod shell served to support water pressure was never widely accepted prior to its unequivocal demonstration in modern cephalopods in the 1960s.

Since the early 1960s there has been a concerted effort to examine and test precise hypotheses of function in cephalopod paleobiology. This includes not only the work of Denton and Gilpin-Brown and Bidder in the early 1960s, but also a subsequent dramatic increase in activity, with workers such as Westermann, Ward, Saunders, and Chamberlain devoting their careers to testing and interpreting cephalopod functional morphology. This is a dramatic change from the previous 40 years when taxonomic and stratigraphic questions, albeit with broad implications, were the primary focus of cephalopod paleontology. The interest in neontologic studies of cephalopods during the preceding 40 years was also limited. This period in neontologic studies is best exemplified by the reductionist focus on the squid giant axon.

The post-1960 period is also distinct from the turn-of-the-century period when laws involving recapitulation and the teleological operation of ontogenetic forces in the evolutionary process were in vogue. The emphasis placed on these laws tended to exclude the testing of functional morphological questions. Observations of ontogeny were the primary objective of natural historical investigation, and they were directed toward an understanding of phylogeny via recapitulation.

The post-1960 interest in functional morphological issues most closely resembles the cephalopod studies engaged in by Owen and Buckland. Owen directly called for tests of functional questions. In the case of gas generation within the shell, tests were not conducted until the early 1960s work of Denton and Gilpin-Brown. Buckland's Paleyan adaptationist program involves experimentation on the buoyancy of *Nautilus,* foreshadowing the work of Ward (1979, 1980, 1982), Saunders (Saunders and Wehman, 1977), and Chamberlain (Chamberlain and Moore, 1982; Chamberlain and Chamberlain, 1985, 1986). In addition, Buckland's (1836) approach to the structural question closely parallels that of Westermann (1956, 1971, 1973, 1975, 1977) and Jacobs (1990) among others. Thus, the use of adaptation, whether Paleyan or neo-Synthetic, resulted in similarities of methodology and scientific reasoning as well as a similarity of the questions addressed. These similarities were not shared by late 19th and early 20th century workers, who focused on the reconstruction of phylogeny via ontogeny and placed less emphasis on adaptation.

Acknowledgments

This research was conducted at Virginia Polytechnic Institute and State University and represents a portion of my Ph.D. dissertation. I would like to thank my committee, R. Bambach, E. Benfield, N. Gilinsky, N. Landman, and D. Porter, for their support. I would also like to thank B. Bennington, R. Burian, M. Grene, D. Lindberg, B. Wallace, E. Yochelson, and an anonymous reviewer for reading and commenting on the manuscript.

REFERENCES

Arkell, W. J., 1949, Jurassic ammonites in 1949, *Sci. Prog.* **147**:401–417.

Arkell, W. J., 1956, *The Jurassic Geology of the World,* Oliver & Boyd, London.

Arkell, W. J., 1957*a,* Sutures and septa in Jurassic ammonite systematics, *Geol. Mag.* **94**:235–248.

Arkell, W. J., 1957*b,* Introduction to Mesozoic ammonoidea, in: *Treatise on Invertebrate Paleontology, Part L* (R. C. Moore, ed.), pp. 89–129, University of Kansas Press, Lawrence, Kansas.

Arkell, W. J., and Moy-Thomas, J. A., 1940, Palaeontology and the taxonomic problem, in: *The New Systematics* (J. Huxley, ed.), pp. 395–410, Oxford University Press, Oxford.

Bandel, K., and von Boletzky, S., 1979, A comparative study of the structure, development, and morphologic relationships of chambered cephalopod shells, *Veliger* **21**:313–354.

Bather, F. A., 1893, The recapitulation theory in palaeontology, *Nat. Sci.* **3**:275–281.

Bather, F. A., 1897, The growth of the cephalopod shell, *Geol. Mag.* **4**:446–449.

Beecher, C. E., 1889, On the development of some Silurian Brachiopoda, *Mem. N. Y. State Mus.* **1**:1–95.

Beecher, C. E., 1890, On the development of the genus *Tornoceras* Hyatt, *Am. J. Sci.* **40**:71–75.

Beecher, C. E., 1891*a*, Development of the Brachiopoda. I. Introduction, *Am. J. Sci.* **41**:71–75.

Beecher, C. E., 1891*b*, The development of Paleozoic porferous corals, *Trans. Conn. Acad. Sci.* **8**:207–214.

Beecher, C. E., 1892, Development of the Brachiopoda. II. Classification of the stages of growth and decline, *Am. J. Sci.* **44**:133–152.

Beecher, C. E., 1893, Larval forms of trilobites in the Lower Helderberg Group, *Am. J. Sci.* **46**:142–147.

Beecher, C. E., 1895, The larval stages of trilobites, *Am. Geol.* **16**:166–197.

Beecher, C. E., 1897, Development of the Brachiopoda. III. Morphology of the brachia, *U. S. Geol. Surv. Bull.* **87**:105–112.

Beecher, C. E., 1901, *Studies in Evolution,* Scribner, New York.

Bert, P., 1867, Mémoire sur la physiologie de la Seiche, *Mem. Soc. Sci. Phys. Nat. Bordeaux* **5**:114–138.

Bidder, A. M., 1962, Use of the tentacles, swimming and buoyancy control in the Pearly Nautilus, *Nature* **196**:451–454.

Bowler, P. J., 1976, *Fossils and Progress: Paleontology and the Idea of Progressive Evolution in the Nineteenth Century,* Science History Publications, New York.

Bruun, A. F., 1943, The biology of *Spirula spirula* (L.), *Dana Rep.* **4**:1–46.

Bruun, A. F., 1950, New light on the biology of *Spirula,* a mesopalagic cephalopod, in: *Essays on the natural sciences in honour of Captain Allan Hancock,* pp. 61–72, University of Southern California Press, Los Angeles, California.

Buckland, W., 1836, *Geology and Mineralogy Considered with Reference to Natural Theology,* Vol. 1, William Pickering, London.

Buckman, S. S., 1909–1912, *Type Ammonites,* Vol. I, Buckman, London.

Buckman, S. S., 1919–1921, *Type Ammonites,* Vol. III, Buckman, London.

Buckman, S. S., 1922–1923, *Type Ammonites,* Vol. IV, Buckman, London.

Chamberlain, J. A., Jr., and Chamberlain, R. B., 1985, Septal fracture in Nautilus: Implications for cephalopod paleobathymetry, *Lethaia* **18**:261–270.

Chamberlain, J. A., Jr., and Chamberlain, R. B., 1986, Is cephalopod septal strength index an index of cephalopod septal strength?, *Alcheringa* **10**:85–97.

Chamberlain, J. A., Jr., and Moore, W. A., Jr., 1982, Rupture strength and flow rate of Nautilus siphuncular tube, *Paleobiology* **8**:408–425.

Checa, A., 1986, Interrelated structural variations in Physoderoceratinae (Aspidoceratidae, Ammonitina), *Neues Jahrb. Geol. Paläontol. Mitt.* **1986**:16–26.

Collins, D. H., and Minton, P., 1967, Siphuncular tube of Nautilus, *Nature* **216**:916–917.

Cummings, E. R., 1903, The morphogenesis of Platystrophia. A study of the evolution of a Paleozoic brachiopod, *Am. J. Sci.* **15**:1–48, 121–146.

Cummings, E. R., 1904, Development of some Paleozoic Bryozoa, *Am. J. Sci.* **17**:49–78.

Cummings, E. R., 1905, Development of Fenestella, *Am. J. Sci.* **20**:169–177.

Cummings, E. R., 1909, Paleontology and the recapitulation theory, *Proc. Indiana Acad. Sci. 25th Annu. Mtg.* **1909**:305–340.

Currey, J. D., 1976, Further studies on the mechanical properties of mollusc shell material, *J. Zool. Lond.* **180**:445–453.

Currey, J. D., and Taylor, J. D., 1974, The mechanical behavior of some molluscan hard tissues, *J. Zool. Lond.* **173**:395–406.

Davis, R. A., 1987, *Nautilus*—The first twenty-two centuries, in: *Nautilus: The Biology and*

Paleobiology of a Living Fossil (W. B. Saunders and N. H. Landman, eds.), pp. 3–24, Plenum Press, New York.

De Beer, G., 1958, *Embryos and Ancestors,* Clarendon Press, Oxford.

Denton, E. J., 1974, On the buoyancy and the lives of modern and fossil cephalopods, *Proc. R. Soc. Lond.* **185**:273–299.

Denton, E. J., and Gilpin-Brown, J. B., 1961*a,* The buoyancy of the cuttlefish *Sepia officinalis* (L.), *J. Mar. Biol. Assoc. U. K.* **41**:319–342.

Denton, E. J., and Gilpin-Brown, J. B., 1961*b,* The distribution of gas and liquid within the cuttlebone, *J. Mar. Biol. Assoc. U. K.* **41**:365–381.

Denton, E. J., and Gilpin-Brown, J. B., 1961*c,* The effect of light on the buoyancy of the cuttlefish, *J. Mar. Biol. Assoc. U. K.* **41**:343–350.

Denton, E. J., and Gilpin-Brown, J. B., 1966, On the buoyancy of pearly Nautilus, *J. Mar. Biol. Assoc. U. K.* **46**:365–381.

Denton, E. J., and Gilpin-Brown, J. B., 1973, Floatation mechanisms in modern and fossil cephalopods, *Adv. Mar. Biol.* **11**:197–268.

Denton, E. J., Gilpin-Brown, J. B., and Howarth, J. V., 1961, The osmotic mechanism of the cuttlebone, *J. Mar. Biol. Assoc. U. K.* **41**:351–364.

Denton, E. J., Gilpin-Brown, J. B., and Howarth, J. V., 1967, On the buoyancy of *Spirula spirula, J. Mar. Biol. Assoc. U. K.* **47**:181–191.

Derham, W., 1726, *Philisophical Experiments and Observations of the late Eminent Dr. Robert Hooke,* Derham, London.

Diamond, J. M., and Bossert, W. H., 1967, Standing gradient osmotic flow. A mechanism for coupling water and solute transport in epitheilia, *J. Gen. Physiol.* **50**:2061–2083.

Dommergues, J., Cariou, E., Contini, D., Hantzpergue, P., Marchand, D., Meister, C., and Thierry, J., 1989, Homéomorphies et canalisations évolutives: Le Rôle del'ontogenèse. Quelques examples pris chez les ammonites du Jurassique, *Geobios* **22**:5–48.

Eimer, G. H. T., 1898, *On Orthogenesis and the Impotence of Natural Selection in Species-formation,* Open Court, Chicago, Illinois.

Garstang, W., 1922, The theory of recapitulation: A critical restatement of the biogenetic law, *Linn. Soc. J. Zool.* **35**:81–101.

Garstang, W., 1966, *Larval Forms with Other Zoological Verses,* Blackwell, Oxford.

George, T. N., 1933, Palingenesis and palaeontology, *Biol. Rev.* **7**:108–135.

Goodrich, E. S., 1930, Edwin Ray Lankester—1947–1927, *Philos. Trans. R. Soc.* **218**:x–xv.

Gould, S. J., 1989, *Wonderful Life,* Norton, New York.

Gould, S. J., and Lewontin, R. C., 1979, The spandrels of San Marco and the Panglossian paradigm: A critique of the adaptationist programme, *Proc. R. Soc. B* **205**:581–598.

Grabau, A. W., 1902, Studies of Gastropoda, *Am. Nat.* **36**:917–945.

Grabau, A. W., 1903, Studies of Gastropoda II. Fulgar and Sycotypus, *Am. Nat.* **37**:515–539.

Grabau, A. W., 1906, Notes on the development of the biserial arm in certain crinoids, *Am. J. Sci.* **21**:289–300.

Grabau, A. W., 1907, Studies of Gastropoda III. On orthogenetic variation in Gastropoda, *Am. Nat.* **41**:607–646.

Greenwald, K. P., Cook, C. B., and Ward, P., 1982, The structure of the chambered *Nautilus* siphuncle: The siphuncular epithelium, *J. Morphol.* **172**:5–22.

Hewitt, R. A., 1988. Nautiloid shell taphonomy: Interpretations based on water pressure, *Palaeogeogr. Palaeoclimatol. Palaeoecol.* **63**:15–25.

Hewitt, R. A., and Westermann, G. E. G., 1986, Function of complexly fluted septa in ammonoid shells I. Mechanical principles and functional models, *Neves Jahrb. Geol. Paläontol.* **172**:47–69.

Hewitt, R. A., and Westermann, G. E. G., 1987*a,* Function of complexly fluted septa in Ammonoid shells II. Septal evolution and conclusions, *Neues Jahrb. Geol. Paläontol.* **174:**135–169.

Hewitt, R. A., and Westermann, G. E. G., 1987*b,* Nautilus shell architecture, in: *Nautilus: The Biology and Paleobiology of a Living Fossil* (W. B. Saunders and N. H. Landman, eds.), pp. 435–461, Plenum Press, New York.

Hewitt, R. A., and Westermann, G. E. G., 1988, Nautiloidseptal strength: Revisited and revised concepts, *Alcheringa* **12:**123–128.

Hurst, C. H., 1893, The recapitulation theory, *Nat. Sci.* **2:**195–200, 364–369.

Hyatt, A., 1889, *Genesis of the Arietidae,* Smithsonian Contributions to Knowledge No. 673, Washington, D. C.

Hyatt, A., 1894, The phylogeny of an acquired characteristic, *Proc. Am. Philos. Soc.* **32:**350–647.

Jackson, R. T., 1890, Phylogeny of the Pelecypoda, Aviculidae and their allies, *Mem. Bost. Soc. Nat. Hist.* **4:**277–400.

Jackson, R. T., 1899, Localized stages in the development of plants and animals, *Mem. Bost. Soc. Nat. Hist.* **5:**89–154.

Jacobs, D. K., 1990, Sutural pattern and shell stress in *Baculites* with implications for other cephalopod shell morphologies, *Paleobiology* **16:**336–348.

Kanie, Y., Fukuda, Y., Nakayama, H., Seli, K., and Hattori, M., 1980, Implosion of living *Nautilus* under increased pressure, *Paleobiology* **6:**44–47.

Kennedy, W. J., and Cobban, W. A., 1976, Aspects of ammonite biology, biogeography and biostratigraphy, *Spec. Pap. Palaeontol.* **17:**1–94.

Kling, G. W., Clark, M. A., Compton, H. R., Devine, J. D., Evans, W. C., Humphrey, A. M., Koenigsberg, E. J., Lockwood, J. P., Tuttle, M. L., and Wagner, G. W., 1987, The 1986 Lake Nyos gas disaster in Cameroon, West Africa, *Science* **236:**169–175.

Landman, N. H., and Waage, K. M., 1982, Terminology of structures in embryonic shells of Mesozoic ammonites, *J. Paleontol.* **56:**1293–1295.

Lang, W. D., 1919*a,* Old age and extinction in fossils, *Proc. Geol. Assoc.* **30:**102–113.

Lang, W. D., 1919*b,* The evolution of ammonites, *Proc. Geol. Assoc.* **30:**49–65.

Lankester, E. R., 1876, Contributions to the developmental history of the mollusca, *Philos. Trans. R. Soc.* **165:**1–48.

Lankester, E. R., 1890, Degeneration a Chapter in Darwinism, in: E. R. Lankester, *The Advancement of Science,* pp. 1–61, British Society for the Advancement of Science.

Mayr, E., 1963, *Animal Species and Evolution,* Harvard University Press, Cambridge, Massachusetts.

Meister, C., 1988, Ontogenèse et évolution des Amaltheidae (Ammonoidea), *Eclogae Geol. Helv.* **81:**763–841.

Moreau, A., 1876, Recherches expérimentales sur la fonctions de la vessie natatoire, *Ann. Sci. Nat. Ser. 6* **4:**1–85.

Needham, J., 1933, On the dissociability of the fundamental processes in ontogenesis, *Biol. Rev.* **8:**180–223.

Oppenheimer, J. M., 1967, *Essays in the History of Embryology and Biology,* MIT Press, Cambridge, Massachusetts.

Owen, R., 1832, *Memoir on the Pearly Nautilus,* Royal College of Surgeons, London.

Owen, R., 1878, On the relative positions to their construction of the chambered shells of cephalopods, *Proc. Zool. Soc. Lond.* **1878:**955–975.

Pavlov, A. P., 1901, Le Crétace inferieur de la Russie et sa faune, *Nouv. Mem. Soc. Imp. Nat. Mosc.* **16:**87.

Pfaff, E., 1911, Über Form und Bau der Ammonitensepten und ihre Beziehungen zur Suturlinie, *Jahrb. Nieder. Geol. Ver. Hann.* **1911:**207–223.

Raup, D. M., 1966, Geometric analysis of shell coiling: General problems, *J. Paleontol.* **40**:1178–1190.
Raup, D. M., 1967, Geometric analysis of shell coiling: Coiling in ammonoids, *J. Paleontol.* **41**:43–65.
Raup, D. M., and Takahashi, T., 1966, Experiments on strength of cephalopod shells in: *Geological Society of America Annual Meeting 1966,* Abstract 173.
Raymond, P. E., 1904, Developmental change in some common Devonian brachiopods, *Am. J. Sci.* **17**:276–301.
Raymond, P. E., 1914, Notes on the ontogeny of *Isotelus gigas* Dekay, *Bull. Mus. Comp. Zool.* **58**:247–263.
Saunders, W. B., and Landman, N. H., (eds.), 1987, *Nautilus: The Biology and Paleobiology of a Living Fossil,* Plenum Press, New York.
Saunders, W. B., and Wehman, D. A., 1977, Shell strength of *Nautilus* as a depth limiting factor, *Paleobiology* **3**:83–89.
Sedgewick, A. C., 1909, The influence of Darwin on the study of animal embryology, in: *Darwin and Modern Science* (A. C. Seward, ed.), pp. 171–184, Cambridge University Press, Cambridge.
Simpson, G. G., 1944, *Tempo and Mode in Evolution,* Columbia University Press, New York.
Smith, J. P., 1897, The development of *Glyphioceras* and the phylogeny of the Glyphioceratidae, *Proc. Calif. Acad. Sci.* **1**:105–128.
Smith, J. P., 1898, The development of *Lytoceras* and *Phylloceras, Proc. Calif. Acad. Sci.* **1**:129–152.
Smith, J. P., 1900*a,* The development and phylogeny of *Placenticeras, Proc. Calif. Acad. Sci.* **3**:181–232.
Smith, J. P., 1900*b,* The biogenic law from the standpoint of paleontology, *J. Geol.* **8**:413–425.
Spath, L. F., 1914, On the development of *Tragophylloceras loscombi, Q. J. Geol. Soc.* **70**:336–362.
Spath, L. F., 1919, Notes on ammonites, *Geol. Mag.* **56**:26–58, 65–74, 115–122, 170–177, 220–225.
Spath, L. F., 1933, Evolution of the cephalopoda, *Biol. Rev.* **8**:418–462.
Spath, L. F., 1938, *The Ammonites of the Liassic Family Liparoceratidae,* British Museum (Natural History), London.
Swinnerton, H. H., 1938, Development and evolution, *Br. Assoc. Adv. Sci. Annu. Mtg.* **1938**:57–84.
Swinnerton, H. H., 1947, *Outlines of Palaeontology,* Edward Arnold, London.
Swinnerton, H. H., and Trueman, A. E., 1918, The morphology and development of the ammonite septum, *Q. J. Geol. Soc.* **73**:26–58.
Trueman, A. E., 1919, The evolution of the *Lipoceratidae, Q. J. Geol. Soc.* **74**:247–298.
Vermeij, G. J., 1987, *Evolution and Escalation An Ecological History of Life,* Princeton University Press, Princeton, New Jersey.
Verrill, A. E., 1896, The molluscan archetype considered as a veliger-like form, with discussion of certain points in molluscan morphology, *Am. J. Sci.* **2**:91–136.
Ward, P. D., 1979, Cameral liquid in *Nautilus* and ammonites, *Paleobiology* **5**:40–49.
Ward, P., 1980, Comparative shell shape distributions in Jurassic–Cretaceous ammonites and Jurassic–Tertiary nautilids, *Paleobiology* **6**:32–43.
Ward, P., 1982, The relationship of siphuncle size to emptying rates in chambered cephalopods: Implications for cephalopod paleobiology, *Paleobiology* **8**:426–433.
Ward, P. D., 1988, *In Search of Nautilus,* Simon & Schuster, New York.
Ward, P. D., and von Boletzky, S., 1984, Shell implosion depth and implosion morphologies in three species of *Sepia* (cephalopoda) from the Mediterranean Sea, *J. Mar. Biol. Assoc. U. K.* **64**:955–966.

Ward, P., Greenwald, L., and Fougerie, F., 1980, Shell implosion depth for living *Nautilus macromphalus* in New Caledonia, *Lethaia* **13**:182.

Ward, P., Greenwald, L., and Magnier, Y., 1981, The chamber formation cycle in *Nautilus macromphalus, Paleobiology* **7**:481–493.

Westermann, G. E. G., 1956, Phylogenie der Stephanocerataceae und Perisphinctaceae des Dogger, *Neues Jahrb. Geol. Paläontol.* **103**:233–279.

Westermann, G. E. G., 1958, The significance of septa and sutures in Jurassic ammonite systematics, *Geol. Mag.* **95**:441–455.

Westermann, G. E. G., 1971, Form, structure and function of shell and siphuncle in coiled mesozoic ammonoids, *Life Sci. Contrib. R. Ont. Mus.* **78**:1–39.

Westermann, G. E. G., 1973, Strength of concave septa and depth limits of fossil cephalopods, *Lethaia* **6**:383–403.

Westermann, G. E. G., 1975, Model for origin, function and fabrication of fluted cephalopod septa, *Paläontol. Z.* **49**:235–253.

Westermann, G. E. G., 1977, Form and function of orthoconic cephalopod shells with concave septa, *Paleobiology* **3**:300–321.

Westermann, G. E. G., 1982, The connecting rings of *Nautilus* and Mesozoic ammonoids: Implications for ammonite bathymetry, *Lethaia* **15**:323–334.

Westermann, G. E. G., 1985*a,* Exploding *Nautilus* camerae does not test septal strength index, *Lethaia* **18**:348.

Westermann, G. E. G., 1985*b,* Post-mortem descent with septal implosion in Silurian nautiloids, *Paläontol. Z.* **59**:79–97.

Willey, A., 1895, In the home of the *Nautilus, Nat. Sci.* **6**:405–414.

Willey, A., 1897, Zoological observations in the South Pacific, *Q. J. Microscop. Soc.* **39**:219–231.

Willey, A., 1902, *Contributions to the Natural History of the Pearly Nautilus: A. Willey's zoological results,* Cambridge University Press, Cambridge.

9

Inversion Polymorphism in Island Species of *Drosophila*

MARVIN WASSERMAN and FLORENCE WASSERMAN

INTRODUCTION

Chromosomal mutations, such as inversions, are major, observable changes in the *Drosophila* genome, and therefore have been extensively and intensively studied for many years (Bush *et al.*, 1977; Carson and Yoon, 1982; Dobzhansky, 1970; Sperlich and Pfriem, 1986; Wasserman, 1982*a,b*). The factors involved in their origin, survival in the heterozygous condition, and fixation must be almost as diverse as the species in which they occur. Moreover, those elements that tend to promote polymorphism are almost certainly antagonistic to those that lead to the fixation of these mutations. It should not be surprising, then, to find that there is no single evolutionary mechanism that can fully explain the chromosomal variability found in nature.

Chromosomal mutations, in general, are alterations in the genome which change recombination frequencies and which may show position effects. In addition, a negative heterosis is exhibited in polymorphic populations because the heterokaryotypes produce at least some duplication-deficiency offspring. This negative heterosis is frequency dependent: To chose a simple situation, suppose the two homokaryotypes and the heterokaryotype are equally fit. Then, if mating is at random, the proportion of the number of a mutation present in the homozygous condition, and thereby

MARVIN WASSERMAN • Biology Department, Queens College of the City University of New York, Flushing, New York 11367. FLORENCE WASSERMAN • Department of Natural Sciences, York College of the City University of New York, Jamaica, New York 11451.

Evolutionary Biology, Volume 26, edited by Max K. Hecht *et al.* Plenum Press, New York, 1992.

not selected against, is equal to *p*, the frequency of the mutation. Selection acts against the chromosomal mutants that are present in the heterozygous condition, $1 - p$. Therefore, the relative fitness of the chromosomal mutation is directly proportional to its frequency. Selection will operate against a newly arisen chromosomal rearrangement until, and unless, it becomes the most frequent gene order in the population. In inversion heterokaryotes, this negative heterosis is manifest when an odd number of crossovers occurs within the inversion loop. It follows, then, that if chromosomal polymorphism is maintained in a population for very long periods of time, there must be other factors that actively counteract the negative heterosis. Such long-term stability may be relatively common in some groups of organisms. In the *Drosophila repleta* group, 15 of the 178 polymorphic paracentric inversions have been maintained in the heterozygous condition for a period of time longer than that needed for speciation (Wasserman, 1982*a*, 1992). Two are shared heterozygous inversions, polymorphic in both of two daughter species: 3g in *D. mercatorum* and *D. paranaensis;* and $2n^3$ in *D. pictura* and *D. pictilis.* Thirteen inversions are polymorphic in one species and fixed in a sister species (Table I). Only some form of balancing selection can account for this stability.

The reduction in the frequency of recombination within and around the inversion during the meiosis of the heterokaryote allows for the development of supergenes by the accumulation of cooperating *cis*-located alleles. The supergenes do not produce obvious phenotypic deviants, although there is strong selection acting upon the various karyotypes (Dobzhansky, 1970) and

TABLE I. Stable Inversion Polymorphisms in *Drosophila*

Paracentric inversion	Polymorphic species	Homozygous species
$2l^2$	*D. fulvimacula*	*D. fulvimaculoides*
$2k^5$	*D. melanopalpa*	*D. neorepleta*
5p	*D. melanopalpa*	*D. neorepleta*
2e	*D. paranaensis*	*D. mercatorum*
$2v^3$	*D. mercatorum*	*D. paranaensis*
2w	*D. leonis*	*D. anceps*
2p	*D. ritae*	*D. mathisi*
$2h^4$	*D. propachuca*	*D. pachuca*
$2i^4$	*D. propachuca*	*D. pachuca*
$2f^2$	*D. starmeri*	*D. martensis*
$2t^6$	*D. starmeri*	*D. uniseta*
$2n^8$	*D. straubae*	*D. parisiena*
$2l^8$	*D. straubae*	*D. parisiena*

chromosomal rearrangements are believed to play important roles in the adaptation and evolution of populations.

Within variable species, centrally located populations are usually cytologically more polymorphic than marginal and peripheral populations. This has been well documented by Brussard (1984) in at least 14 different *Drosophila* species. He discussed several possible, but not necessarily exclusive, explanations. These seem to fall into two broad categories: (1) The number of chromosomal arrangements that can survive may be correlated with the number of ecological niches; there usually are more niches in the central part of the species range, while the niches in marginal regions are more limited in number and, perhaps, transient; and (2) heterokaryotes are heterotic and have a superior fitness over homokaryotes. Heterokaryotes are, therefore, favored under the crowded conditions found in the center of the species range where intraspecific competition is important despite the fact that they produce many poorly adapted homokaryotypic offspring. On the other hand, in a marginal habitat, with few breeding sites, selection may favor the few homokaryotypic individuals that find the rare host and that can produce complete broods of individuals of average fitness.

Chromosomal rearrangements have also been indicted as being important contributing factors in the speciation processes. Speciation is supposed to involve a "cytological revolution" and even closely related, sister species often differ cytologically from each other (Bush, 1975; Bush *et al.*, 1977; White, 1968; Yoon, 1989). Moreover, wide-ranging, continentally distributed species usually show a high level of polymorphism, while their marginal, island-inhabiting sister species are essentially monomorphic (Blair, 1950; Carson, 1959; Eldredge and Gould, 1972; Heed and Russell, 1971).

Here we present the results of a cytological study of the *Drosophila mulleri* cluster including new data obtained from material from the Caribbean Islands. These results are interesting in that they seem to refute the generalities described above. An explanation is presented which may account for the *mulleri* cluster data.

When last reviewed by Wasserman (1982*a,b*), the *Drosophila mulleri* cluster of the *D. repleta* species group consisted of four species, *D. wheeleri, D. aldrichi, D. mulleri,* and an undescribed species, "from Venezuela." Fontdevila *et al.* (1990) have since described and formally named "from Venezuela" as *D. nigrodumosa* and also described a new *D. mulleri* cluster species from Peru, *D. huaylasi.* The *mulleri* cluster species breed in cactus and, as a group, are found in the arid and semiarid regions from Nebraska and California, south to northern South America, and out to Australia. These species are unusual in that they are homosequential, being monomorphic for identical gene orders. When the work was begun, the gene order present in *Drosophila repleta* was chosen as the standard for the *repleta* species group (Wasserman, 1954). The gene sequences in all of the other species were compared back to that found in *repleta* and the changes were interpreted as having arisen by simple two-break inversions. We now believe that the ancestral gene order of the species group differs from that of *repleta* by at least six inversions, Xa, Xb, Xc, 2a, 2b, and 3b. The putative ancestral gene order of the *D. repleta* species group, PRIMITIVE I, is, therefore, Xabc;2ab;3b;4;5;6 (Wasserman, 1982*a*). The homosequential *mulleri* cluster species differ from this PRIMITIVE I by an additional five inversions, 2c, 2f, 2g, 3a, and 3c; their cytological formula is Xabc;2abcfg;3abc;4;5;6. The presence of this homosequentiality demonstrates that speciation has occurred among these five species without any obvious cytological changes, i.e., no "cytological revolutions."

Recent collections on the Caribbean Islands have yielded data on four *D. mulleri* cluster species. These include *D. mulleri,* and three species that were unknown in 1982, *D. mayaguana,* a species described by Vilela (1983), and two species described by Heed and Grimaldi (1991), *D. straubae* and *D. parisiena. D. mayaguana* is homosequential with the other monomorphic species. However, *D. straubae* and *D. parisiena,* despite limited, peripheral distributions, show a considerable amount of chromosomal evolution, being fixed or polymorphic for advanced, derived chromosomal mutations. The distributions, ecological specializations, and chromosomal evolution of the eight forms in the *D. mulleri* cluster, with special emphasis on the new data on the Caribbean Island forms, are described here. Of particular interest are the new karyotype data and their influence on our understanding of the mechanisms involved in cytological evolution.

KARYOTYPE EVOLUTION IN THE *D. MULLERI* COMPLEX

Table II lists the Caribbean localities for *D. mayaguana* and *D. mulleri.* Both species are widely distributed throughout the western Caribbean. A total of 45 strains from 12 localities of *D. mayaguana* and 28 strains from 11 localities of *D. mulleri* were analyzed cytologically.

Tables III and IV list the collection localities of the cytologically analyzed strains of *D. straubae* and *D. parisiena,* while Table V lists the cacti found at each of these localities. These two species are sympatric on Hispaniola and Cuba. However, only *D. straubae* is present on Navassa, while only *D. parisiena* has been found on Jamaica.

Hispaniola has three dry cactus regions separated by wet mountain ranges: (a) north-central Hispaniola, represented by a sample from Montecristi, Dominican Republic; (b) northwestern Hispaniola, represented by the

TABLE II. Cytologically Analyzed Localities of *Drosophila mayaguana* and *Drosophila mulleri*[a]

Locality number	Locality	*Drosophila mayaguana*	*Drosophila mulleri*
920	Conception Island, Bahamas	—	X*
921	Great Inagua, Bahamas	X*	X*
909	Tortola, Virgin Islands	X	—
927	Cayman Brac, Cayman Islands	—	X*
928	Grand Cayman, Cayman Islands	X*	—
901	Fond Parisien, Haiti	X*	X*
942	Gonaives, Haiti	X	X*
943	Port-au-Prince, Haiti	X	X
905	Hatillo, Dominican Republic	X	—
940, 950	Montecristi, Dominican Republic	X	X*
941, 951	Barahona, Dominican Republic	X	X
923	Discovery Bay, Jamaica	—	X*
925	Port Henderson, Jamaica	X	—
945	Kingston, Jamaica	—	X
946	Hellshire, Jamaica	X	X
960	Guantanamo Bay, Cuba	X	—
	Total localities	12	10

[a] X, locality sampled cytologically; —, locality not sampled; asterisk indicates the strain used in intraspecific mating tests.

TABLE III. Chromosomal Constitution of *Drosophila straubae*[a]

Locality number	Locality	Additional inversions				
		2ST	$2n^8$	$2r^8$	$2n^8r^8$	$2n^8l^8$
922	Navassa Island	0	12	0	0	0
950	Montecristi, Dominican Republic	2	0	1	0	11
942	Gonaives, Haiti	1	0	2	0	5
901	Fond Parisien, Haiti	0	0	0	1	1
941	Barahona, Dominican Republic	0	8	0	0	28
Total	Hispaniola	3	8	3	1	45
959	Guantanamo Bay, Cuba	0	26	0	0	0
961	Sigua Beach, Cuba	0	120	0	0	0
Total	Cuba	0	146	0	0	0

[a] Standard sequence: Xabc;2abcfg;3abc;4;5.

TABLE IV. Chromosomal Constitution of *Drosophila parisiena*[a]

Locality number	Locality	Additional inversions						
		XST	Xz	2ST	$2o^8$	$2s^8$	3ST	$3n^2$
940, 950	Montecristi, Dominican Republic	118	0	118	0	0	0	118
902	Gonaives, Haiti	4	0	2	2	0	0	4
901	Fond Parisien, Haiti	4	0	2	2	0	0	4
903	Mirebalais, Haiti	2	0	1	1	0	0	2
943	Port-au-Prince, Haiti	6	0	3	2	1	0	6
905	Hatillo, Dominican Republic	2	0	2	0	0	0	2
941, 951	Barahona, Dominican Republic	154	0	136	10	8	3	151
Total	Hispaniola	290	0	264	17	9	3	287
925	Port Henderson, Jamaica	0	2	2	0	0	2	0
945	Kingston, Jamaica	0	17	21	1	0	22	0
946	Hellshire, Jamaica	0	37	40	0	0	39	1
Total	Jamaica	0	56	63	1	0	63	1
959	Guantanamo Bay, Cuba	0	4	8	0	0	0	6
961	Sigua Beach, Cuba	0	33	44	0	0	1	43
Total	Cuba	0	37	52	0	0	1	49

[a] Standard Sequence: Xabc,2abcfgl^{8}m^{8}n^8;3abc;4;5r.

TABLE V. Locality Data

Locality number	Strain number	Locality	Cactus[a]
1	940, 950	Montecristi, Dominican Republic	O.m., S
2	902	Gonaives, Haiti	O.m., O.st., O.l., S, P
3	903	Mirebalais, Haiti	S
	901	Fond Parisien, Haiti	O.m., O.st., S, P
	943	Port-au-Prince, Haiti	O.m., O.st., O.l., S, P
	905	Hatillo, Dominican Republic	O.m., O.st., O.l., S, P
4	941, 951	Barahona, Dominican Republic	O.st., O.c., S, N, P
5[b]	925, 945, 946	Kingston area, Jamaica	O.st., O.sp., S, P, M, Hy
5[c]	922	Navassa	O.st., O.l., M
6	959	Guantanamo Bay, Cuba	S, P
	961	Sigua Beach, Cuba	O.d., S, P, C, Ha

[a] O.m., *Opuntia moniliformis;* O.st., *Opuntia sticta;* O.l., *Opuntia leptocellis;* O.c., *Opuntia caribae;* O.sp., *Opuntia spinosissima;* O.d., *Opuntia delayni;* S, *Stenocereus;* P, *Pelosocereus;* N, *Neoabbotta;* M., *Melocactus;* Hy, *Hylocereus;* C, *Consolea;* Ha, *Harisia.*
[b] Only *Drosophila parisiena* present.
[c] Only *Drosophila straubae* present.

Gonaives, Haiti, collection; and (c) a southwestern cactus region, sampled at Mirebalais, Fond Parisien, and Port-au-Prince, Haiti, and Hatillo and Barahona, Dominican Republic.

The Cuban collections were from the region east of Santiago de Cuba, the principal arid region of a relatively wet island, at Sigua Beach, and in the Guantanamo Bay area. In addition to the wild cacti listed in Table V, a cactus garden containing exotic species was present at Sigua Beach. The Guantanamo collection was kindly sent to us by Dr. W. B. Heed.

Navassa, a small island between Haiti and Jamaica, has a small *Opuntia* cactus patch. The Jamaican localities were in the relatively dry Kingston region.

D. straubae and *D. parisiena* are extremely difficult to separate morphologically (Heed and Grimaldi, 1991). They are, however, ecologically and cytologically distinct. Flies emerging from cactus have thus far always proven to be monospecific. *D. straubae* breeds on *Opuntia, Cephalocereus,* and *Pelosocereus,* while *D. parisiena* is limited to *Stenocereus* (Heed and Grimaldi, 1991). Three methods were used to obtain cytological data. In most instances, each strain was derived from a single collected female. The salivary gland chromosomes of a single larva were sampled from each strain; usually this larva was the F_1 of the collected female. The Cuban data were exceptions to this. Some strains were initiated by mass cultures of a number of females and males. The Guantanamo Bay samples came from two mass cultures of *D. straubae* and one mass culture of *D. parisiena,* each of which originated from a number of flies bred from cactus and proved to be conspecific. The Sigua Beach *D. straubae* sample, all F_1 larvae, came from 16 isofemale lines and 14 mass cultures containing both species. These mixed mass cultures were initiated with 6–30 females and 12–30 males of the two species. The number of wild genomes of *D. straubae* sampled from this locality was between 60 and 120; the actual number depends upon how many females gave rise to the larvae sampled from the mass cultures. The Sigua Beach *D. parisiena* sample, all F_1 larvae, came from 11 isofemale lines and 7 of the mass cultures. The number of its wild genomes sampled lies between 36 and 44. In both cases, the actual number of genomes sampled is most likely closer to the larger number.

The Homosequential Species

The Standard salivary gland chromosome sequence of the *Drosophila repleta* species group is that found in the monomorphic *Drosophila repleta* and figured by Wharton (1942). The putative ancestor of the species group, the PRIMITIVE I sequence, differs from that of *D. repleta* by at least six

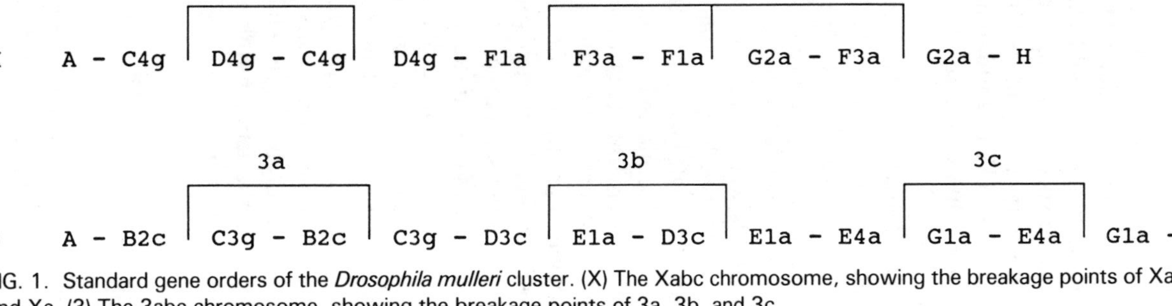

FIG. 1. Standard gene orders of the *Drosophila mulleri* cluster. (X) The Xabc chromosome, showing the breakage points of Xa, Xb, and Xc. (3) The 3abc chromosome, showing the breakage points of 3a, 3b, and 3c.

inversions, Xa, Xb, Xc, 2a, 2b, and 3b (Wasserman, 1982*a*). Using Wharton's (1942) map regions, we give the approximate breakage points of the PRIMITIVE I inversions on the X chromosome and chromosome 3 in Fig. 1; those of chromosome 2 are shown in Fig. 2A. For the latter, two letters represent each segment. The key to the segments is given in the legend. Two letters are used for each segment so that the orientation of the segment can be determined. The *D. mulleri* cluster species differ from PRIMITIVE I by at least five inversions, 3a and 3c (Fig. 1) and 2g, 2c, and 2f (Fig. 2B). Figure 3 diagrams the *D. mulleri* cluster Standard X chromosome (Xabc), the Standard chromosome 3 (3abc), and the Standard chromosome 5, which appears to be identical to the *D. repleta* chromosome 5. Figure 4A shows the *D.*

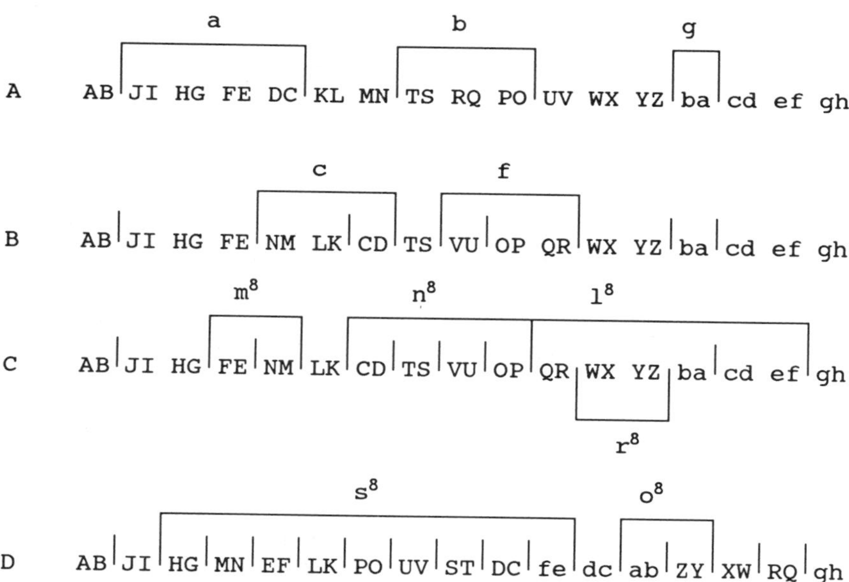

FIG. 2. Evolution of chromosome 2 in the *Drosophila mulleri* cluster. (A) The 2abg chromosome, the Standard gene order of the *D. mulleri* complex, showing the breakage points of 2a, 2b, and 2g. (B) The 2abgcf chromosome, the Standard gene order of the *D. mulleri* cluster, showing the breakage points of 2c and 2f. (C) The 2abgcf chromosome, showing the breakage points of 2m^{8}, 2r^{8}, 2n^{8}, and 2l^{8}; the latter two inversions seem to share a common breakage point. (D) The 2abgcfm^{8}n^{8}l^{8} chromosome, the Standard of *D. parisiena,* showing the breakage points of 2o^{8} and 2s^{8}. The key to the segments (letter pair = salivary segment) of the chromosomes of the *D. repleta* map of Wharton (1942) are as follows: AB, A–A3a; CD, A3a–B1b; EF, B1b–B1f; GH, B1f–B2d; IJ, B2d–C1a; KL, C1a–C2a; MN, C2a–C6a; OP, C6a–C6-; QR, C6--D1a; ST, D1a–D1g; UV, D1g–D3a; WX, D3a–D5a; YZ, D5a–E6a; ab, E6a–F6a; cd, F6a–F6g; ef, F6g–G1a; gh, G1a–H.

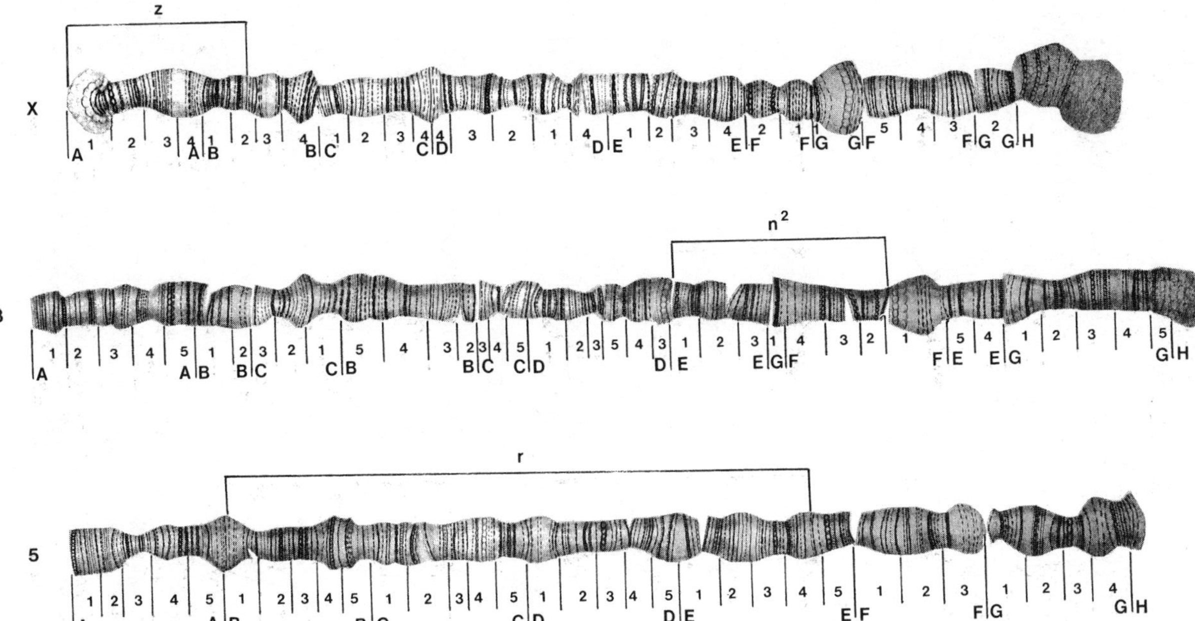

FIG. 3. The salivary gland chromosome maps of the *Drosophila mulleri* cluster. (X) The Xabc chromosome, the Standard X of the *D. mulleri* complex, showing the breakage points of Xz, a terminal inversion present in *D. parisiena*. (3) The 3abc chromosome, the Standard 3 of the *D. mulleri* cluster, showing the breakage points of $3n^2$, an inversion present in *D. parisiena*. (5) The Standard 5 chromosome of the *D. repleta* species group, showing the breakage points of 5r, the Standard chromosome of *D. parisiena*.

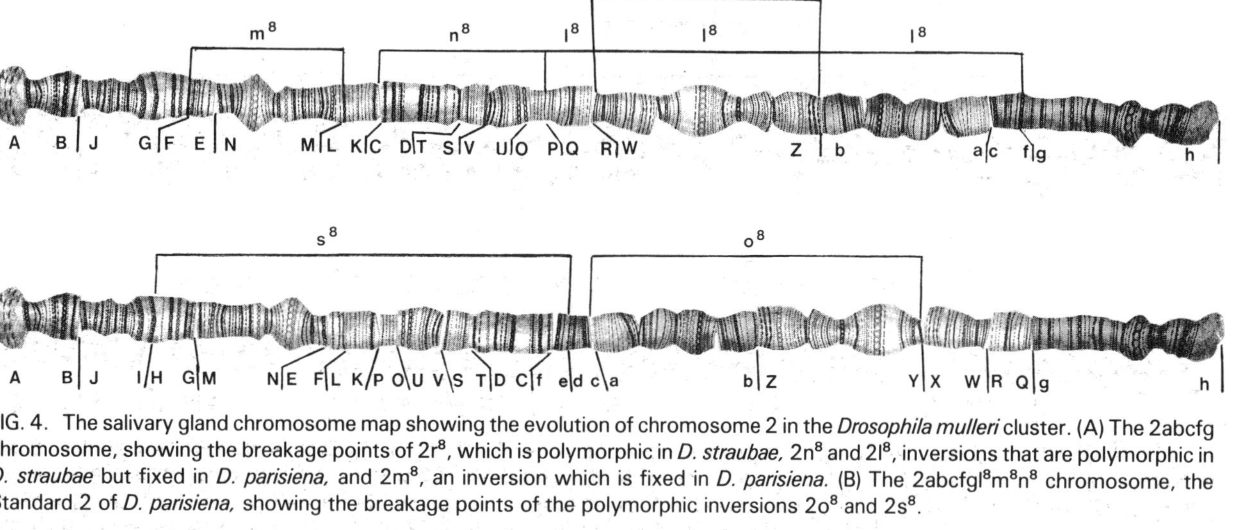

FIG. 4. The salivary gland chromosome map showing the evolution of chromosome 2 in the *Drosophila mulleri* cluster. (A) The 2abcfg chromosome, showing the breakage points of $2r^8$, which is polymorphic in *D. straubae,* $2n^8$ and $2l^8$, inversions that are polymorphic in *D. straubae* but fixed in *D. parisiena,* and $2m^8$, an inversion which is fixed in *D. parisiena.* (B) The $2abcfgl^8m^8n^8$ chromosome, the Standard 2 of *D. parisiena,* showing the breakage points of the polymorphic inversions $2o^8$ and $2s^8$.

mulleri cluster Standard chromosome 2 (2abcfg). No changes have been observed in chromosome 4, and, therefore, it will not be discussed further.

The salivary gland chromosomes of at least one larva were examined from 28 strains from eleven localities of *D. mulleri* and from 45 strains from 12 localities of *D. mayaguana* (Table II). All larvae were homozygous for Xabc;2abcfg;3abc;4;5;6, the Standard chromosomes of the cluster.

To ascertain whether the Caribbean *D. mulleri* species is, in fact, the same species as the *D. mulleri* of Texas, 10 replicas of reciprocal crosses in mass using 10 pairs for each replica were made between *D. mulleri* from Lake Travis, Texas, and strains from the seven localities indicated in Table II. The strains originated in the Bahamas, the Caymans, the Dominican Republic, Haiti, and Jamaica. All crosses produced fully fertile offspring.

Seventy-one crosses using single pairs per vial between strains of *D. mayaguana* from the Great Inagua Island of the Bahamas, the Grand Cayman Island, and Fond Parisien, Haiti, were made. After 2 weeks, 63 females produced F_1 larvae, 4 others had sperm in their ventral recepticles, 2 showed a vaginal reaction mass indicating that they had been inseminated, while only 2 failed to show any indication of having been mated. These numbers are well within the range one might expect from conspecific crosses. An F_2 was obtained from the F_1. Comparable interspecific tests between *D. mayaguana* and *D. parisiena* yielded no matings in 39 attempts.

Table VI summarizes the results of attempts to obtain interspecific hybrids between homosequential members of the *D. mulleri* cluster using matings in mass with no choice (Wasserman, 1982*a*). The table includes our new data as well as those accumulated over a period of years by many workers. Fertile F_1 males have never been produced. Fertile F_1 females are obtained only in crosses between species that are allopatric.

TABLE VI. Results of Crosses in the *Drosophila mulleri* Cluster[a]

	Males					
Females	*wheeleri*	*aldrichi*	*mulleri*	*nigrodumosa*	*huaylasi*	*mayaguana*
wheeleri	—	FF; SM	FF; SM	NO	—	—
aldrichi	FF; SM	—	NO	NO	NO	NO
mulleri	FF; SM	SF; SM	—	FF; SM	NO	NO
nigrodumosa	NO	SM	SF; SM	—	NO	NO
huaylasi	—	SM	FF; SM	FF; SM	—	—
mayaguana	NO	NO	NO	NO	—	—

[a] NO, no offspring; FF, fertile F_1 females; FM, fertile F_1 males; SF, sterile F_1 females; SM, sterile F_1 males.

The Polymorphic Species

A total of nine new inversions have occurred and survived during the history of the two polymorphic species. Of the two, *D. straubae* is cytologically more primitive. Its chromosomal constitution is given in Table III. A total of 109 larvae from 69 strains of *D. straubae* were analyzed. The Standard sequence of this species is identical to that of the six homosequential species. The species is also polymorphic for three inversions, $2n^8$, $2r^8$, and $2l^8$, whose approximate breakage points are shown in Figs. 2C and 4. Inversions $2n^8$ and $2r^8$ are independent inversions which assort from each other. Inversion $2l^8$ seems to share one breakage point with $2n^8$, is only found in chromosomes which have $2n^8$, and, therefore, is almost certainly more advanced.

There is some conspecific cytological differentiation between island populations of *D. straubae* (Table III; Fig. 5). Both the Navassa population (Fig. 5: population 5) and the Cuban populations (Fig. 5: population 6) appear to be monomorphic, being homozygous for the $2n^8$ inversion. Navassa is a small island with a very small cactus patch; homozygosity on such an island in an otherwise polymorphic species is not unexpected. The sampled cactus region in Cuba is a relatively large one. Yet, if there is cytological polymorphism present, it is at a low level and was not detected in our examination of at least 60 genomes. The Hispaniola populations are clearly distinct cytologically from those on the other two islands (Table III). The most common sequence on the island is the most derived chromosome containing $2n^8$ and $2l^8$. This chromosome has not been found on the other two islands. Despite the small samples, each locality on Hispaniola is polymorphic, containing at least two different chromosome 2 gene orders (Table III). Although the data must be considered preliminary, there is a hint of some intraisland differentiation within *D. straubae* (Table III). The northern localities, Montecristi (Fig. 5: population 1) and Gonaives (Fig. 5: population 2), still preserve the Standard chromosomes which are characteristic of the homosequential species but are not found elsewhere in *D. straubae.* Inversion $2n^8$ has not yet been found in these northern localities, while $2r^8$ has. The southern, rift, populations from Fond Parisien and Barahona (Fig. 5: populations 3 and 4, respectively), lack the Standard and $2r^8$, but have $2n^8$ and a chromosome with both $2n^8$ and $2r^8$.

D. parisiena evolved from *D. straubae* by fixing two of the three inversions which are polymorphic within *D. straubae,* $2n^8$ and $2l^8$, and by fixing both a new, independent, distally located inversion, $2m^8$ (Figs. 2C and 4) and a new inversion in chromosome 5, 5r (Fig. 3). The basic formula for this species is Xabc;2abcfg$l^8m^8n^8$;3abc;4;5r;6 (Table IV). The Hispaniola populations also have two newer inversions which occur at low frequencies, $2o^8$, a

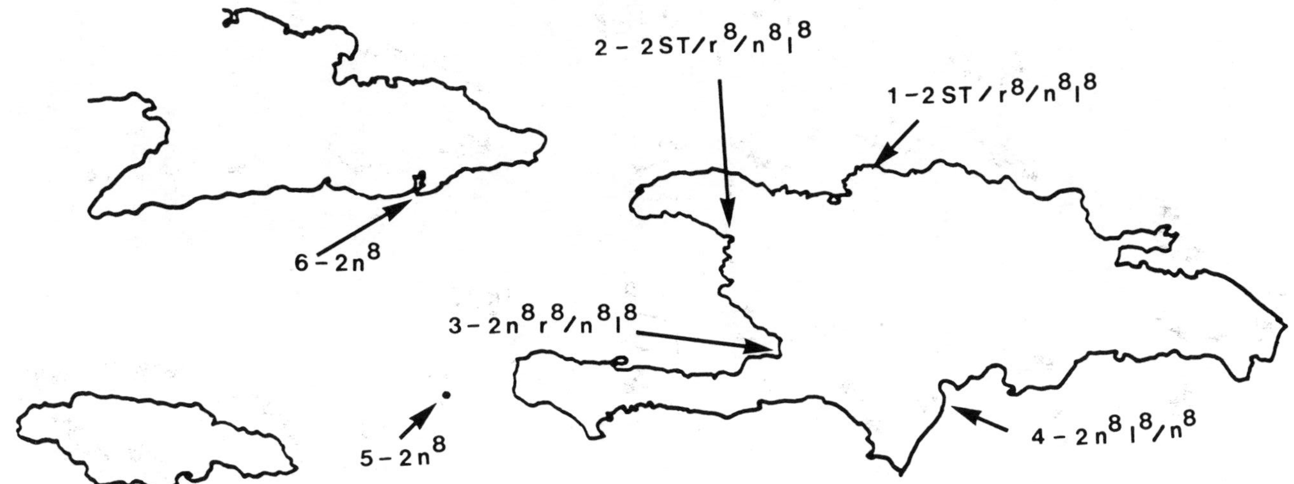

FIG. 5. Geographical distribution of cytolcgically sampled populations of *Drosophila straubae*. 1, Montecristi, Dominican Republic; 2, Gonaives, Haiti; 3, Fond Parisien, Haiti; 4, Barahona, Dominican Republic; 5, Navassa; 6, Sigua Beach and Guantanamo Bay, Cuba.

proximal inversion included within $2l^8$, and $2s^8$, a distal inversion which includes $2m^8$ and overlaps $2l^8$ (Figs. 2D and 4). The Standard chromosome 3 has been essentially replaced by one which contains a new inversion, $3n^2$ (Figs. 1 and 3). There are three isolated cactus regions in Hispaniola, each of which has populations which appear to be cytologically distinct (Table IV). Our samples from Montecristi, Dominican Republic (Fig. 6: population 1) are homozygous for the Standard of the species, $2abcfgn^8l^8m^8$, and homozygous for $3n^2$; the population from Gonaives, Haiti (Fig. 6: population 2) is polymorphic, having both the Standard and $2o^8$; while the populations in the strip which runs from Port-au-Prince, Haiti, to Barahona, Dominican Republic (Fig. 6: populations 3 and 4, respectively) are polymorphic, having both the $2o^8$ and $2s^8$ as well as the Standard chromosome 2. The Barahona population also has the Standard chromosome 3 at a low frequency.

The Jamaican populations of *D. parisiena* (Fig. 6: population 5) differ considerably from the Hispaniola populations in that they are homozygous for a new terminally situated X chromosome inversion, Xz (Fig. 3), which is absent in Hispaniola, and they are essentially homozygous for the Standard chromosome 3 which has practically been replaced by $3n^2$ in Hispaniola. Although the $2s^8$ inversion has not yet been found in Jamaica, our sample sizes are probably too small to determine whether there are qualitative and/or quantitative differences in the chromosome 2 polymorphisms between the two islands (Table IV).

Cuban *D. parisiena* were homozygous for Xz, $2abcfgl^8m^8n^8$, and 5r (Table IV and Fig. 6: population 6). Of the 17 larvae sampled, one was heterozygous for $3n^2$/ST, the other 16 being homozygous for $3n^2$. Thus, the Cuban flies have the X chromosome of Jamaica and the chromosome 3 of Hispaniola.

Interspecific Hybridization on Hispaniola

D. straubae and *D. parisiena* adults are virtually indistinguishable in the field. Each strain was established by adding one or more field-collected males to a single field-collected female. To obtain quantitative data on chromosome frequencies, the salivary gland chromosomes of one larva were analyzed from each strain. Of the 173 analyzed larvae from Hispaniola, where the species are sympatric, 168 resulted from conspecific matings, 3 larvae were F_1 interspecific hybrids, while 2 were F_2 or backcross offspring. The 3 F_1 larvae could have resulted from matings that had taken place in the laboratory after their parents had been collected and are therefore not compelling evidence for the occurrence of interspecific hybridization in nature. However, the two F_2 or backcross offspring could only have resulted from two

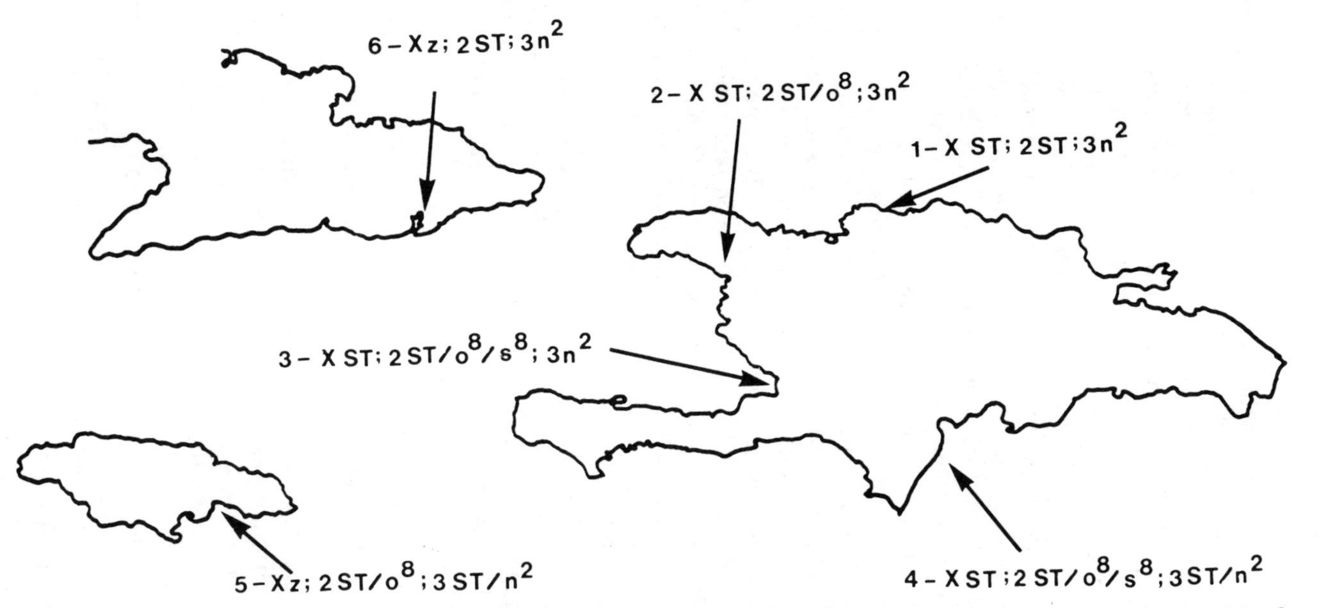

FIG. 6. Geographical distribution of cytologically sampled populations of *Drosophila parisiena*. 1, Montecristi, Dominican Republic; 2, Gonaives, Haiti; 3, Fond Parisien, Mirebalais, Port-au-Prince, Haiti, and Hatillo, Dominican Republic, combined; 4, Barahona, Dominican Republic; 5, various Jamaican populations; 6, Sigua Beach and Guantanamo Bay, Cuba.

different interspecific hybridizations that had occurred in nature prior to the time of collection. This is particularly interesting since the F_1 interspecific hybrids are fertile and, under laboratory conditions, we have obtained an F_4 without any indication of either inviability or sterility.

DISCUSSION

Evolution of the Homosequential Species

Only a total of nine inversions have occurred and survived during the history of the eight *D. mulleri* cluster species. Taken together, the six species homosequential for the primitive sequence of the cluster are widely distributed throughout a major part of the New World, while the two cytologically variable species are limited to a few Caribbean islands. In this cluster, most of the speciations have occurred without any detectable cytological changes, much less "cytological revolutions." Speciation independent of cytological changes may be the common mode of evolution in the genus *Drosophila*. Of the 70 *D. repleta* group species analyzed, 20 share identical standard chromosomes with at least one other species (Wasserman, 1982*a*, 1992). Heed and Russell (1971) report that 7 of the 16 species in the *Drosophila cardini* species group are homosequential with at least one other species. Other examples are found in the Hawaiian *Drosophila* (Carson and Yoon, 1982). Moreover, the presence of cytological differences between species does not necessarily mean that the rearrangements have been instrumental in the isolation leading to speciation. The degree of genetic isolation between species is not related to the degree of cytological differentiation (Wasserman, 1982*a*). For example, in the *D. mulleri* cluster, interspecific crosses between homosequential species may produce fertile F_1 females, but only sterile F_1 males (Table VI). However, crosses between *D. straubae* and *D. parisiena* produce fertile F_1 females and males with no evidence of genetic incompatability. A hybrid strain was carried in the lab for four generations before it was discarded. This, despite the fact that the two species differ from each other by a minimum of two inversions on two chromosomes. In some crosses, the two species may differ by up to eight inversions spread over four of the five major chromosomes. In addition, there is no evidence in the genus *Drosophila* for asserting that those genes which are involved in isolating species are localized in cytologically differentiated chromosomes (Coyne, 1984, 1985; Naveira and Fontdevila, 1986; Naveira *et al.*, 1984; Vigneault and Zouros, 1986; Zouros, 1981*a,b;* Zouros *et al.*, 1988). In fact, Dodd and Powell (1985), in founder-flush experiments, obtained a greater degree of sexual isolation in their homokaryotypic strain of *Drosophila pseudoobscura* than in their heterokaryotypic stains, indicating that free recombination may enhance rapid evolution. Thus, although chromosomal rearrangements can be used as phylogenetic determinants, for *Drosophila* at least, there is no compelling evidence that these mutations have been a driving force in speciation.

The lack of chromosomal variability among the homosequential species precludes us from using cytology as a basis for identifying the species, much less determining phylogenetic relationships among the species within this cluster. We must, therefore, look at distribution data, crossability, and allozyme studies.

Five of the six monomorphic forms are distinct species whose known geographical distributions are as follows. *D. mulleri* is found from Nebraska, south into Texas, and into northern Mexico in Tamaulipas. Our new data show that it ranges into the Caribbean, where it occurs on the southern tip of Florida, Great Inagua in the Bahamas, Cayman Brac in the Cayman Islands, Jamaica, and Hispaniola. *D. wheeleri* is only known from the southern part of California; *D. mayaguana* is widely distributed throughout the western Caribbean, having been found on Great Inagua of the Bahamas, Grand Cayman, Jamaica, Cuba, Hispaniola, and Tortola. Two species are known only from local areas in South America, *D. nigrodumosa* collected in the Andes, west of Merida, Venezuela, and *D. huaylasi* from Peru. The extreme limitation in their distributions may be due to a lack of collection data from most of the Andean Mountains of South America.

D. aldrichi, described from Texas, undoubtedly consists of a number of cryptic, sibling species which, being morphologically and cytologically identical, can only be distinguished by mating tests. This sibling array has a Y-shaped distribution, ranging south along the low-land coast of eastern Mexico across the Isthmus of Tehuantepec and north along the western coast of Mexico to Sonora. From the Isthmus of Tehuantepec, it ranges south through Central America to Colombia. It has also been reported to be in Brazil (Vilela *et al.,* 1983) and in Australia (Barker and Mulley, 1976). The Australian, and perhaps the Brazilian, populations were undoubtedly introduced into these locations along with *Opuntia* by man. Crosses among nine strains using mass cultures with no choice failed to yield data that would be consistent with those which one should get if *D. aldrichi* were a single species. Our unpublished data, along with the published information, can be summarized here. Reciprocal crosses between two isofemale lines from Tuxtla Gutierrez, Chiapas, Mexico, produced no offspring. Females of one of these strains, when crossed to a strain from Jojutla, Morelia, Mexico, yielded fertile females and sterile males. The reciprocal cross produced a few F_2, but no F_3. Reciprocal crosses between the Morelia strain and a strain from Zumpango, Guerrero, Mexico, yielded fertile females and sterile males. Wasserman and Wilson (1957) reported that a strain from Colombia was fully fertile with one from El Salvador. Reciprocal crosses between a Mexican Standard and the Colombian strain produced only an F_1, while the Mexican females crossed to the El Salvador males yielded a total F_2 of two females and one male. Ruiz and Fontdevila (1981) found a strain from Venezuela which proved to be fully fertile with another southern Mexico strain from Tehuantepec. Obviously, the *D. aldrichi* category is a mixed bag of homosequential forms. A systematic study of the morphology and mating behavior of the various forms should be undertaken, not only to determine the nature of the *D. aldrichi* category, but to understand the evolution of the species cluster. Unfortunately, thus far it has been impossible to do this because the strains usually die out before newer ones are obtained.

Heed *et al.* (1990), studying allozyme variability, divided the cluster into three sections: (1) *D. aldrichi* and *D. wheeleri;* (2) *D. mulleri, D. huaylasi,* and *D. nigrodumosa;* and (3) *D. mayaguana, D. straubae,* and *D. parisiena.* The results of tests of matings in mass with no choice are consistant with this division (Table VI). Thus, *D. wheeleri* can exchange genes with *D. aldrichi, s. s.,* while *D. mulleri, D. nigrodumosa,* and *D. huaylasi* can exchange genes among themselves. *D. mulleri* can also exchange genes with *D. wheeleri.*

One cannot interpret the evolution within the *D. mulleri* cluster with a great degree of confidence without knowing what the name *D. aldrichi* signifies. However, a few suggestions seem reasonable. The ancestral population probably had a Y-shaped distribution, the center being in southern Mexico. The northeastern tip split off and formed *D. mulleri;* the northwestern tip formed *D. wheeleri;* while the remainder became *D. aldrichi. D. wheeleri* is allopatric to the other species and under laboratory conditions, is still capable of exchanging genes with *D. mulleri* and *D. aldrichi. D. aldrichi* and *D. mulleri* are now sympatric in Texas and Tamaulipas. The sympatric *D. aldrichi* females will not mate with *D. mulleri* males. However, *D. mulleri* females will accept *D. aldrichi* males, producing sterile F_1 females and males, the latter being detectable due to their abnormal testes. Patterson (1947) and Heed (1957) found evidence for such interspecific hybridization in nature.

The affinities of the South American *D. nigrodumosa* and *D. huaylasi* are not clear. Morphologically, they are most similar to *D. aldrichi* (Fontdevila *et al.,* 1990). They could, therefore, have split off from a South American *D. aldrichi.* However, the allozyme work of Heed *et al.* (1990) and the production of fertile offspring in laboratory crosses (Table VI) indicate that these two species are more closely related to *D. mulleri* than to *D. aldrichi.* These South American forms may have come from the Caribbean *D. mulleri;* on the continent, this latter species is not known to exist south of Tamaulipas, in northeastern Mexico. The above must be considered highly speculative since South American *D. aldrichi* forms have not been studied and there are large cactus regions in South America that have not yet been sampled.

The relationship between the *D. mayaguana* and the other homosequential species is not known. Very little interisland sexual differentiation has taken place within *D. mayaguana.* Single-pair mating tests between strains from different islands showed a high degree of cross-fertility (97% fertile) and yielded an F_2 with no indication of reduced fertility in the F_1. Heed *et al.* (1990) calculated interisland genetic distances from their allozyme work, and found a rough correlation between the genetic distances and the geographical distances between islands. Their work also indicates that *D. mayaguana* is the ancestor of the two polymorphic species.

Chromosomal Evolution of the Polymorphic Species

Chromosomal evolution is limited to the two peripherally distributed species, *D. straubae* and *D. parisiena.* These species have not been found in the Bahamas, the Lesser Antilles, Puerto Rica, or the Caymans; they appear to be limited to Jamaica, Hispaniola, Navassa, Cuba, and perhaps some other small unsampled islands in the immediate neighborhood. Moreover, the two species show interisland, and, in the case of *D. parisiena,* intraisland, cytological differentiation.

The overall cytological evolution of the polymorphic species is shown in Fig. 7. The two species probably have recently evolved and are continuing to evolve rapidly. Evidence for this are two seemingly conflicting observations: all intermediate cytological steps are still present, yet most of the populations are becoming, or are already, fixed for their gene orders.

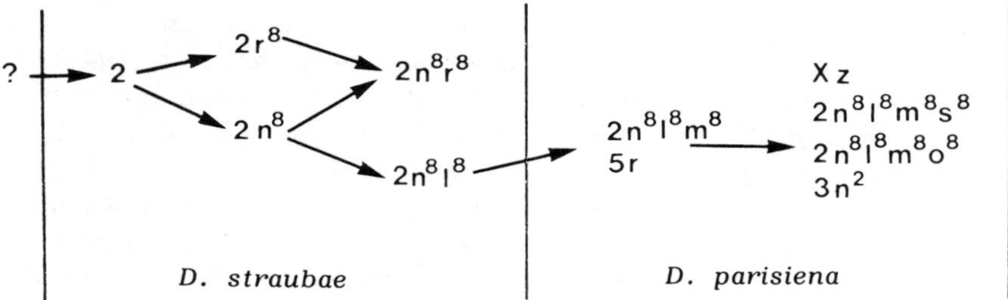

FIG. 7. Cytological evolution of the *Drosophila mulleri* cluster. The evolution proceeds from left to right. Five gene orders, including the Standard 2 chromosome of the cluster, are present in *D. straubae.* In addition to the Standard 2 of the cluster, *D. parisiena* is homozygous for $2l^8$, $2m^8$, $2n^8$, and 5r, and is polymorphic for Xz, $2s^8$, $2o^8$, and $3n^2$.

Our sample sizes for most localities of *D. straubae* are very small and, therefore, the apparent absence of a gene order from a locality may not reflect the true situation. Accepting this *caveat,* the simplest interpretation of the data is that *D. straubae* arose on Hispaniola. All five *D. straubae* chromosome 2 types are found there, including the Standard, the primitive sequence of the species cluster. This gene order, which is present in the six homozygous homosequential species, appears to be limited to the northern populations (Fig. 5: populations 1 and 2) and is relatively rare, comprising only 5% of our sample from this island. The most common (75%) is the advanced sequence, $2n^8l^8$, which is found in all Hispaniola localities. Our data indicate that only the intermediate sequence, $2n^8$, approximately 13% of the Hispaniola sample, has succeeded in becoming established in Navassa and Cuba. Navassa, a small island with a small cactus patch, probably is monomorphic. The principal arid region of Cuba, sampled at Guantanamo Bay and Sigua Beach, yielded only a single genotype despite sample sizes larger than those from Hispaniola. It appears that Cuba is also, essentially, if not actually, monomorphic for the intermediate sequence $2n^8$. Why has only $2n^8$ succeeded in Navassa and Cuba? It is not a particularly common sequence: of the 60 chromosomes tested from Hispaniola, only eight were $2n^8$, and these were found near Barahona, some distance from the Haitian coast (Fig. 5: population 4). Inversion $2n^8$ may have been the first rearrangement in the ancestral *D. straubae* population and emigration from Hispaniola may have occurred before $2l^8$ arose and when $2n^8$ was more common. This implies that *D. straubae* has been able to invade Navassa and Cuba only once. Perhaps a combination of a relatively poor ability to disperse—*straubae* has not been found on Jamaica and the Bahamas despite the presence of its host species—combined with the prior adaptation of the local *straubae* populations have prevented successful incursions of the advanced gene orders into Navassa and Cuba.

D. parisiena arose from *D. straubae* by fixing 5r and adding $2m^8$ to the $2n^8l^8$ chromosome (Fig. 7). Thus, it is readily distinguishable cytologically from *D. straubae.* The new $2n^8l^8m^8$ is the Standard chromosome 2 (2ST) of this species. *D. parisiena* is also variable for Xz, $2o^8$, $2s^8$, and $3n^2$ (Table IV). The three islands are cytologically differentiated, showing a high level of monomorphy for their chromosomes. There is no obvious center of origin for the species. Cuba and Jamaica are fixed for the advanced X chromosome, Xz, while Hispaniola is homozygous for the primitive XST; Cuba and Hispaniola are virtually fixed for the advanced $3n^2$ (98–99%), while Jamaica is almost homozygous for 3ST (99%); all islands have the 2ST at a high frequency, but Hispaniola has the greatest frequency of advanced sequences (10%). Within Hispaniola, the *D. parisiena* populations have diverged cytologically. Montecristi, Dominican Republic, appears to be fixed for 2ST

(118/118); approximately 88% (138/158) of chromosome 2 are 2ST in the southern Dominican Republic localities, Barahona and Hatillo; while only 50% (8/16) are 2ST in Haiti. The Haitian populations, of which we have the smallest samples, seem to be the most variable and may prove to be the cytological center of distribution for the species. This indicates that Hispaniola may be the site of origin of *D. parisiena* as well as *D. straubae*. (This does not imply sympatric speciation, since geographical divergence is certainly possible within a large island dissected by mountain chains.)

An alternative case can be made for Jamaica as the site of origin for *parisiena*. If so, a second group of *straubae* migrants carrying $2l^8$ as well as $2n^8$ invaded Jamaica. Allopatric speciation involving cytological differentiation occurred on Jamaica. Further movement from Jamaica to Hispaniola involved *parisiena* that were polymorphic for the $3n^2$ and the inversions in chromosome 2, and were fixed, or essentially fixed, for the standard X chromosome. Migrations, at a later date, from Jamaica to Cuba would have occurred after Xz was fixed in Jamaica. The interisland differentiation indicates a paucity of successful interisland migrations. Whatever movement occurred in the past, clearly there is little, if any, occurring now. The chromosomal trail does not allow us to chose between the two alternative scenarios presented above. It is hoped that behavioral and/or molecular data will disclose which if either is correct.

Chromosomal Polymorphism or Monomorphism

We would like to understand why chromosomal variability in the *mulleri* cluster seems to differ from that in other species. In general, the degree of chromosomal variability is believed to be correlated with the age of the species and its relative success in utilizing the environment. Chromosomal polymorphism is usually thought to be associated with successful populations and species. The populations presumably consist of many heterotic individuals and occupy multiple habitats under favorable conditions (Brussard, 1984; Heed, 1981).

The evidence for overdominant gene heterosis, where the heterozygote is superior to both homozygotes, as the mechanism underlying cytological polymorphism in nature is weak (Wasserman, 1968; Wasserman and Koepfer, 1975). More likely is either dominance heterosis, where the heterozygote masks the presence of recessive harmful, and even lethal, alleles, or some kind of interaction between internally balanced supergenes. Dominance heterosis is the most reasonable explanation for the rigid stable polymorphisms such as those found in the *Drosophila mesophragmatica* species group (Brncic, 1957*a,b*) and in small laboratory populations (Levene and

Dobzhansky, 1958; Carson, 1987). Other examples of fixed heterozygosity are discussed by Carson (1967). The stability of inversion polymorphisms for exceptionally long periods of time as seen in some *repleta* group species (Table I) may also involve some kind of heterotic interaction, but more likely between internally balanced supergenes rather than recessive alleles.

The number of chromosomal arrangements that can survive may be correlated with the number of ecological niches available. These, in turn, may be influenced by the amount of competition presented by other species: the weaker the competition, the greater the number of available niches, and, therefore, the greater the amount of polymorphism that can be present. These generalities do not seem to hold for the *mulleri* cluster species. There is no evidence that ecological diversity influences the degree of polymorphism among the *D. mulleri* cluster species. *D. straubae* is most variable on Hispaniola, being essentially monomorphic on Navassa and Cuba. It is sympatric on Hispaniola and Cuba with three other *D. mulleri* cluster species, *D. parisiena, D. mayaguana,* and *D. mulleri,* and with *D. stalkeri,* a distantly related *D. mulleri* subgroup species. *D. straubae* is ecologically distinct from *D. parisiena.* Heed and Grimaldi (1991) report that *D. straubae* coexists and has been reared with *D. mulleri* and *D. stalkeri* from rots of *Opuntia moniliformis* and *Opuntia stricta.* It has also been reared from *Cephalocereus. D. mayaguana,* they state, although not yet reared together from the same rots with *D. straubae,* also breeds in *Opuntia stricta* and *Cephalocereus.* A small, monomorphic population of *D. straubae* exists alone without competitors on Navassa. Larger, monomorphic (in Cuba) and polymorphic (in Hispaniola) populations coexist with the three competitors. *D. straubae,* then, does not appear to be exploiting new or more habitats than do the sympatric homosequential species, nor does the presence or absence of possible competitors appear to influence the degree of polymorphism in *D. straubae.*

Unlike *D. straubae, D. parisiena* is ecologically isolated from the other species. Heed and Grimaldi (1991) report that it breeds only in *Stenocereus hystrix,* and is in fact the only species reared from this cactus in this region. However, having eluded competitors by specializing, *D. parisiena* did not increase, but rather decreased, the number of its habitats and the number of islands available to itself. *Stenocereus*—and therefore *D. parisiena*—is not present in the Bahamas or the Cayman Islands. Thus, the increase in the number of gene orders in *D. parisiena* seems to have occurred with a concomitant decrease in the number of niches.

Monomorphism, as opposed to polymorphism, is considered to be a depauperate state found in populations, or species, that can barely eke out a living under marginal conditions (Brussard, 1984). Is monomorphism associated with struggling populations living in a stressful environment? One might argue that the *D. mulleri* cluster homosequential species are really marginal populations despite their overall wide distribution. The limited distributions of the two South American species may be due, at least in part, to a lack of collections in the South American Andes. The homosequential species do have characteristics similar to those of marginal populations. They inhabit arid and semiarid regions, localities that are considered stressful. *Opuntia* appears to be the primitive habitat of the *D. mulleri* subgroup, a taxon that contains about 45 species. For the most part, the homosequential continental species of the *D. mulleri* cluster have maintained this breeding site, although *D. huaylasi* also breeds in the columnar cactus *Cephalocereus* (Fontdevila *et al.*, 1990). With the exception of the sympatric Texas species *D. mulleri* and *D. aldrichi,* the continentally distributed homosequential species are allopatric with each other. However, these species coexist with other *Opuntia*-breeding *D. repleta* group species many of which are also monomorphic or have a relatively low degree of polymorphism (Wasserman 1982*a*). The homosequential island forms, *D. mayaguana* and *D. mulleri,* are sympatric and widely distributed throughout the western Caribbean (Heed and Grimaldi, 1991). One might suppose that the paucity in the number of ecological habitats combined with interspecific competition in the stressful environment of arid and semiarid regions have placed limitations on the degree of cytological polymorphism permissible to these species. But it has not limited the degree of success in their exploitation of the environment. *D. mulleri, D. aldrichi,* and *D. mayaguana* are not merely holding on in the "stressful" environments, but are widely distributed forms, with large population sizes at least in some regions of their distributions. After extensive and intensive collecting by the University of Texas group, Patterson (1943) found that *D. mulleri* and *D. aldrichi* together make up the third largest population of *Drosophila* in Texas, only being surpassed by the combined populations of *D. melanogaster* and *D. simulans,* and by *D. hydei.* The homosequential species, *D. mulleri* and *D. aldrichi,* are not poor, struggling forms, but should be considered to be highly successful species. Parenthetically, *D. hydei,* a cosmopolitan *D. repleta* group species, is polymorphic for only a single ubiquitous inversion.

Even within the polymorphic *mulleri* cluster species, the monomorphic populations may be highly successful. *Stenocereus hystrix,* the host of *D. parisiena,* is quite common in Jamaica, where *D. parisiena* shows little variability; it is less common in Hispaniola, where *D. parisiena* is most polymorphic. In Jamaica, monomorphic *D. parisiena* makes up 70.3% (6168/8765) of the *D. repleta* species, and 52.2% (6168/11813) of the 17 *Drosophila* species collected with banana baits. In Hispaniola, *D. straubae* and *D. parisiena,* both polymorphic, taken together only make up 59.3% (1116/1882) of the *D. repleta* species, and 4.7% (1116/24,955) of the 18 *Drosophila* species collected there. These values, of course, are merely rough estimates of the frequencies of the different species present, but they do show that the monomorphic Jamaican *D. parisiena* is a highly successful dominant population, whereas, even taken together, the polymorphic species on Hispaniola are less so.

It is conceivable that monomorphic species lack inversions because they are too young to have been exposed to these rare mutations, while the polymorphic species have been able to accumulate inversions because they are older. The argument for this is as follows. During speciation, populations go through "bottlenecks" losing cytological as well as genetic variability. Then, through time, they accumulate new chromosomal mutations. Thus, new species are supposed to be essentially monomorphic, while older, ancestral species should show a greater amount of chromosomal evolution including polymorphism.

The underlying genetic assumption is that chromosomal rearrangements are essentially neutral. The negative heterosis that one finds in heterozygotes for chromosomal mutations results in a frequency-dependent selection that tends to eliminate newly arisen inversions when meiotic crossing over occurs in the heterokaryotype. In *Drosophila,* this is alleviated, at least in part, by special meiotic mechanisms that tend to prevent lethal crossover gametes from being formed. In *Drosophila,* at least, chromosomal mutations *per se* approach being neutral and, therefore, one might expect them to accumulate through time and become fixed at rates related their mutation rate, but inversely to the population size. The mutation rates, although generally relatively low, can under special circumstances be quite high, particularly if transposable elements are involved. P factors in *D. melanogaster* can produce up to 10% chromosomal rearrangements per chromosome arm per generation in dysgenic hybrid males (Berg *et al.,* 1980; Engels and Preston, 1984). In some mutator strains, the spontaneous rate of all chromosomal rearrangements may be on the order of 10^{-3}–10^{-4} per gamete (Lande, 1979, 1984; Yamaguchi and Mukai, 1974; Yamaguchi *et al.,* 1976). Of these, a relatively high number are inversions (Imai *et al.,* 1986). Interspecific hybridization may also lead to a high incidence of chromosomal mutations in the F_2 and backcrosses (Naveira and Fontdevila, 1985). Clearly, whatever the cause, under certain circumstances, the spontaneous mutation rate for chromosomal mutations may be relatively high. Therefore, given a reasonable mutation rate of neutral chromosomal mutations, one would expect them to accumulate through time.

However, the alleged correlation between the age of a species and its level of cytological polymorphism does not hold true for the *D. mulleri* cluster, and very likely does not apply to most other taxa either. Assuming an age–polymorphism correlation, one must conclude that *D. parisiena* with eight inversions, although ecologically the most advanced species, is the oldest, *D. straubae* with three inversions is next in age, while the homosequential species are the most recent forms. This sequence must be incorrect for at least two specific reasons: (1) The outgroup (the rest of the *D. repleta* species group along with several other species groups) is cytologically and ecologically more closely related to the homosequential species; if *D. parisiena* is the oldest member of this species cluster, it is also the oldest member of a major section of the genus *Drosophila,* which includes a half a dozen other species groups. (2) The use of this alleged correlation often leads to other irreconcilable inconsistencies when one expands the scope of the study. To cite one example, *D. mojavensis,* a closely related *D. mulleri* complex species, is polymorphic for eight inversions (Ruiz *et al.,* 1990), and, therefore, should be older than the homosequential *mulleri* cluster species; *D. straubae* and *D. parisiena* are polymorphic and, therefore, should also be older than the homosequential species; however, the homosequential species are cytologically intermediate between *D. mojavensis* and the *D. straubae–D. parisiena* siblings, and, therefore, cannot be younger than both phylads. (However, they may be older.) Indeed, one of the generalities that have come from studies such as that on the *D. repleta* group is that there is no correlation between time and the rate of chromosomal evolution: some older species continue to be monomorphic for old primitive gene orders, while other, derived species became polymorphic and even fixed for whole series of newer chromosomal rearrangements. Chromosomal mutations, even in *Drosophila,* are clearly not neutral, but if preserved in the population, must be maintained by selection either through heterosis or by the creation of internally balanced supergenes (Dobzhansky, 1970). Thus, any attempt to use them as a cytological clock would be a mistake.

It should be apparent that the cytological variability found in the *mulleri* cluster does not readily fit into the scenarios proposed for other *Drosophila* species. These species differ, however, in two critical ways: they are relatively young species and they are island species. A reasonable scenario for these island forms might be the following: Migrants from populations of *D. mayaguana* arrived in Hispaniola. With no competitors, and perhaps through a series of flush–crash incidences as proposed by Carson and Templeton (1984), the Hispaniola population built up a store of cytological mutations concurrently with its geographical speciation giving rise to *D. straubae.* Of primary importance was the lack of closely related competitors—this allowed for the loosening of the ecological and genetic constraints placed on the colony and resulted in an accumulation of cytological as well as genetic variability during differentiation. Unlike Carson, we do not believe that a population need go through a period of "disorganization." If one or more previously allopatric close relatives are already present in the new region, there may be a reduction in the genetic variability and a reinforcement of the isolation between the migrants and the endemic forms. On the other hand, if close relatives are not present (as we believe was the situation with *D. straubae*) a new colony may be able to increase its permissible genetic variability, including cytological variability, during the period when selection pressure is reduced. Subsequent contact with competitors could reduce this genetic variability. Further evolution led to the origin of *D. parisiena,* perhaps as outlined above. Part of the differentiation of *D. parisiena* involved specialization to a new host plant and thus escape from competitors allowing for a new round of chromosomal evolution.

This type of cytological release when a migrant reaches a new, virgin territory has also been found in *D. mojavensis* (Ehrman and Wasserman, 1987). During the evolution of *D. mojavensis,* migrants from an essentially monomorphic, continentally distributed species invaded Baja California. With no competitors, they were able to adapt to a new host, *Agria* cactus, and they became fixed for three new inversions and polymorphic for eight others. Subsequent reinvasion of Sonora by the Baja form brought it into secondary contact with the original mainland form and through sympatric reinforcement of the sexual isolation initiated in allopatry, the speciation process was completed. The original mainland forms are now named *D. arizonae,* while the migrant populations comprise *D. mojavensis* (Ruiz *et al.,* 1990). Besides the sexual behavioral change, the new Sonoran *D. mojavensis* populations, in competition with several *D. repleta* group species including *D. arizonae,* lost much of their chromosomal variability and are now polymorphic for only one inversion.

Other examples of cytological release occurred among the Hawaiian *Drosophila* (Carson and Yoon, 1982). In the *D. adiastola* subgroup, the 12 species living on the older islands and therefore presumed to be older in age show no chromosomal polymorphism. The two youngest species, living on the island of Hawaii, *D. setosimentum* and *D. ochrobasis,* taken together, are homozygous for 11 new inversions and polymorphic for an additional 15 inversions. A somewhat similar situation exists in the *D. planitibia* subgroup-II, where the ancestral species *D. differens* and *D. planitibia* are monomorphic and have the same standard sequences that are found in the younger species, *D. heteroneura* and *D. sylvestris. D. sylvestris* is polymorphic for 12 inversions, one of which is also heterozygous in *D. heteroneura.*

To conclude, the invasion of a new, virgin territory, geographical or ecological, where there is no intraspecific or interspecific competition may result in a cytological release where an increase in chromosomal as well as genic variability is tolerated. On the other hand, the invasion of territories where competitors are present will prevent this cytological and genetic release and may, in fact, lead to the reduction in the amount of variability allowed. Homokaryotypic populations and species may be highly successful forms which are closely adapted to their environment.

SUMMARY

The distribution, ecology, and chromosomal evolution of the eight species of the *D. mulleri* cluster of the *D. repleta* group, with special emphasis on the new data on the Caribbean Island forms, are described. Six are homosequential species, being monomorphic for identical gene orders while two species, *D. straubae* and *D. parisiena,* are homozygous or polymorphic for nine inversions. The homosequential species are widely distributed, successful forms; the polymorphic species, although also successful, are limited in distribution to several islands in the Caribbean Sea. Both polymorphic species exhibit interisland, and perhaps, intraisland, differentiation including monomorphic as well as polymorphic populations. The monomorphic populations are successful dominant populations where they occur. The chromosomal evolution is explained by a combination of founder–flush events and the absence of interspecific competition at critical periods during the species evolution.

Acknowledgments

This work was supported by grant No. BSR-8508457 from the National Science Foundation and CUNY-PSC grant No. 669172. The authors thank Dr. Alfredo Ruiz for his hospitality and for the use of his laboratory during their stay at the Universidad Autonoma de Barcelona, Spain.

REFERENCES

Barker, J. S. F., and Mulley, J. C., 1976, Isozyme variation in natural populations of *Drosophila buzzatii, Evolution* **30:**213–233.

Berg, R. L., Engels, W. R., and Kreber, R. A., 1980, Site-specific X-chromosome rearrangements from hybrid dysgenesis in *Drosophila melanogaster, Science* **210:**427–429.

Blair, W. F., 1950, Ecological factors in speciation of *Peromyscus, Evolution* **4:**253–275.

Brncic, D., 1957*a,* A comparitive study of chromosomal variation in species of the *mesophragmatica* group of *Drosophila, Genetics* **42:**798–805.

Brncic, D., 1957*b*, Chromosomal polymorphism in natural populations of *Drosophila pavani, Chromosoma* **8:**699–708.

Brussard, P. F., 1984, Geographical patterns and environmental gradients: The central—marginal model of *Drosophila* revisited, *Annu. Rev. Ecol. Syst.* **15:**25–64.

Bush, G. L., 1975, Modes of animal speciation, *Annu. Rev. Ecol. Syst.* **6:**339–364.

Bush, G. L., Case, S. M., Wilson, A. C., and Patton, J. L., 1977, Rapid speciation and chromosomal evolution in mammals, *Proc. Natl. Acad. Sci. USA* **74:**3942–3946.

Carson, H. L., 1959, Genetic conditions which promote or retard the formation of species, *Cold Spring Harbor Symp. Quant. Biol.* **24:**87–105.

Carson, H. L., 1967, Permanent heterozygosity, in: *Evolutionary Biology,* Vol. 1 (Th. Dobzhansky, M. K. Hecht, and W. C. Steere, eds.), pp. 143–168, Appleton-Centrury-Crofts, New York.

Carson, H. L., 1987, High fitness of heterokaryotypic individuals segregating naturally within a long-standing laboratory population of *Drosophila sylvestris, Genetics* **116:**415–422.

Carson, H. L., and Templeton, A. R., 1984, Genetic revolution in relation to speciation phenomena: The founding of new populations, *Annu. Rev. Ecol. Syst.* **15:**97–131.

Carson, H. L., and Yoon, J. S., 1982, Genetics and evolution of Hawaiian *Drosophila,* in: *Genetics and Biology of Drosophila,* Vol. 3b (M. Ashburner, H. L. Carson, and J. N. Thompson, Jr., eds.), pp. 298–344, Academic Press, London.

Coyne, J. A., 1984, Genetic basis of male sterility in hybrids between two closely related species of *Drosophila, Proc. Natl. Acad. Sci. USA* **84:**4444–4447.

Coyne, J. A., 1985, The genetic basis of Haldane's rule, *Nature* **314:**736–738.

Dobzhansky, Th., 1970, *Genetics of the Evolutionary Process,* Columbia University Press, New York.

Dodd, D. M. B., and Powell, J. R., 1985, Founder-flush speciation: An update of experimental results with *Drosophila, Evolution* **39:**1388–1392.

Ehrman, L., and Wasserman, M., 1987, The significance of asymmetrical sexual isolation, in: *Evolutionary Biology,* Vol. 21 (M. K. Hecht, B. Wallace, and G. T. Prance, eds.), pp. 1–20, Plenum Press, New York.

Eldredge, N., and Gould, S. J., 1972, Punctuated equilibria: An alternative to phyletic gradualism, in: *Models in Paleobiology* (T. J. M. Schopf, ed.), pp. 82–115, Freeman, Cooper and Co., San Francisco.

Engels, W. R., and Preston, C. R., 1984, Formation of chromosome rearrangements by P factors in *Drosophila, Genetics* **107:**657–678.

Fontdevila, A., Wasserman, M., Pla, C., Pilares, L., de Armengol, R., Suyo, M. P., Sanchez, A., Vasquez, J., Ruiz, A., and Garcia, J. L., 1990, Description and evolutionary relationships of two species of the *Drosophila mulleri* cluster (Diptera: Drosophilidae), *Ann. Entomol. Soc. Am.* **83:**444–452.

Heed, W. B., 1957, An attempt to detect hybrid mating between *D. mulleri* and *D. aldrichi* under natural conditions, *Univ. Tex. Pub.* **5721:**182–185.

Heed, W. B., 1981, Central and marginal populations revisited, *Drosophila Information Service* **56:**60–61.

Heed, W. B., and Grimaldi, D., 1991, Revision of the morphocryptic, Caribbean *mayaguana* species subcluster, in the *Drosophila repleta* group (Diptera: Drosophilidae), *Am. Mus. Novit.* **2999:**1–15.

Heed, W. B., and Russell, J. S., 1971, Phylogeny and population structure in island and continental species of the *cardini* group of *Drosophila* studied by inversion analysis, *Univ. Tex. Pub.* **7103:**91–130.

Heed, W. B., Sanchez, A., Armengol, R., and Fontdevila, A., 1990, Genetic differentiation among island populations and species of cactophilic *Drosophila* in the West Indies, in:

Ecology and Evolutionary Genetics of Drosophila (J. S. F. Barker, R. MacIntyre, and W. T. Starmer, eds.), pp. 447–489, Plenum Press, New York.

Imai, H. T., Maruyama, T., Gojobori, T., Inoue, Y., and Crozier, R., 1986, Theoretical bases for karyotype evolution. I. The minimum-interaction hypothesis, *Am. Nat.* **128:**900–920.

Lande, R., 1979, Effective deme sizes during long-term evolution estimated from rates of chromosomal rearrangement, *Evolution* **33:**234–251.

Lande, R., 1984, The expected fixation rate of chromosomal inversions, *Evolution* **38:**743–752.

Levene, H., and Dobzhansky, Th., 1958, New evidence of heterosis in naturally occurring inversion heterozygotes in *Drosophila pseudoobscura, Heredity* **12:**37–49.

Naveira, H., and Fontdevila, A., 1985, The evolutionary history of *Drosophila buzzatii.* IX. High frequencies of new chromosomal rearrangements induced by introgressive hybridization, *Chromosoma* **91:**87–94.

Naveira, H., and Fontdevila, A., 1986, The evolutionary history of *Drosophila buzzatii.* XII. The genetic basis of sterility in hybrids between *D. buzzatii* and its sibling *D. serido* from Argentina, *Genetics* **114:**841–857.

Naveira, H., Hauschteck-Jungen, E., and Fontdevila, A., 1984, Spermiogenesis of inversion heterozygotes in backcross hybrids between *Drosophila buzzatii* and *D. serido, Genetica* **65:**205–214.

Patterson, J. T., 1943, The Drosophilidae of the southwest, *Univ. Tex. Pub.* **4313:**7–216.

Patterson, J. T., 1947, Sexual isolation in the *mulleri* subgroup, *Univ. Tex. Pub.* **4720:**32–40.

Ruiz, A., and Fontdevila, A., 1981, Ecologia y evolucion del subgroup *mulleri* de *Drosophila* en Venezuela y Colombia, *Acta Cient. Venez.* **32:**338–345.

Ruiz, A., Heed, W. B., and Wasserman, M., 1990, Evolution of the *mojavensis* cluster of cactophilic *Drosophila* with descriptions of two new species, *J. Hered.* **81:**30–42.

Sperlich, D., and Pfriem, P., 1986, Chromosomal polymorphism in natural and experimental populations, in: *The Genetics and Biology of Drosophila,* Vol. 3e (M. Ashburner, H. L. Carson, and J. N. Thompson, Jr., eds.), pp. 257–309, Academic Press, London.

Vigneault, G., and Zouros, E., 1986, The genetics of asymmetrical male sterility in *Drosophila mojavensis* and *Drosophila arizonensis* hybrids: Interactions between the Y-chromosome and autosomes, *Evolution* **40:**1160–1170.

Vilela, C. R., 1983, A revision of the *Drosophila repleta* species group (Diptera, Drosophilidae), *Rev. Brasil. Entomol.* **24:**1–114.

Vilela, C. R., Pereira, M. A. Q. R., and Sene, F. M., 1983, Preliminary data on the geographical distribution of *Drosophila* species within morphoclimatic domains of Brazil. II. The *repleta* group, *Cienc. Cult.* **35:**66–70.

Wasserman, M., 1954, Cytological studies on the *repleta* group, *Univ. Tex. Pub.* **5422:**130–152.

Wasserman, M., 1968, Recombination-induced chromosomal heterosis, *Genetics* **58:**125–139.

Wasserman, M., 1982*a,* Evolution and speciation in selected species groups. Evolution of the *repleta* group, in: *The Genetics and Biology of Drosophila,* Vol. 3b (M. Ashburner, H. L. Carson, and J. N. Thompson, Jr.), pp. 61–139, Academic Press, London.

Wasserman, M., 1982*b,* Cytological evolution of the *Drosophila repleta* group, in: *Ecological Genetics and Evolution: The Cactus–Yeast–Drosophila model* (J. S. F. Barker and W. T. Starmer, eds.), pp. 49–64, Academic Press, New York.

Wasserman, M., 1992, Cytological evolution of the *Drosophila repleta* species group, in: *Inversion Polymorphism in Drosophila* (C. Krimbas and J. Powell, eds.), CRC Press, Boca Raton, Florida.

Wasserman, M., and Koepfer, H. R., 1975, Fitness of karyotypes in *Drosophila pseudoobscura, Genetics* **79:**113–126.

Wasserman, M., and Wilson, F. D., 1957, Further studies on the *repleta* group, *Univ. Tex. Pub.* **5721:**132–156.

Wharton, L. T., 1942, Analysis of the *repleta* group of *Drosophila, Univ. Tex. Pub.* **4228:**23–52.
White, M. J. D., 1968, Models of speciation, *Science* **159:**1065–1070.
Yamaguchi, O., and Mukai, T., 1974, Variation of spontaneous occurrence rates of chromosomal aberrations in the second chromosomes of *Drosophila melanogaster, Genetics* **78:**1209–1221.
Yamaguchi, O., Cardellino, R. A., and Mukai, T., 1976, High rates of occurrence of spontaneous chromosome aberrations in *Drosophila melanogaster, Genetics* **83:**409–422.
Yoon, J. S., 1989, Chromosomal evolution and speciation in Hawaiian *Drosophila,* in: *Genetics, Speciation and the Founder Principle* (L. V. Giddings, K. Y. Kaneshiro, and W. W. Anderson, eds.), pp. 129–147, Oxford University Press, Oxford.
Zouros, E., 1981*a,* The chromosomal basis of sexual isolation in two sibling species of *Drosophila: Drosophila arizonensis* and *Drosophila mojavensis, Genetics* **97:**703–718.
Zouros, E., 1981*b,* The chromosomal basis of viability in interspecific hybrids between *Drosophila arizonensis* and *Drosophila mojavensis, Can. J. Genet. Cytol.* **23:**65–72.
Zouros, E., Lofdahl, K., and Martin, P. A., 1988, Male hybrid sterility in *Drosophila:* Interactions between autosomes and sex chromosomes in crosses of *D. mojavensis* and *D. arizonensis, Evolution* **42:**1321–1331.

Index